Mathematical Handbook of Formulas and Tables

Mathematical Handbook of Formulas and Tables

Third Edition

Murray R. Spiegel, PhD

Former Professor and Chairman
Mathematics Department
Rensselaer Polytechnic Institute
Hartford Graduate Center

Seymour Lipschutz, PhD

Mathematics Department
Temple University

John Liu, PhD

Mathematics Department
University of Maryland

Schaum's Outline Series

New York Chicago San Francisco
Lisbon London Madrid Mexico City
Milan New Delhi San Juan
Seoul Singapore Sydney Toronto

The McGraw-Hill Companies

SEYMOUR LIPSCHUTZ is on the faculty of Temple University and formerly taught at the Polytechnic Institute of Brooklyn. He received his PhD from New York University and is one of Schaum's most prolific authors. In particular he has written, among others, *Linear Algebra, Probability, Discrete Mathematics, Set Theory, Finite Mathematics,* and *General Topology*.

JOHN LIU is presently a professor of mathematics at University of Maryland, and he formerly taught at Temple University. He received his PhD from the University of California, and he has held visiting positions at New York University, Princeton University, and Berkeley. He has published many papers in applied mathematics, including the areas of partial differential equations and numerical analysis.

The late **MURRAY R. SPIEGEL** received the MS degree in Physics and the PhD degree in Mathematics from Cornell University. He had positions at Harvard University, Columbia University, Oak Ridge, and Rensselaer Polytechnic Institute, and served as a mathematical consultant at several large companies. His last position was Professor and Chairman of Mathematics at the Rensselaer Polytechnic Institute, Hartford Graduate Center. He was interested in most branches of mathematics, especially those that involve applications to physics and engineering problems. He was the author of numerous journal articles and 14 books on various topics in mathematics.

Schaum's Outline of MATHEMATICAL HANDBOOK of FORMULAS and TABLES

2 3 4 5 6 7 8 9 0 CUS CUS 0 1 4 3 2 1 0 9 8

ISBN 978-0-07-154855-7
MHID 0-07-154855-6

Sponsoring Editor: Charles Wall
Production Supervisor: Tama L. Harris
Editing Supervisor: Maureen Walker
Interior Designer: Jane Tenenbaum
Project Manager: Madhu Bhardwaj

Library of Congress Cataloging-in-Publication Data is on file with the Library of Congress.

Preface

This handbook supplies a collection of mathematical formulas and tables which will be valuable to students and research workers in the fields of mathematics, physics, engineering, and other sciences. Care has been taken to include only those formulas and tables which are most likely to be needed in practice, rather than highly specialized results which are rarely used. It is a "user-friendly" handbook with material mostly rooted in university mathematics and scientific courses. In fact, the first edition can already be found in many libraries and offices, and it most likely has moved with the owners from office to office since their college times. Thus, this handbook has survived the test of time (while most other college texts have been thrown away).

This new edition maintains the same spirit as the second edition, with the following changes. First of all, we have deleted some out-of-date tables which can now be easily obtained from a simple calculator, and we have deleted some rarely used formulas. The main change is that sections on Probability and Random Variables have been expanded with new material. These sections appear in both the physical and social sciences, including education.

Topics covered range from elementary to advanced. Elementary topics include those from algebra, geometry, trigonometry, analytic geometry, probability and statistics, and calculus. Advanced topics include those from differential equations, numerical analysis, and vector analysis, such as Fourier series, gamma and beta functions, Bessel and Legendre functions, Fourier and Laplace transforms, and elliptic and other special functions of importance. This wide coverage of topics has been adopted to provide, within a single volume, most of the important mathematical results needed by student and research workers, regardless of their particular field of interest or level of attainment.

The book is divided into two main parts. Part A presents mathematical formulas together with other material, such as definitions, theorems, graphs, diagrams, etc., essential for proper understanding and application of the formulas. Part B presents the numerical tables. These tables include basic statistical distributions (normal, Student's t, chi-square, etc.), advanced functions (Bessel, Legendre, elliptic, etc.), and financial functions (compound and present value of an amount, and annuity).

McGraw-Hill wishes to thank the various authors and publishers—for example, the Literary Executor of the late Sir Ronald A. Fisher, F.R.S., Dr. Frank Yates, F.R.S., and Oliver and Boyd Ltd., Edinburgh, for Table III of their book *Statistical Tables for Biological, Agricultural and Medical Research*—who gave their permission to adapt data from their books for use in several tables in this handbook. Appropriate references to such sources are given below the corresponding tables.

Finally, I wish to thank the staff of the McGraw-Hill Schaum's Outline Series, especially Charles Wall, for their unfailing cooperation.

SEYMOUR LIPSCHUTZ
Temple University

Contents

Part A FORMULAS 1

Section I Elementary Constants, Products, Formulas 3

1. Greek Alphabet and Special Constants 3
2. Special Products and Factors 5
3. The Binomial Formula and Binomial Coefficients 7
4. Complex Numbers 10
5. Solutions of Algebraic Equations 13
6. Conversion Factors 15

Section II Geometry 16

7. Geometric Formulas 16
8. Formulas from Plane Analytic Geometry 22
9. Special Plane Curves 28
10. Formulas from Solid Analytic Geometry 34
11. Special Moments of Inertia 41

Section III Elementary Transcendental Functions 43

12. Trigonometric Functions 43
13. Exponential and Logarithmic Functions 53
14. Hyperbolic Functions 56

Section IV Calculus 62

15. Derivatives 62
16. Indefinite Integrals 67
17. Tables of Special Indefinite Integrals 71
18. Definite Integrals 108

Section V Differential Equations and Vector Analysis 116

19. Basic Differential Equations and Solutions 116
20. Formulas from Vector Analysis 119

Section VI Series 134

21. Series of Constants 134
22. Taylor Series 138
23. Bernoulli and Euler Numbers 142
24. Fourier Series 144

Section VII Special Functions and Polynomials **149**

25. The Gamma Function 149
26. The Beta Function 152
27. Bessel Functions 153
28. Legendre and Associated Legendre Functions 164
29. Hermite Polynomials 169
30. Laguerre and Associated Laguerre Polynomials 171
31. Chebyshev Polynomials 175
32. Hypergeometric Functions 178

Section VIII Laplace and Fourier Transforms **180**

33. Laplace Transforms 180
34. Fourier Transforms 193

Section IX Elliptic and Miscellaneous Special Functions **198**

35. Elliptic Functions 198
36. Miscellaneous and Riemann Zeta Functions 203

Section X Inequalities and Infinite Products **205**

37. Inequalities 205
38. Infinite Products 207

Section XI Probability and Statistics **208**

39. Descriptive Statistics 208
40. Probability 217
41. Random Variables 223

Section XII Numerical Methods **227**

42. Interpolation 227
43. Quadrature 231
44. Solution of Nonlinear Equations 233
45. Numerical Methods for Ordinary Differential Equations 235
46. Numerical Methods for Partial Differential Equations 237
47. Iteration Methods for Linear Systems 240

Part B TABLES **243**

Section I Logarithmic, Trigonometric, Exponential Functions **245**

1. Four Place Common Logarithms $\log_{10} N$ or $\log N$ 245
2. Sin x (x in degrees and minutes) 247
3. Cos x (x in degrees and minutes) 248
4. Tan x (x in degrees and minutes) 249

5. Conversion of Radians to Degrees, Minutes, and Seconds or Fractions of Degrees 250
6. Conversion of Degrees, Minutes, and Seconds to Radians 251
7. Natural or Napierian Logarithms $\log_e x$ or $\ln x$ 252
8. Exponential Functions e^x 254
9. Exponential Functions e^{-x} 255
10. Exponential, Sine, and Cosine Integrals 256

Section II **Factorial and Gamma Function, Binomial Coefficients** **257**

11. Factorial n 257
12. Gamma Function 258
13. Binomial coefficients 259

Section III **Bessel Functions** **261**

14. Bessel Functions $J_0(x)$ 261
15. Bessel Functions $J_1(x)$ 261
16. Bessel Functions $Y_0(x)$ 262
17. Bessel Functions $Y_1(x)$ 262
18. Bessel Functions $I_0(x)$ 263
19. Bessel Functions $I_1(x)$ 263
20. Bessel Functions $K_0(x)$ 264
21. Bessel Functions $K_1(x)$ 264
22. Bessel Functions $\mathrm{Ber}(x)$ 265
23. Bessel Functions $\mathrm{Bei}(x)$ 265
24. Bessel Functions $\mathrm{Ker}(x)$ 266
25. Bessel Functions $\mathrm{Kei}(x)$ 266
26. Values for Approximate Zeros of Bessel Functions 267

Section IV **Legendre Polynomials** **268**

27. Legendre Polynomials $P_n(x)$ 268
28. Legendre Polynomials $P_n(\cos\theta)$ 269

Section V **Elliptic Integrals** **270**

29. Complete Elliptic Integrals of First and Second Kinds 270
30. Incomplete Elliptic Integral of the First Kind 271
31. Incomplete Elliptic Integral of the Second Kind 271

Section VI **Financial Tables** **272**

32. Compound amount: $(1+r)^n$ 272
33. Present Value of an Amount: $(1+r)^{-n}$ 273
34. Amount of an Annuity: $\frac{(1+r)^n-1}{r}$ 274
35. Present Value of an Annuity: $\frac{1-(1+r)^{-n}}{r}$ 275

Section VII **Probability and Statistics** **276**

36. Areas Under the Standard Normal Curve 276
37. Ordinates of the Standard Normal curve 277
38. Percentile Values (t_p) for Student's t Distribution 278
39. Percentile Values (χ_p^2) for χ^2 (Chi-Square) Distribution 279
40. 95th Percentile Values for the F distribution 280
41. 99th Percentile Values for the F distribution 281
42. Random Numbers 282

Index of Special Symbols and Notations **283**

Index **285**

FORMULAS

Section I: Elementary Constants, Products, Formulas

1 GREEK ALPHABET and SPECIAL CONSTANTS

Greek Alphabet

Greek name	Greek letter	
	Lower case	Capital
Alpha	α	A
Beta	β	B
Gamma	γ	Γ
Delta	δ	Δ
Epsilon	ϵ	E
Zeta	ζ	Z
Eta	η	H
Theta	θ	Θ
Iota	ι	I
Kappa	κ	K
Lambda	λ	Λ
Mu	μ	M

Greek name	Greek letter	
	Lower case	Capital
Nu	ν	N
Xi	ξ	Ξ
Omicron	o	O
Pi	π	Π
Rho	ρ	P
Sigma	σ	Σ
Tau	τ	T
Upsilon	υ	Υ
Phi	ϕ	Φ
Chi	χ	X
Psi	ψ	Ψ
Omega	ω	Ω

Special Constants

1.1. $\pi = 3.14159\ 26535\ 89793\ \ldots$

1.2. $e = 2.71828\ 18284\ 59045\ \ldots = \lim_{n\to\infty}\left(1+\frac{1}{n}\right)^n$

= natural base of logarithms

1.3. $\gamma = 0.57721\ 56649\ 01532\ 86060\ 6512\ \ldots =$ *Euler's constant*

$$= \lim_{n\to\infty}\left(1+\frac{1}{2}+\frac{1}{3}+\cdots+\frac{1}{n}-\ln n\right)$$

1.4. $e^{\gamma} = 1.78107\ 24179\ 90197\ 9852\ \ldots$ [see 1.3]

1.5. $\sqrt{e} = 1.64872\ 12707\ 00128\ 1468\ \ldots$

1.6. $\sqrt{\pi} = \Gamma(\frac{1}{2}) = 1.77245\ 38509\ 05516\ 02729\ 8167\ \ldots$
where Γ is the *gamma function* [see 25.1].

1.7. $\Gamma(\frac{1}{3}) = 2.67893\ 85347\ 07748\ \ldots$

1.8. $\Gamma(\frac{1}{4}) = 3.62560\ 99082\ 21908\ \ldots$

1.9. $1 \text{ radian} = 180°/\pi = 57.29577\ 95130\ 8232\ \ldots°$

1.10. $1° = \pi/180 \text{ radians} = 0.01745\ 32925\ 19943\ 29576\ 92\ \ldots \text{ radians}$

2 SPECIAL PRODUCTS and FACTORS

2.1. $(x+y)^2 = x^2 + 2xy + y^2$

2.2. $(x-y)^2 = x^2 - 2xy + y^2$

2.3. $(x+y)^3 = x^3 + 3x^2y + 3xy^2 + y^3$

2.4. $(x-y)^3 = x^3 - 3x^2y + 3xy^2 - y^3$

2.5. $(x+y)^4 = x^4 + 4x^3y + 6x^2y^2 + 4xy^3 + y^4$

2.6. $(x-y)^4 = x^4 - 4x^3y + 6x^2y^2 - 4xy^3 + y^4$

2.7. $(x+y)^5 = x^5 + 5x^4y + 10x^3y^2 + 10x^2y^3 + 5xy^4 + y^5$

2.8. $(x-y)^5 = x^5 - 5x^4y + 10x^3y^2 - 10x^2y^3 + 5xy^4 - y^5$

2.9. $(x+y)^6 = x^6 + 6x^5y + 15x^4y^2 + 20x^3y^3 + 15x^2y^4 + 6xy^5 + y^6$

2.10. $(x-y)^6 = x^6 - 6x^5y + 15x^4y^2 - 20x^3y^3 + 15x^2y^4 - 6xy^5 + y^6$

The results 2.1 to 2.10 above are special cases of the *binomial formula* [see 3.3].

2.11. $x^2 - y^2 = (x-y)(x+y)$

2.12. $x^3 - y^3 = (x-y)(x^2 + xy + y^2)$

2.13. $x^3 + y^3 = (x+y)(x^2 - xy + y^2)$

2.14. $x^4 - y^4 = (x-y)(x+y)(x^2 + y^2)$

2.15. $x^5 - y^5 = (x-y)(x^4 + x^3y + x^2y^2 + xy^3 + y^4)$

2.16. $x^5 + y^5 = (x+y)(x^4 - x^3y + x^2y^2 - xy^3 + y^4)$

2.17. $x^6 - y^6 = (x-y)(x+y)(x^2 + xy + y^2)(x^2 - xy + y^2)$

2.18. $x^4 + x^2y^2 + y^4 = (x^2 + xy + y^2)(x^2 - xy + y^2)$

2.19. $x^4 + 4y^4 = (x^2 + 2xy + 2y^2)(x^2 - 2xy + 2y^2)$

Some generalizations of the above are given by the following results where n is a positive integer.

2.20. $$x^{2n+1} - y^{2n+1} = (x - y)(x^{2n} + x^{2n-1}y + x^{2n-2}y^2 + \cdots + y^{2n})$$
$$= (x - y)\left(x^2 - 2xy\cos\frac{2\pi}{2n+1} + y^2\right)\left(x^2 - 2xy\cos\frac{4\pi}{2n+1} + y^2\right)$$
$$\cdots\left(x^2 - 2xy\cos\frac{2n\pi}{2n+1} + y^2\right)$$

2.21. $$x^{2n+1} + y^{2n+1} = (x + y)(x^{2n} - x^{2n-1}y + x^{2n-2}y^2 - \cdots + y^{2n})$$
$$= (x + y)\left(x^2 + 2xy\cos\frac{2\pi}{2n+1} + y^2\right)\left(x^2 + 2xy\cos\frac{4\pi}{2n+1} + y^2\right)$$
$$\cdots\left(x^2 + 2xy\cos\frac{2n\pi}{2n+1} + y^2\right)$$

2.22. $$x^{2n} - y^{2n} = (x - y)(x + y)(x^{n-1} + x^{n-2}y + x^{n-3}y^2 + \cdots)(x^{n-1} - x^{n-2}y + x^{n-3}y^2 - \cdots)$$
$$= (x - y)(x + y)\left(x^2 - 2xy\cos\frac{\pi}{n} + y^2\right)\left(x^2 - 2xy\cos\frac{2\pi}{n} + y^2\right)$$
$$\cdots\left(x^2 - 2xy\cos\frac{(n-1)\pi}{n} + y^2\right)$$

2.23. $$x^{2n} + y^{2n} = \left(x^2 + 2xy\cos\frac{\pi}{2n} + y^2\right)\left(x^2 + 2xy\cos\frac{3\pi}{2n} + y^2\right)$$
$$\cdots\left(x^2 + 2xy\cos\frac{(2n-1)\pi}{2n} + y^2\right)$$

3 THE BINOMIAL FORMULA and BINOMIAL COEFFICIENTS

Factorial n

For $n = 1, 2, 3, \ldots$, *factorial n* or *n factorial* is denoted and defined by

3.1. $n! = n(n-1)\cdot\cdots\cdot 3\cdot 2\cdot 1$

Zero factorial is defined by

3.2. $0! = 1$

Alternately, n factorial can be defined recursively by

$$0! = 1 \quad \text{and} \quad n! = n\cdot(n-1)!$$

EXAMPLE: $4! = 4\cdot 3\cdot 2\cdot 1 = 24$,
$5! = 5\cdot 4\cdot 3\cdot 2\cdot 1 = 5\cdot 4! = 5(24) = 120$,
$6! = 6\cdot 5! = 6(120) = 720$

Binomial Formula for Positive Integral n

For $n = 1, 2, 3, \ldots$,

3.3. $(x+y)^n = x^n + nx^{n-1}y + \dfrac{n(n-1)}{2!}x^{n-2}y^2 + \dfrac{n(n-1)(n-2)}{3!}x^{n-3}y^3 + \cdots + y^n$

This is called the *binomial formula.* It can be extended to other values of n, and also to an infinite series [see 22.4].

EXAMPLE:

(a) $(a-2b)^4 = a^4 + 4a^3(-2b) + 6a^2(-2b)^2 + 4a(-2b)^3 + (-2b)^4 = a^4 - 8a^3b + 24a^2b^2 - 32ab^3 + 16b^4$

Here $x = a$ and $y = -2b$.

(b) See Fig. 3-1a.

Binomial Coefficients

Formula 3.3 can be rewritten in the form

3.4. $(x+y)^n = x^n + \binom{n}{1}x^{n-1}y + \binom{n}{2}x^{n-2}y^2 + \binom{n}{3}x^{n-3}y^3 + \cdots + \binom{n}{n}y^n$

where the coefficients, called *binomial coefficients,* are given by

3.5. $$\binom{n}{k} = \frac{n(n-1)(n-2)\cdots(n-k+1)}{k!} = \frac{n!}{k!(n-k)!} = \binom{n}{n-k}$$

EXAMPLE: $\binom{9}{4} = \frac{9\cdot 8\cdot 7\cdot 6}{1\cdot 2\cdot 3\cdot 4} = 126$, $\binom{12}{5} = \frac{12\cdot 11\cdot 10\cdot 9\cdot 8}{1\cdot 2\cdot 3\cdot 4\cdot 5} = 792$, $\binom{10}{7} = \binom{10}{3} = \frac{10\cdot 9\cdot 8}{1\cdot 2\cdot 3} = 120$

Note that $\binom{n}{r}$ has exactly r factors in both the numerator and the denominator.

The binomial coefficients may be arranged in a triangular array of numbers, called Pascal's triangle, as shown in Fig. 3.1b. The triangle has the following two properties:

(1) The first and last number in each row is 1.

(2) Every other number in the array can be obtained by adding the two numbers appearing directly above it. For example

$$10 = 4 + 6, \qquad 15 = 5 + 10, \qquad 20 = 10 + 10$$

Property (2) may be stated as follows:

3.6. $$\binom{n}{k} + \binom{n}{k+1} = \binom{n+1}{k+1}$$

$$(a+b)^0 = 1$$
$$(a+b)^1 = a + b$$
$$(a+b)^2 = a^2 + 2ab + b^2$$
$$(a+b)^3 = a^3 + 3a^2b + 3ab^2 + b^3$$
$$(a+b)^4 = a^4 + 4a^3b + 6a^2b^2 + 4ab^3 + b^4$$
$$(a+b)^5 = a^5 + 5a^4b + 10a^3b^2 + 10a^2b^3 + 5ab^4 + b^5$$
$$(a+b)^6 = a^6 + 6a^5b + 15a^4b^2 + 20a^3b^3 + 15a^2b^4 + 6ab^5 + b^6$$
....

(*a*)

1
1 1
1 2 1
1 3 3 1
1 4 6 4 1
1 5 10 10 5 1
1 6 15 20 15 6 1
....

(*b*)

Fig. 3-1

Properties of Binomial Coefficients

The following lists additional properties of the binomial coefficients:

3.7. $$\binom{n}{0} + \binom{n}{1} + \binom{n}{2} + \cdots + \binom{n}{n} = 2^n$$

3.8. $$\binom{n}{0} - \binom{n}{1} + \binom{n}{2} - \cdots (-1)^n \binom{n}{n} = 0$$

3.9. $$\binom{n}{n} + \binom{n+1}{n} + \binom{n+2}{n} + \cdots + \binom{n+m}{n} = \binom{n+m+1}{n+1}$$

3.10. $\binom{n}{0}+\binom{n}{2}+\binom{n}{4}+\cdots=2^{n-1}$

3.11. $\binom{n}{1}+\binom{n}{3}+\binom{n}{5}+\cdots=2^{n-1}$

3.12. $\binom{n}{0}^2+\binom{n}{1}^2+\binom{n}{2}^2+\cdots+\binom{n}{n}^2=\binom{2n}{n}$

3.13. $\binom{m}{0}\binom{n}{p}+\binom{m}{1}\binom{n}{p-1}+\cdots+\binom{m}{p}\binom{n}{0}=\binom{m+n}{p}$

3.14. $(1)\binom{n}{1}+(2)\binom{n}{2}+(3)\binom{n}{3}+\cdots+(n)\binom{n}{n}=n2^{n-1}$

3.15. $(1)\binom{n}{1}-(2)\binom{n}{2}+(3)\binom{n}{3}-\cdots(-1)^{n+1}(n)\binom{n}{n}=0$

Multinomial Formula

Let $n_1, n_2, \ldots, n_r$ be nonnegative integers such that $n_1+n_2+\cdots+n_r=n$. Then the following expression, called a *multinomial coefficient,* is defined as follows:

3.16. $\binom{n}{n_1, n_2, \ldots, n_r}=\dfrac{n!}{n_1!\,n_2!\cdots n_r!}$

EXAMPLE: $\binom{7}{2,\ 3,\ 2}=\dfrac{7!}{2!3!2!}=210,\quad \binom{8}{4,\ 2,\ 2,\ 0}=\dfrac{8!}{4!2!2!0!}=420$

The name multinomial coefficient comes from the following formula:

3.17. $(x_1+x_2+\cdots+x_p)^n=\sum\binom{n}{n_1, n_2, \ldots, n_r}x_1^{n_1}x_2^{n_2}\cdots x_r^{n_r}$

where the sum, denoted by Σ, is taken over all possible multinomial coefficients.

4 COMPLEX NUMBERS

Definitions Involving Complex Numbers

A complex number z is generally written in the form

$$z = a + bi$$

where a and b are real numbers and i, called the *imaginary unit,* has the property that $i^2 = -1$. The real numbers a and b are called the *real* and *imaginary parts* of $z = a + bi$, respectively.

The *complex conjugate* of z is denoted by $\bar{z}$; it is defined by

$$\overline{a + bi} = a - bi$$

Thus, $a + bi$ and $a - bi$ are conjugates of each other.

Equality of Complex Numbers

4.1. $a + bi = c + di \quad$ if and only if $\quad a = c$ and $b = d$

Arithmetic of Complex Numbers

Formulas for the addition, subtraction, multiplication, and division of complex numbers follow:

4.2. $(a + bi) + (c + di) = (a + c) + (b + d)i$

4.3. $(a + bi) - (c + di) = (a - c) + (b - d)i$

4.4. $(a + bi)(c + di) = (ac - bd) + (ad + bc)i$

4.5. $\dfrac{a + bi}{c + di} = \dfrac{a + bi}{c + di} \cdot \dfrac{c - di}{c - di} = \dfrac{ac + bd}{c^2 + d^2} + \left(\dfrac{bc - ad}{c^2 + d^2}\right)i$

Note that the above operations are obtained by using the ordinary rules of algebra and replacing i^2 by -1 wherever it occurs.

EXAMPLE: Suppose $z = 2 + 3i$ and $w = 5 - 2i$. Then

$$z + w = (2 + 3i) + (5 - 2i) = 2 + 5 + 3i - 2i = 7 + i$$

$$zw = (2 + 3i)(5 - 2i) = 10 + 15i - 4i - 6i^2 = 16 + 11i$$

$$\bar{z} = \overline{2 + 3i} = 2 - 3i \text{ and } \bar{w} = \overline{5 - 2i} = 5 + 2i$$

$$\frac{w}{z} = \frac{5 - 2i}{2 + 3i} = \frac{(5 - 2i)(2 - 3i)}{(2 + 3i)(2 - 3i)} = \frac{4 - 19i}{13} = \frac{4}{13} - \frac{19}{13}i$$

Complex Plane

Real numbers can be represented by the points on a line, called the *real line*, and, similarly, complex numbers can be represented by points in the plane, called the *Argand diagram* or *Gaussian plane* or, simply, the *complex plane*. Specifically, we let the point (a, b) in the plane represent the complex number $z = a + bi$. For example, the point P in Fig. 4-1 represents the complex number $z = -3 + 4i$. The complex number can also be interpreted as a vector from the origin O to the point P.

The *absolute value* of a complex number $z = a + bi$, written $|z|$, is defined as follows:

4.6. $|z| = \sqrt{a^2 + b^2} = \sqrt{z\bar{z}}$

We note $|z|$ is the distance from the origin O to the point z in the complex plane.

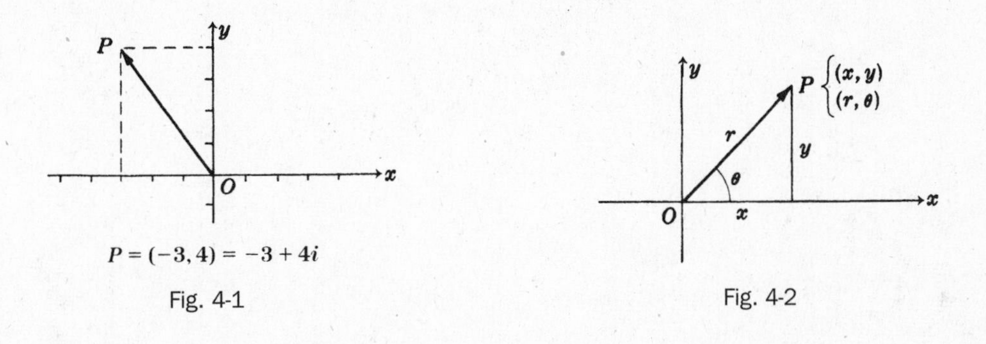

Fig. 4-1

Fig. 4-2

Polar Form of Complex Numbers

The point P in Fig. 4-2 with coordinates (x, y) represents the complex number $z = x + iy$. The point P can also be represented by *polar coordinates* (r, θ). Since $x = r\cos\theta$ and $y = r\sin\theta$, we have

4.7. $z = x + iy = r(\cos\theta + i\sin\theta)$

called the *polar form* of the complex number. We often call $r = |z| = \sqrt{x^2 + y^2}$ the *modulus* and θ the *amplitude* of $z = x + iy$.

Multiplication and Division of Complex Numbers in Polar Form

4.8. $[r_1(\cos\theta_1 + i\sin\theta_1)][r_2(\cos\theta_2 + i\sin\theta_2)] = r_1 r_2[\cos(\theta_1 + \theta_2) + i\sin(\theta_1 + \theta_2)]$

4.9. $\dfrac{r_1(\cos\theta_1 + i\sin\theta_1)}{r_2(\cos\theta_2 + i\sin\theta_2)} = \dfrac{r_1}{r_2}[\cos(\theta_1 - \theta_2) + i\sin(\theta_1 - \theta_2)]$

De Moivre's Theorem

For any real number p, De Moivre's theorem states that

4.10. $[r(\cos\theta + i\sin\theta)]^p = r^p(\cos p\theta + i\sin p\theta)$

Roots of Complex Numbers

Let $p = 1/n$ where n is any positive integer. Then 4.10 can be written

4.11. $$[r(\cos\theta + i\sin\theta)]^{1/n} = r^{1/n}\left(\cos\frac{\theta + 2k\pi}{n} + i\sin\frac{\theta + 2k\pi}{n}\right)$$

where k is any integer. From this formula, all the nth roots of a complex number can be obtained by putting $k = 0, 1, 2, \ldots, n - 1$.

5 SOLUTIONS of ALGEBRAIC EQUATIONS

Quadratic Equation: $ax^2 + bx + c = 0$

5.1. Solutions:

$$x = \frac{-b \pm \sqrt{b^2 - 4ac}}{2a}$$

If a, b, c are real and if $D = b^2 - 4ac$ is the *discriminant*, then the roots are

(i) real and unequal if $D > 0$
(ii) real and equal if $D = 0$
(iii) complex conjugate if $D < 0$

5.2. If x_1, x_2 are the roots, then $x_1 + x_2 = -b/a$ and $x_1 x_2 = c/a$.

Cubic Equation: $x^3 + a_1 x^2 + a_2 x + a_3 = 0$

Let

$$Q = \frac{3a_2 - a_1^2}{9}, \quad R = \frac{9a_1 a_2 - 27a_3 - 2a_1^3}{54},$$

$$S = \sqrt[3]{R + \sqrt{Q^3 + R^2}}, \quad T = \sqrt[3]{R - \sqrt{Q^3 + R^2}}$$

where $ST = -Q$.

5.3. Solutions:

$$\begin{cases} x_1 = S + T - \frac{1}{3}a_1 \\ x_2 = -\frac{1}{2}(S+T) - \frac{1}{3}a_1 + \frac{1}{2}i\sqrt{3}(S-T) \\ x_3 = -\frac{1}{2}(S+T) - \frac{1}{3}a_1 - \frac{1}{2}i\sqrt{3}(S-T) \end{cases}$$

If a_1, a_2, a_3, are real and if $D = Q^3 + R^2$ is the *discriminant*, then

(i) one root is real and two are complex conjugate if $D > 0$
(ii) all roots are real and at least two are equal if $D = 0$
(iii) all roots are real and unequal if $D < 0$.

If $D < 0$, computation is simplified by use of trigonometry.

5.4. Solutions:

$$\text{if } D < 0: \begin{cases} x_1 = 2\sqrt{-Q}\cos(\frac{1}{3}\theta) - \frac{1}{3}a_1 \\ x_2 = 2\sqrt{-Q}\cos(\frac{1}{3}\theta + 120°) - \frac{1}{3}a_1 \\ x_3 = 2\sqrt{-Q}\cos(\frac{1}{3}\theta + 240°) - \frac{1}{3}a \end{cases}$$

where $\cos\theta = R/\sqrt{-Q^3}$

5.5. $x_1 + x_2 + x_3 = -a_1, \quad x_1x_2 + x_2x_3 + x_3x_1 = a_2, \quad x_1x_2x_3 = -a_3$

where x_1, x_2, x_3 are the three roots.

Quartic Equation: $x^4 + a_1x^3 + a_2x^2 + a_3x + a_4 = 0$

Let y_1 be a real root of the following cubic equation:

5.6. $y^3 - a_2y^2 + (a_1a_3 - 4a_4)y + (4a_2a_4 - a_3^2 - a_1^2a_4) = 0$

The four roots of the quartic equation are the four roots of the following equation:

5.7. $z^2 + \frac{1}{2}\left(a_1 \pm \sqrt{a_1^2 - 4a_2 + 4y_1}\right)z + \frac{1}{2}\left(y_1 \mp \sqrt{y_1^2 - 4a_4}\right) = 0$

Suppose that all roots of 5.6 are real; then computation is simplified by using the particular real root that produces all real coefficients in the quadratic equation 5.7.

5.8.
$$\begin{cases} x_1 + x_2 + x_3 + x_4 = -a_1 \\ x_1x_2 + x_2x_3 + x_3x_4 + x_4x_1 + x_1x_3 + x_2x_4 = a_2 \\ x_1x_2x_3 + x_2x_3x_4 + x_1x_2x_4 + x_1x_3x_4 = -a_3 \\ x_1x_2x_3x_4 = x_4 \end{cases}$$

where x_1, x_2, x_3, x_4 are the four roots.

CONVERSION FACTORS

Length

1 kilometer (km)	= 1000 meters (m)	1 inch (in)	= 2.540 cm
1 meter (m)	= 100 centimeters (cm)	1 foot (ft)	= 30.48 cm
1 centimeter (cm)	= 10^{-2} m	1 mile (mi)	= 1.609 km
1 millimeter (mm)	= 10^{-3} m	1 millimeter	= 10^{-3} in
1 micron (μ)	= 10^{-6} m	1 centimeter	= 0.3937 in
1 millimicron (mμ)	= 10^{-9} m	1 meter	= 39.37 in
1 angstrom (Å)	= 10^{-10} m	1 kilometer	= 0.6214 mi

Area

1 square meter (m^2)	= 10.76 ft^2	1 square mile (mi^2)	= 640 acres
1 square foot (ft^2)	= 929 cm^2	1 acre	= 43,560 ft^2

Volume 1 liter (l) = 1000 cm^3 = 1.057 quart (qt) = 61.02 in^3 = 0.03532 ft^3
1 cubic meter (m^3) = 1000 l = 35.32 ft^3
1 cubic foot (ft^3) = 7.481 U.S. gal = 0.02832 m^3 = 28.32 l
1 U.S. gallon (gal) = 231 in^3 = 3.785 l; 1 British gallon = 1.201 U.S. gallon = 277.4 in^3

Mass 1 kilogram (kg) = 2.2046 pounds (lb) = 0.06852 slug; 1 lb = 453.6 gm = 0.03108 slug
1 slug = 32.174 lb = 14.59 kg

Speed 1 km/hr = 0.2778 m/sec = 0.6214 mi/hr = 0.9113 ft/sec
1 mi/hr = 1.467 ft/sec = 1.609 km/hr = 0.4470 m/sec

Density 1 gm/cm^3 = 10^3 kg/m^3 = 62.43 lb/ft^3 = 1.940 slug/ft^3
1 lb/ft^3 = 0.01602 gm/cm^3; 1 slug/ft^3 = 0.5154 gm/cm^3

Force 1 newton (nt) = 10^5 dynes = 0.1020 kgwt = 0.2248 lbwt
1 pound weight (lbwt) = 4.448 nt = 0.4536 kgwt = 32.17 poundals
1 kilogram weight (kgwt) = 2.205 lbwt = 9.807 nt
1 U.S. short ton = 2000 lbwt; 1 long ton = 2240 lbwt; 1 metric ton = 2205 lbwt

Energy 1 joule = 1 nt m = 10^7 ergs = 0.7376 ft lbwt = 0.2389 cal = 9.481×10^{-4} Btu
1 ft lbwt = 1.356 joules = 0.3239 cal = 1.285×10^{-3} Btu
1 calorie (cal) = 4.186 joules = 3.087 ft lbwt = 3.968×10^{-3} Btu
1 Btu (British thermal unit) = 778 ft lbwt = 1055 joules = 0.293 watt hr
1 kilowatt hour (kw hr) = 3.60×10^6 joules = 860.0 kcal = 3413 Btu
1 electron volt (ev) = 1.602×10^{-19} joule

Power 1 watt = 1 joule/sec = 10^7 ergs/sec = 0.2389 cal/sec
1 horsepower (hp) = 550 ft lbwt/sec = 33,000 ft lbwt/min = 745.7 watts
1 kilowatt (kw) = 1.341 hp = 737.6 ft lbwt/sec = 0.9483 Btu/sec

Pressure 1 nt/m^2 = 10 dynes/cm^2 = 9.869×10^{-6} atmosphere = 2.089×10^{-2} lbwt/ft^2
1 lbwt/in^2 = 6895 nt/m^2 = 5.171 cm mercury = 27.68 in water
1 atm = 1.013×10^5 nt/m^2 = 1.013×10^6 dynes/cm^2 = 14.70 lbwt/in^2
= 76 cm mercury = 406.8 in water

Section II: Geometry

7 GEOMETRIC FORMULAS

Rectangle of Length b and Width a

7.1. Area $= ab$

7.2. Perimeter $= 2a + 2b$

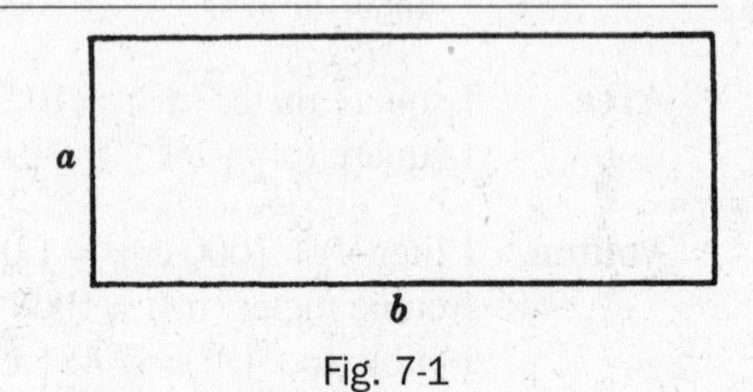

Fig. 7-1

Parallelogram of Altitude h and Base b

7.3. Area $= bh = ab\sin\theta$

7.4. Perimeter $= 2a + 2b$

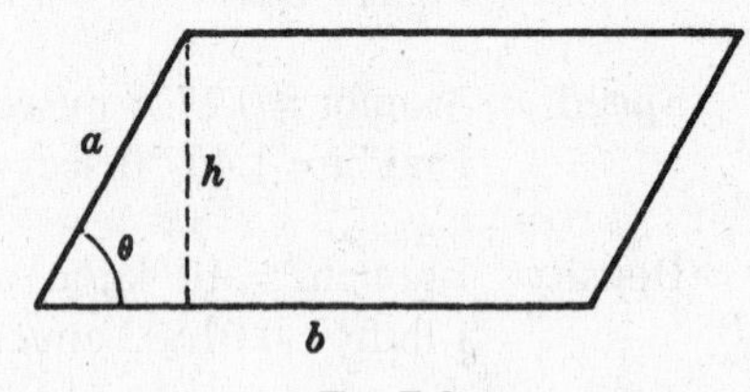

Fig. 7-2

Triangle of Altitude h and Base b

7.5. Area $= \frac{1}{2}bh = \frac{1}{2}ab\sin\theta$

$$= \sqrt{s(s-a)(s-b)(s-c)}$$

where $s = \frac{1}{2}(a+b+c) =$ semiperimeter

Fig. 7-3

7.6. Perimeter $= a + b + c$

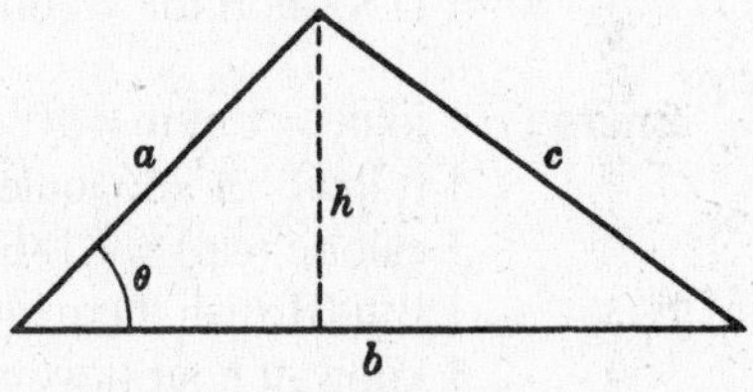

Trapezoid of Altitude h and Parallel Sides a and b

7.7. Area $= \frac{1}{2}h(a+b)$

7.8. Perimeter $= a + b + h\left(\frac{1}{\sin\theta} + \frac{1}{\sin\phi}\right)$

$$= a + b + h(\csc\theta + \csc\phi)$$

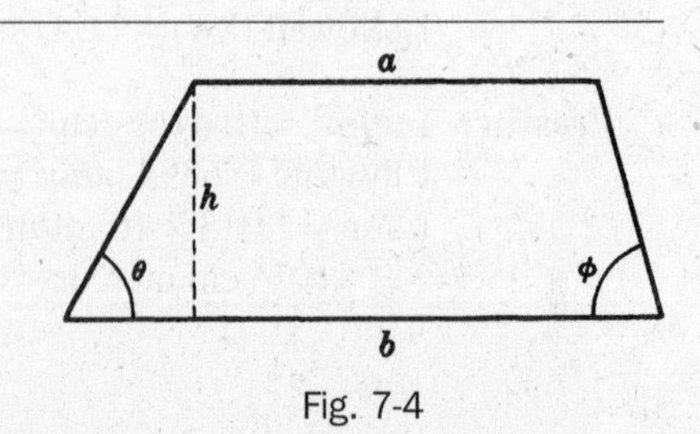

Fig. 7-4

Regular Polygon of n Sides Each of Length b

7.9. Area $= \frac{1}{4}nb^2 \cot\frac{\pi}{n} = \frac{1}{4}nb^2 \frac{\cos(\pi/n)}{\sin(\pi/n)}$

7.10. Perimeter $= nb$

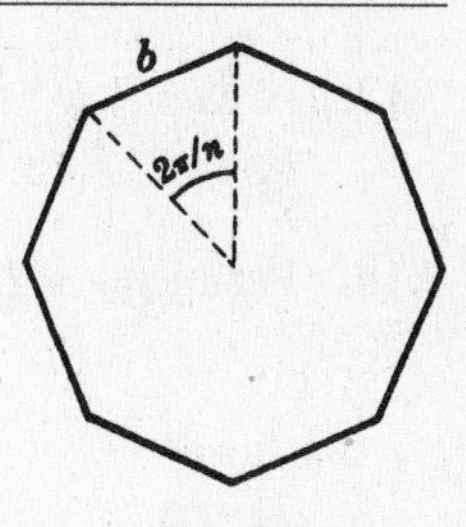

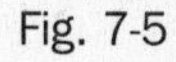

Fig. 7-5

Circle of Radius r

7.11. Area $= \pi r^2$

7.12. Perimeter $= 2\pi r$

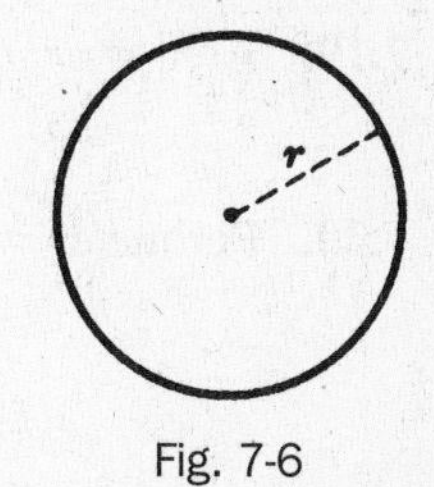

Fig. 7-6

Sector of Circle of Radius r

7.13. Area $= \frac{1}{2}r^2\theta$ [θ in radians]

7.14. Arc length $s = r\theta$

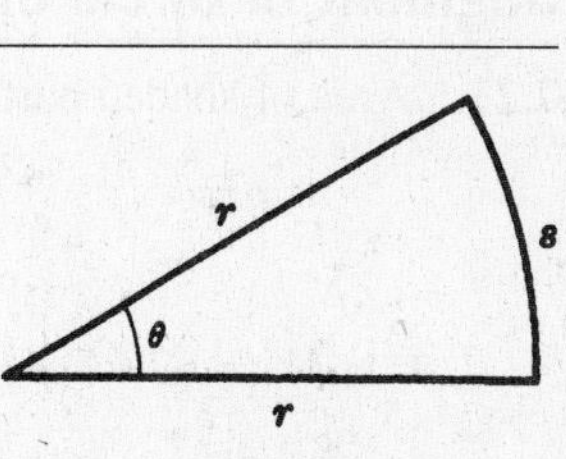

Fig. 7-7

Radius of Circle Inscribed in a Triangle of Sides a, b, c

7.15. $r = \dfrac{\sqrt{s(s-a)(s-b)(s-c)}}{s}$

where $s = \frac{1}{2}(a+b+c) =$ semiperimeter.

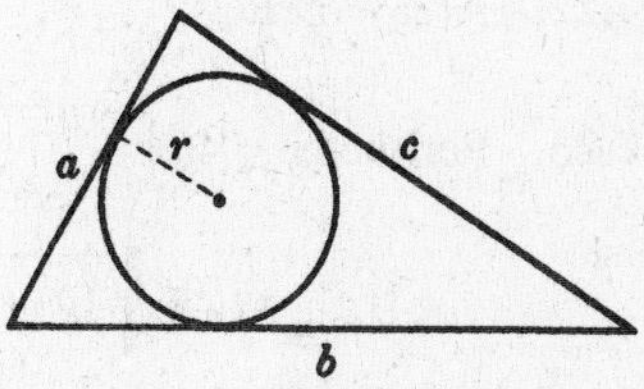

Fig. 7-8

Radius of Circle Circumscribing a Triangle of Sides a, b, c

7.16. $R = \dfrac{abc}{4\sqrt{s(s-a)(s-b)(s-c)}}$

where $s = \frac{1}{2}(a+b+c) =$ semiperimeter.

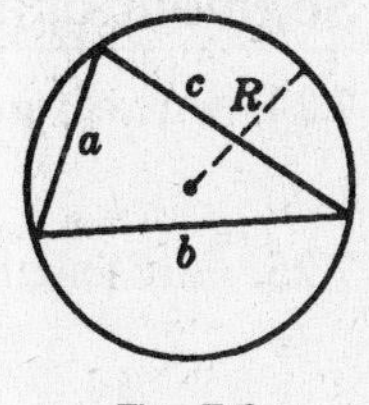

Fig. 7-9

Regular Polygon of n Sides Inscribed in Circle of Radius r

7.17. Area $= \frac{1}{2}nr^2 \sin\dfrac{2\pi}{n} = \frac{1}{2}nr^2 \sin\dfrac{360^\circ}{n}$

7.18. Perimeter $= 2nr \sin\dfrac{\pi}{n} = 2nr \sin\dfrac{180^\circ}{n}$

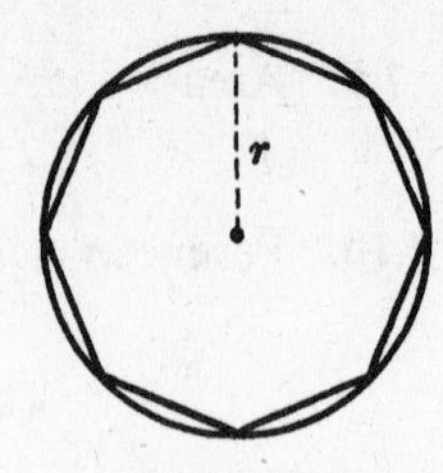

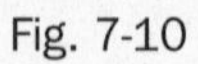

Fig. 7-10

Regular Polygon of n Sides Circumscribing a Circle of Radius r

7.19. Area $= nr^2 \tan\dfrac{\pi}{n} = nr^2 \tan\dfrac{180^\circ}{n}$

7.20. Perimeter $= 2nr \tan\dfrac{\pi}{n} = 2nr \tan\dfrac{180^\circ}{n}$

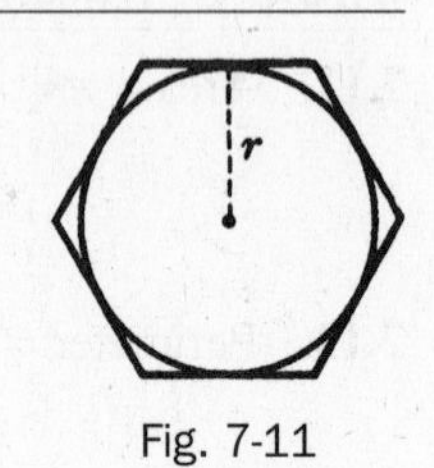

Fig. 7-11

Segment of Circle of Radius r

7.21. Area of shaded part $= \frac{1}{2}r^2(\theta - \sin\theta)$

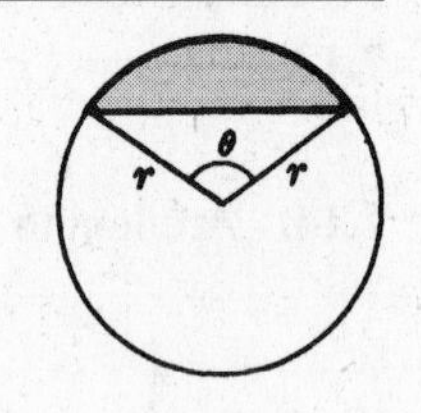

Fig. 7-12

Ellipse of Semi-major Axis a and Semi-minor Axis b

7.22. Area $= \pi ab$

7.23. Perimeter $= 4a\int_0^{\pi/2} \sqrt{1 - k^2 \sin^2\theta}\, d\theta$

$= 2\pi\sqrt{\frac{1}{2}(a^2 + b^2)}$ [approximately]

where $k = \sqrt{a^2 - b^2}/a$. See Table 29 for numerical values.

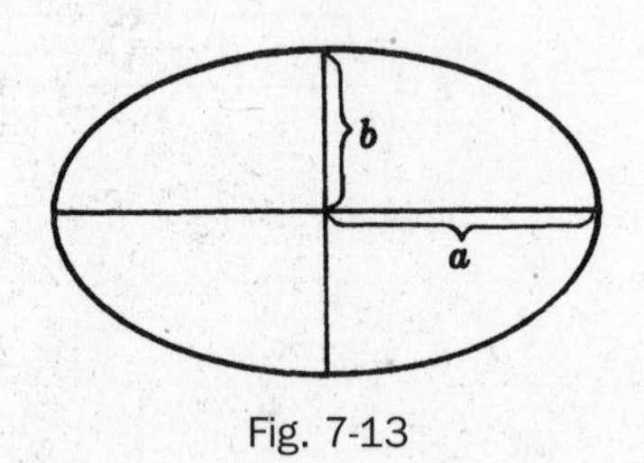

Fig. 7-13

Segment of a Parabola

7.24. Area $= \frac{2}{3}ab$

7.25. Arc length $ABC = \frac{1}{2}\sqrt{b^2 + 16a^2} + \dfrac{b^2}{8a} \ln\left(\dfrac{4a + \sqrt{b^2 + 16a^2}}{b}\right)$

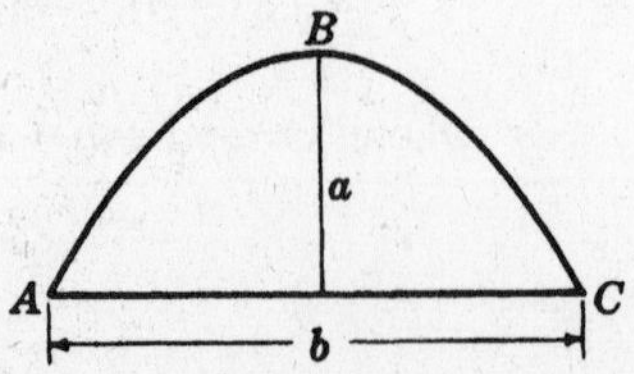

Fig. 7-14

Rectangular Parallelepiped of Length a, Height b, Width c

7.26. Volume $= abc$

7.27. Surface area $= 2(ab + ac + bc)$

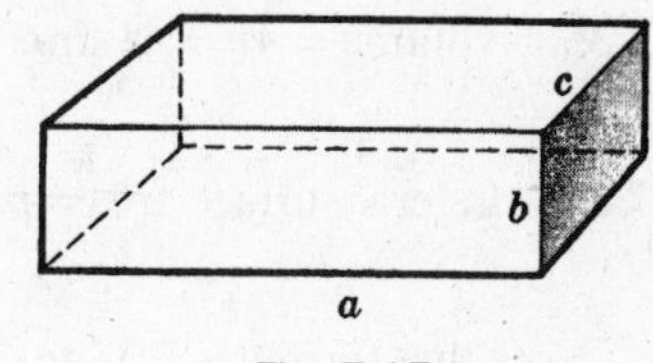

Fig. 7-15

Parallelepiped of Cross-sectional Area A and Height h

7.28. Volume $= Ah = abc \sin\theta$

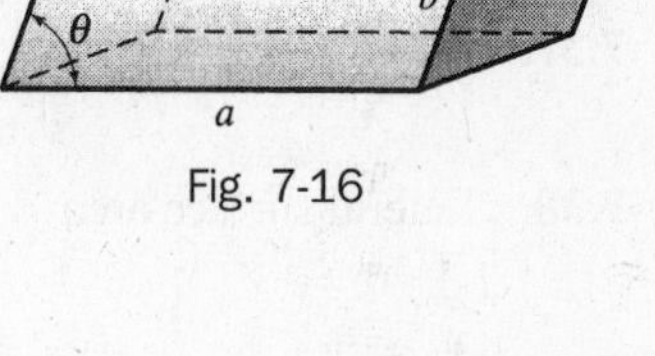

Fig. 7-16

Sphere of Radius r

7.29. Volume $= \frac{4}{3}\pi r^3$

7.30. Surface area $= 4\pi r^2$

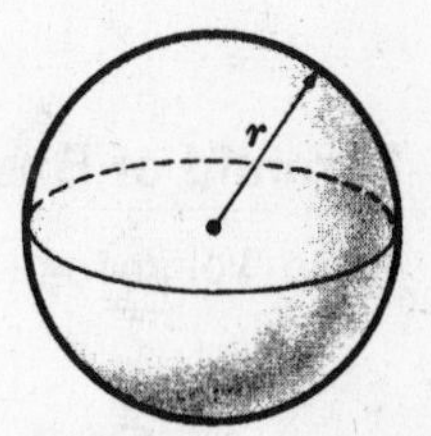

Fig. 7-17

Right Circular Cylinder of Radius r and Height h

7.31. Volume $= \pi r^2 h$

7.32. Lateral surface area $= 2\pi rh$

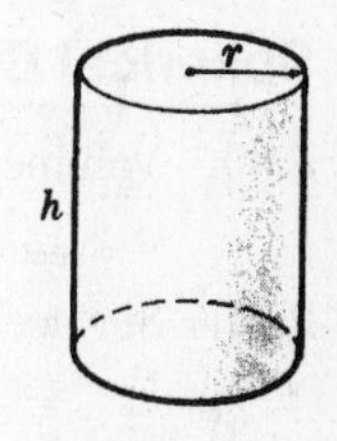

Fig. 7-18

Circular Cylinder of Radius r and Slant Height l

7.33. Volume $= \pi r^2 h = \pi r^2 l \sin\theta$

7.34. Lateral surface area $= 2\pi rl = \frac{2\pi rh}{\sin\theta} = 2\pi rh \csc\theta$

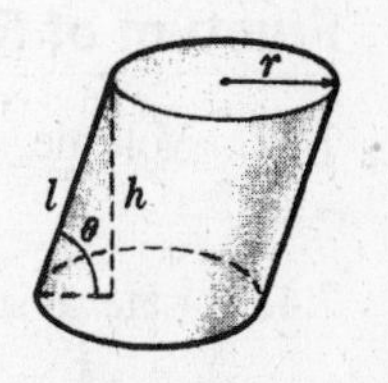

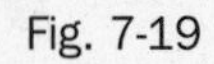

Fig. 7-19

Cylinder of Cross-sectional Area A and Slant Height l

7.35. Volume $= Ah = Al \sin\theta$

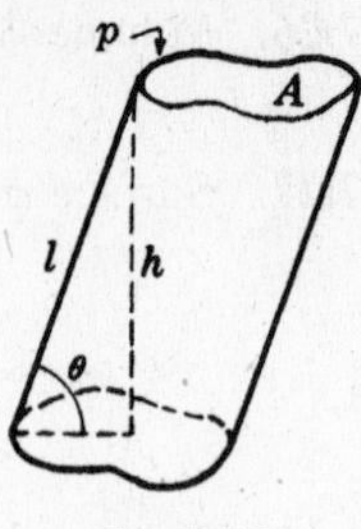

Fig. 7-20

7.36. Lateral surface area $= ph = pl \sin\theta$

Note that formulas 7.31 to 7.34 are special cases of formulas 7.35 and 7.36.

Right Circular Cone of Radius r and Height h

7.37. Volume $= \frac{1}{3}\pi r^2 h$

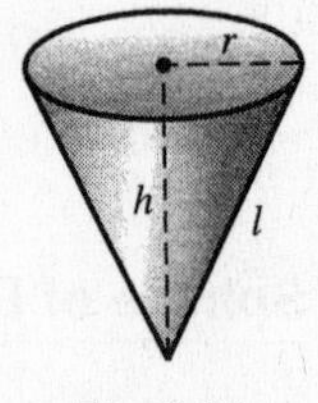

Fig. 7-21

7.38. Lateral surface area $= \pi r\sqrt{r^2 + h^2} = \pi r l$

Pyramid of Base Area A and Height h

7.39. Volume $= \frac{1}{3}Ah$

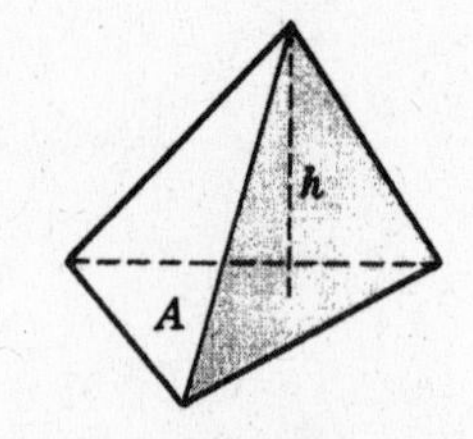

Fig. 7-22

Spherical Cap of Radius r and Height h

7.40. Volume (shaded in figure) $= \frac{1}{3}\pi h^2(3r - h)$

7.41. Surface area $= 2\pi rh$

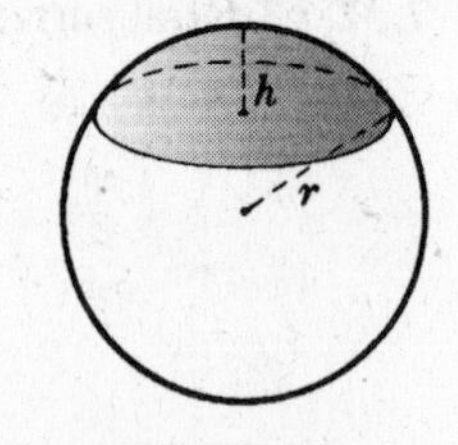

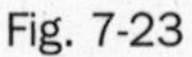

Fig. 7-23

Frustum of Right Circular Cone of Radii a, b and Height h

7.42. Volume $= \frac{1}{3}\pi h(a^2 + ab + b^2)$

7.43. Lateral surface area $= \pi(a+b)\sqrt{h^2 + (b-a)^2}$

$$= \pi(a+b)l$$

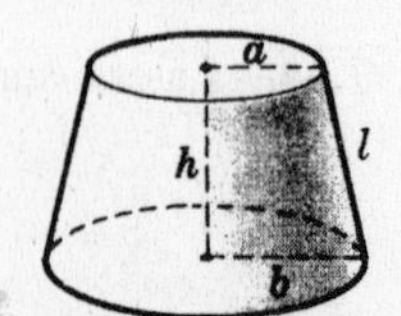

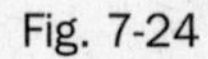

Fig. 7-24

Spherical Triangle of Angles A, B, C on Sphere of Radius r

7.44. Area of triangle $ABC = (A + B + C - \pi)r^2$

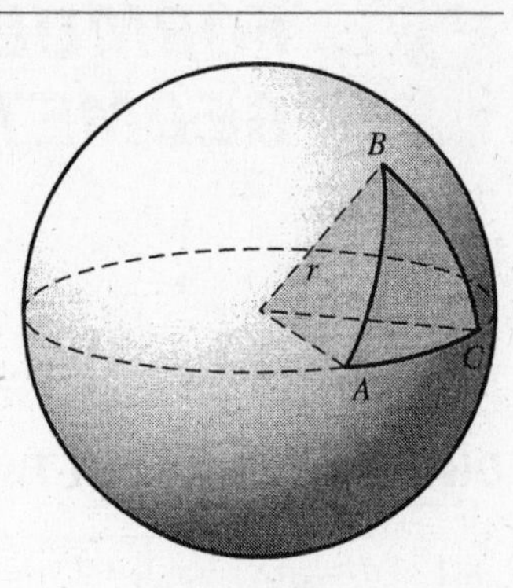

Fig. 7-25

Torus of Inner Radius a and Outer Radius b

7.45. Volume $= \frac{1}{4}\pi^2(a + b)(b - a)^2$

7.46. Surface area $= \pi^2(b^2 - a^2)$

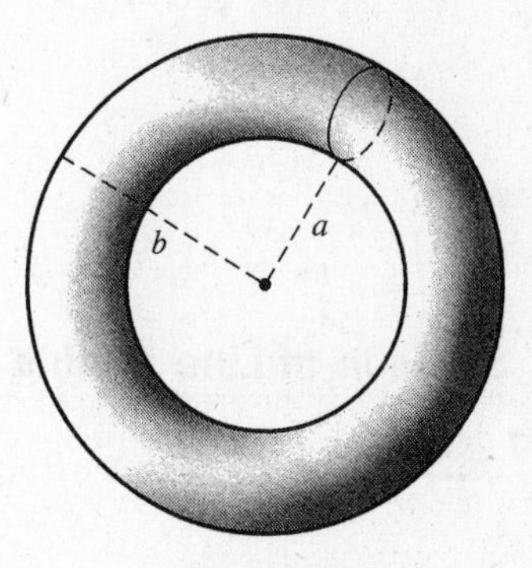

Fig. 7-26

Ellipsoid of Semi-axes a, b, c

7.47. Volume $= \frac{4}{3}\pi abc$

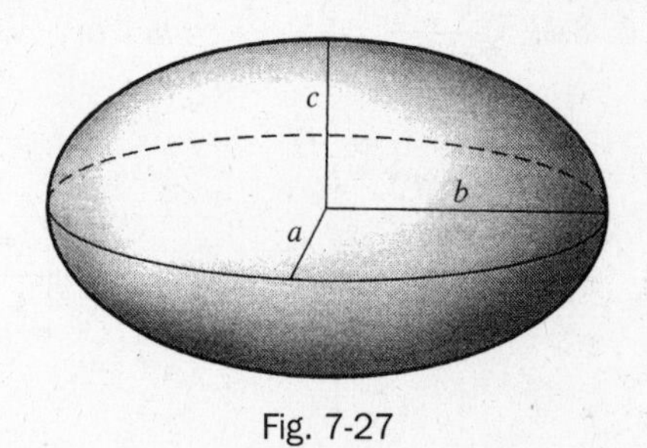

Fig. 7-27

Paraboloid of Revolution

7.48. Volume $= \frac{1}{2}\pi b^2 a$

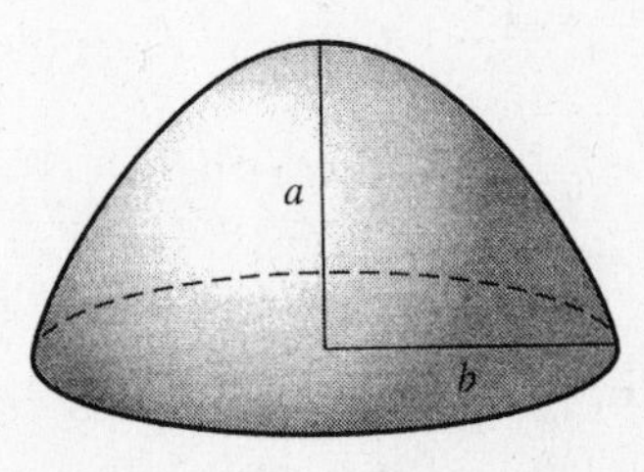

Fig. 7-28

8 FORMULAS from PLANE ANALYTIC GEOMETRY

Distance d Between Two Points $P_1(x_1, y_1)$ and $P_2(x_2, y_2)$

8.1. $d = \sqrt{(x_2 - x_1)^2 + (y_2 - y_1)^2}$

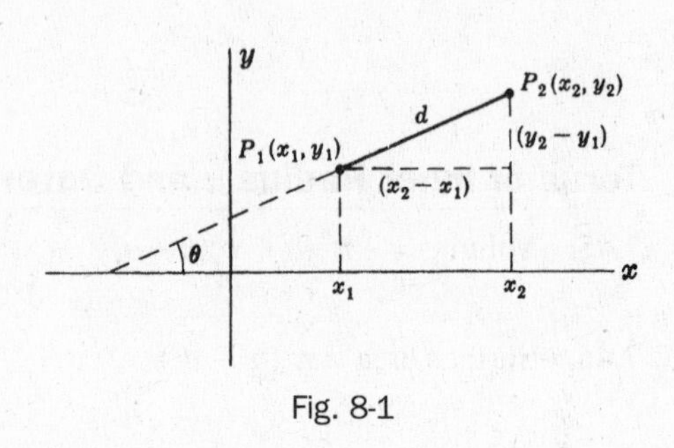

Fig. 8-1

Slope m of Line Joining Two Points $P_1(x_1, y_1)$ and $P_2(x_2, y_2)$

8.2. $m = \dfrac{y_2 - y_1}{x_2 - x_1} = \tan\theta$

Equation of Line Joining Two Points $P_1(x_1, y_1)$ and $P_2(x_2, y_2)$

8.3. $\dfrac{y - y_1}{x - x_1} = \dfrac{y_2 - y_1}{x_2 - x_1} = m$ or $y - y_1 = m(x - x_1)$

8.4. $y = mx + b$

where $b = y_1 - mx_1 = \dfrac{x_2 y_1 - x_1 y_2}{x_2 - x_1}$ is the *intercept* on the y axis, i.e., the y *intercept*.

Equation of Line in Terms of x Intercept $a \neq 0$ and y Intercept $b \neq 0$

8.5. $\dfrac{x}{a} + \dfrac{y}{b} = 1$

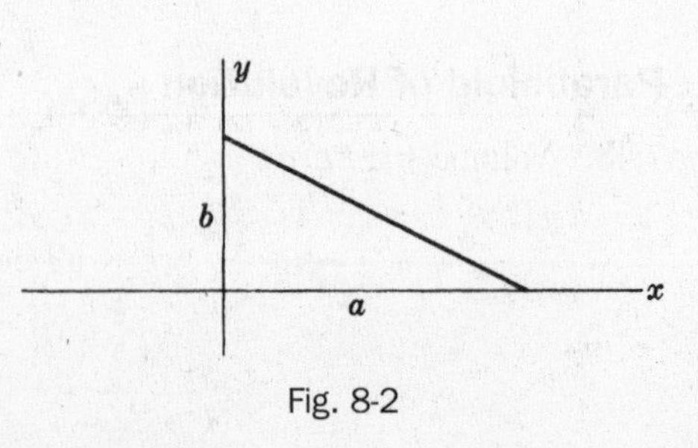

Fig. 8-2

Normal Form for Equation of Line

8.6. $x \cos \alpha + y \sin \alpha = p$

where p = perpendicular distance from origin O to line
and α = angle of inclination of perpendicular with positive x axis.

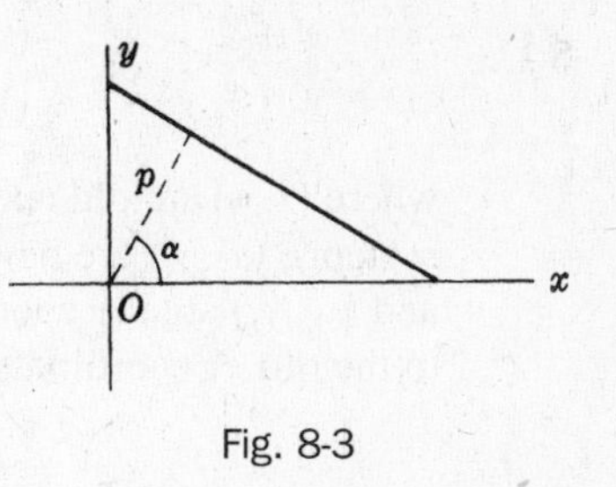

Fig. 8-3

General Equation of Line

8.7. $Ax + By + C = 0$

Distance from Point (x_1, y_1) to Line $Ax + By + C = 0$

8.8. $$\frac{Ax_1 + By_1 + C}{\pm\sqrt{A^2 + B^2}}$$

where the sign is chosen so that the distance is nonnegative.

Angle ψ Between Two Lines Having Slopes m_1 and m_2

8.9. $$\tan \psi = \frac{m_2 - m_1}{1 + m_1 m_2}$$

Lines are parallel or coincident if and only if $m_1 = m_2$.
Lines are perpendicular if and only if $m_2 = -1/m_1$.

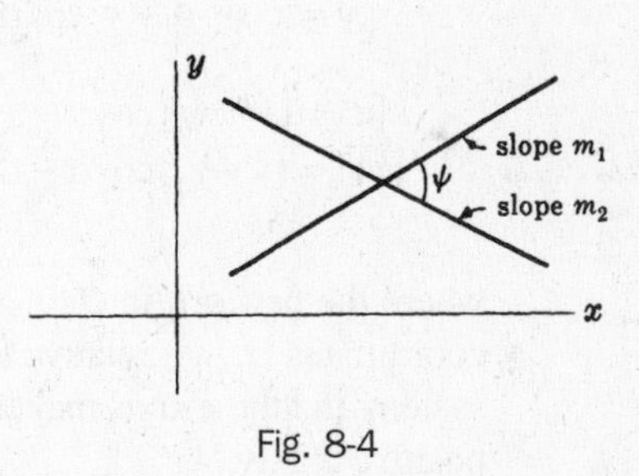

Fig. 8-4

Area of Triangle with Vertices at (x_1, y_1), (x_2, y_2), (x_3, y_3)

8.10. $$\text{Area} = \pm\frac{1}{2}\begin{vmatrix} x_1 & y_1 & 1 \\ x_2 & y_2 & 1 \\ x_3 & y_3 & 1 \end{vmatrix}$$

$$= \pm\frac{1}{2}(x_1 y_2 + y_1 x_3 + y_3 x_2 - y_2 x_3 - y_1 x_2 - x_1 y_3)$$

where the sign is chosen so that the area is nonnegative.
If the area is zero, the points all lie on a line.

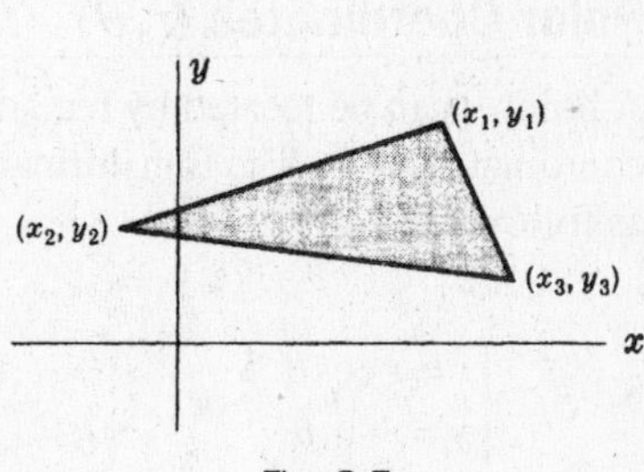

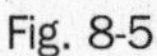
Fig. 8-5

Transformation of Coordinates Involving Pure Translation

8.11. $\begin{cases} x = x' + x_0 \\ y = y' + y_0 \end{cases}$ or $\begin{cases} x' = x - x_0 \\ y' = y - y_0 \end{cases}$

where (x, y) are old coordinates (i.e., coordinates relative to xy system), (x', y') are new coordinates (relative to x', y' system), and (x_0, y_0) are the coordinates of the new origin O' relative to the old xy coordinate system.

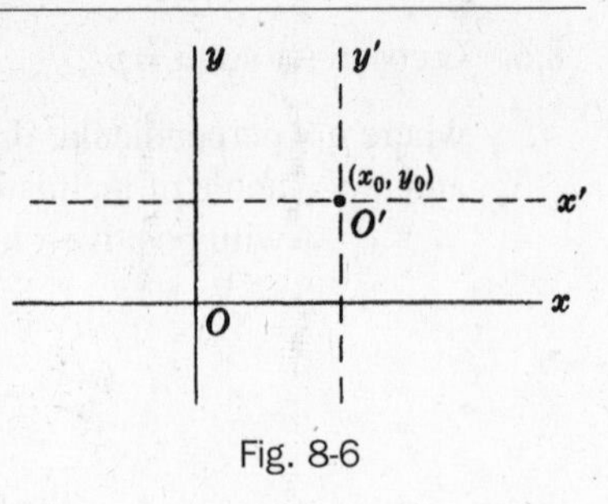

Fig. 8-6

Transformation of Coordinates Involving Pure Rotation

8.12. $\begin{cases} x = x' \cos\alpha - y' \sin\alpha \\ y = x' \sin\alpha + y' \cos\alpha \end{cases}$ or $\begin{cases} x' = x\cos\alpha + y\sin\alpha \\ y' = y\cos\alpha - x\sin\alpha \end{cases}$

where the origins of the old $[xy]$ and new $[x'y']$ coordinate systems are the same but the x' axis makes an angle α with the positive x axis.

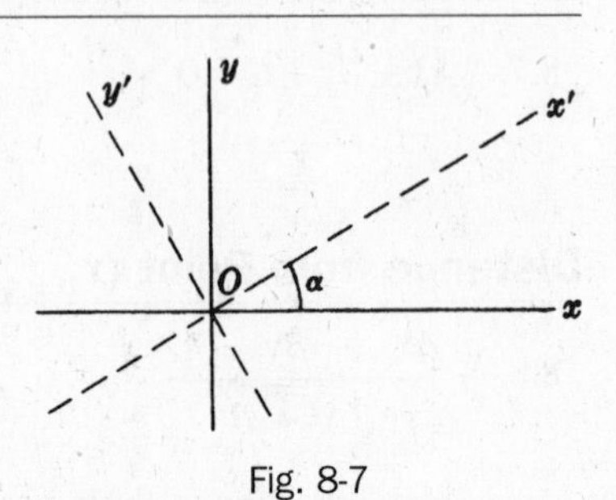

Fig. 8-7

Transformation of Coordinates Involving Translation and Rotation

8.13. $\begin{cases} x = x' \cos\alpha - y' \sin\alpha + x_0 \\ y = x' \sin\alpha + y' \cos\alpha + y_0 \end{cases}$

or $\begin{cases} x' = (x - x_0)\cos\alpha + (y - y_0)\sin\alpha \\ y' = (y - y_0)\cos\alpha - (x - x_0)\sin\alpha \end{cases}$

where the new origin O' of $x'y'$ coordinate system has coordinates (x_0, y_0) relative to the old xy coordinate system and the x' axis makes an angle α with the positive x axis.

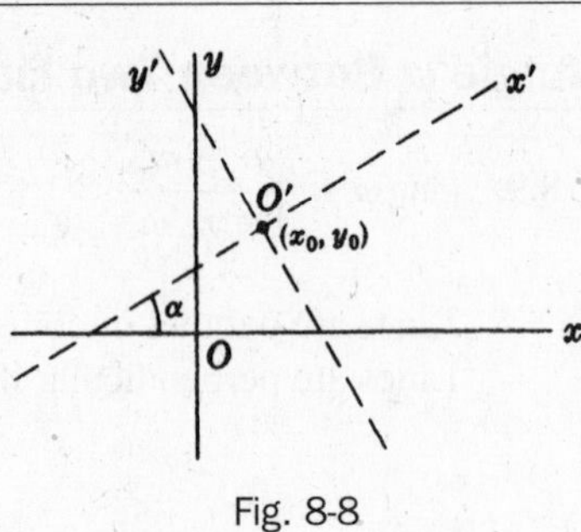

Fig. 8-8

Polar Coordinates (r, θ)

A point P can be located by rectangular coordinates (x, y) or polar coordinates (r, θ). The transformation between these coordinates is as follows:

8.14. $\begin{cases} x = r\cos\theta \\ y = r\sin\theta \end{cases}$ or $\begin{cases} r = \sqrt{x^2 + y^2} \\ \theta = \tan^{-1}(y/x) \end{cases}$

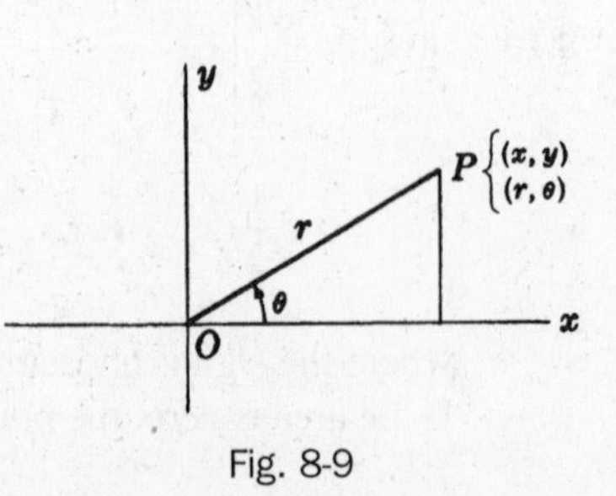

Fig. 8-9

Equation of Circle of Radius R, Center at (x_0, y_0)

8.15. $(x - x_0)^2 + (y - y_0)^2 = R^2$

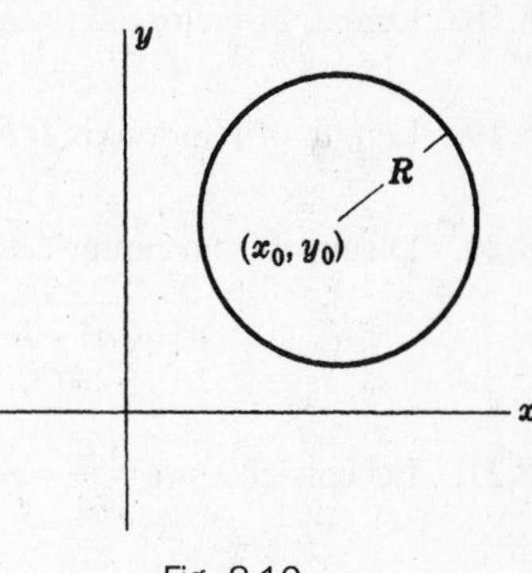

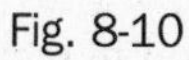
Fig. 8-10

Equation of Circle of Radius R Passing Through Origin

8.16. $r = 2R\cos(\theta - \alpha)$

where (r, θ) are polar coordinates of any point on the circle and (R, α) are polar coordinates of the center of the circle.

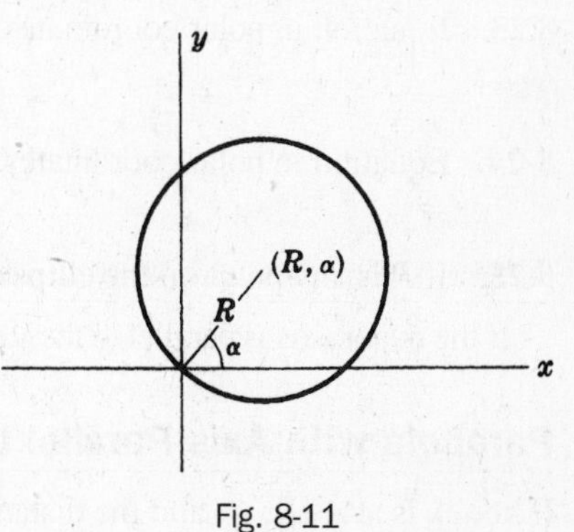

Fig. 8-11

Conics (Ellipse, Parabola, or Hyperbola)

If a point P moves so that its distance from a fixed point (called the *focus*) divided by its distance from a fixed line (called the *directrix*) is a constant ϵ (called the *eccentricity*), then the curve described by P is called a *conic* (so-called because such curves can be obtained by intersecting a plane and a cone at different angles).

If the focus is chosen at origin O, the equation of a conic in polar coordinates (r, θ) is, if $OQ = p$ and $LM = D$ (see Fig. 8-12),

8.17. $$r = \frac{p}{1 - \epsilon\cos\theta} = \frac{\epsilon D}{1 - \epsilon\cos\theta}$$

The conic is

(i) an ellipse if $\epsilon < 1$
(ii) a parabola if $\epsilon = 1$
(iii) a hyperbola if $\epsilon > 1$

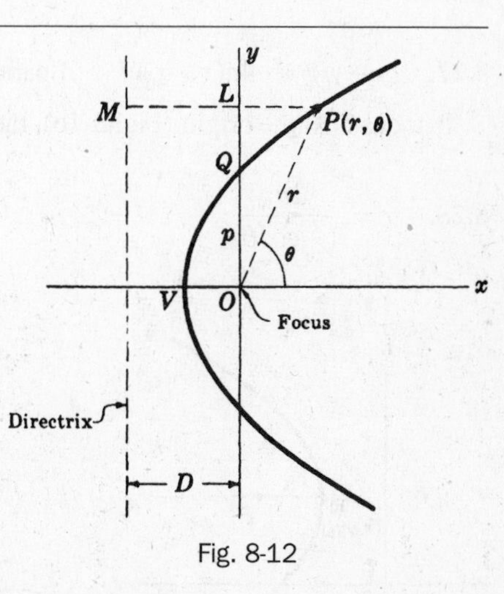

Fig. 8-12

Ellipse with Center $C(x_0, y_0)$ and Major Axis Parallel to x Axis

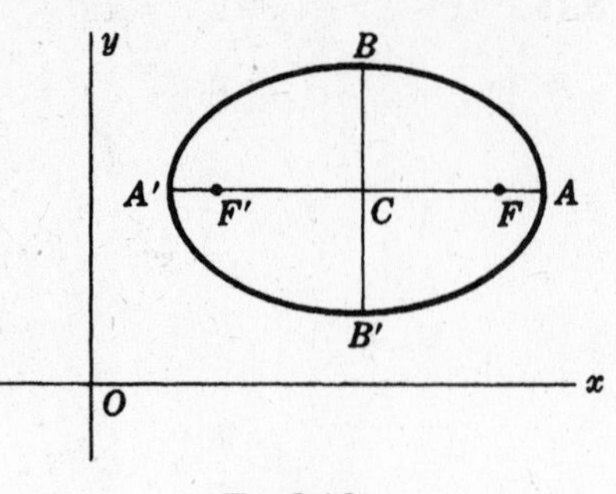

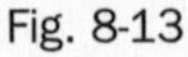
Fig. 8-13

8.18. Length of major axis $A'A = 2a$

8.19. Length of minor axis $B'B = 2b$

8.20. Distance from center C to focus F or F' is

$$c = \sqrt{a^2 - b^2}$$

8.21. Eccentricity $= \epsilon = \dfrac{c}{a} = \dfrac{\sqrt{a^2 - b^2}}{a}$

8.22. Equation in rectangular coordinates:

$$\frac{(x - x_0)^2}{a^2} + \frac{(y - y_0)^2}{b^2} = 1$$

8.23. Equation in polar coordinates if C is at O: $r^2 = \dfrac{a^2 b^2}{a^2 \sin^2\theta + b^2 \cos^2\theta}$

8-24. Equation in polar coordinates if C is on x axis and F' is at O: $r = \dfrac{a(1 - \epsilon^2)}{1 - \epsilon\cos\theta}$

8.25. If P is any point on the ellipse, $PF + PF' = 2a$

If the major axis is parallel to the y axis, interchange x and y in the above or replace θ by $\frac{1}{2}\pi - \theta$ (or $90° - \theta$).

Parabola with Axis Parallel to x Axis

If vertex is at $A\ (x_0, y_0)$ and the distance from A to focus F is $a > 0$, the equation of the parabola is

8.26. $(y - y_0)^2 = 4a(x - x_0)$ if parabola opens to right (Fig. 8-14)

8.27. $(y - y_0)^2 = -4a(x - x_0)$ if parabola opens to left (Fig. 8-15)

If focus is at the origin (Fig. 8-16), the equation in polar coordinates is

8.28. $r = \dfrac{2a}{1 - \cos\theta}$

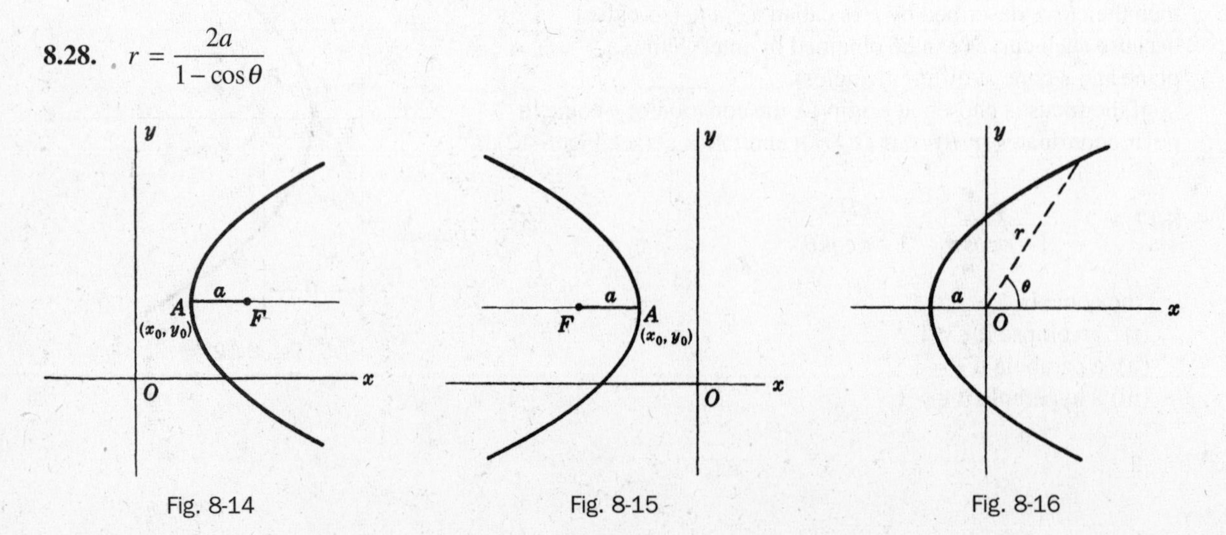

Fig. 8-14 Fig. 8-15 Fig. 8-16

In case the axis is parallel to the y axis, interchange x and y or replace θ by $\frac{1}{2}\pi - \theta$ (or $90° - \theta$).

Hyperbola with Center $C(x_0, y_0)$ and Major Axis Parallel to x Axis

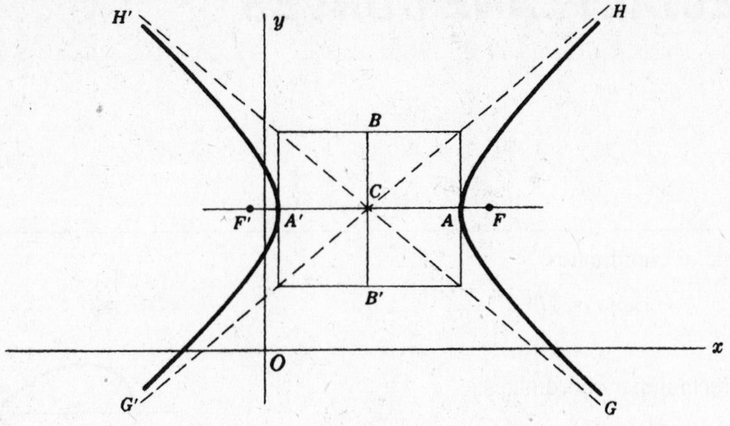

Fig. 8-17

8.29. Length of major axis $A'A = 2a$

8.30. Length of minor axis $B'B = 2b$

8.31. Distance from center C to focus F or $F' = c = \sqrt{a^2 + b^2}$

8.32. Eccentricity $\epsilon = \dfrac{c}{a} = \dfrac{\sqrt{a^2 + b^2}}{a}$

8.33. Equation in rectangular coordinates: $\dfrac{(x - x_0)^2}{a^2} - \dfrac{(y - y_0)^2}{b^2} = 1$

8.34. Slopes of asymptotes $G'H$ and $GH' = \pm \dfrac{b}{a}$

8.35. Equation in polar coordinates if C is at O: $r^2 = \dfrac{a^2 b^2}{b^2 \cos^2 \theta - a^2 \sin^2 \theta}$

8.36. Equation in polar coordinates if C is on x axis and F' is at O: $r = \dfrac{a(\epsilon^2 - 1)}{1 - \epsilon \cos \theta}$

8.37. If P is any point on the hyperbola, $PF - PF' = \pm 2a$ (depending on branch)

If the major axis is parallel to the y axis, interchange x and y in the above or replace θ by $\frac{1}{2}\pi - \theta$ (or $90° - \theta$).

9 SPECIAL PLANE CURVES

Lemniscate

9.1. Equation in polar coordinates:

$$r^2 = a^2 \cos 2\theta$$

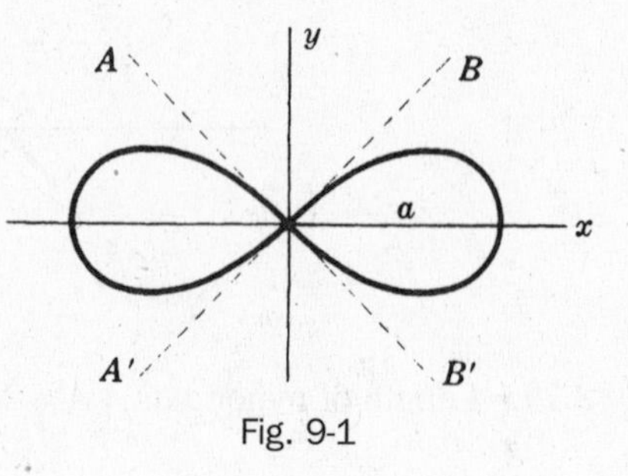

Fig. 9-1

9.2. Equation in rectangular coordinates:

$$(x^2 + y^2)^2 = a^2(x^2 - y^2)$$

9.3. Angle between AB' or $A'B$ and x axis $= 45°$

9.4. Area of one loop $= a^2$

Cycloid

9.5. Equations in parametric form:

$$\begin{cases} x = a(\phi - \sin\phi) \\ y = a(1 - \cos\phi) \end{cases}$$

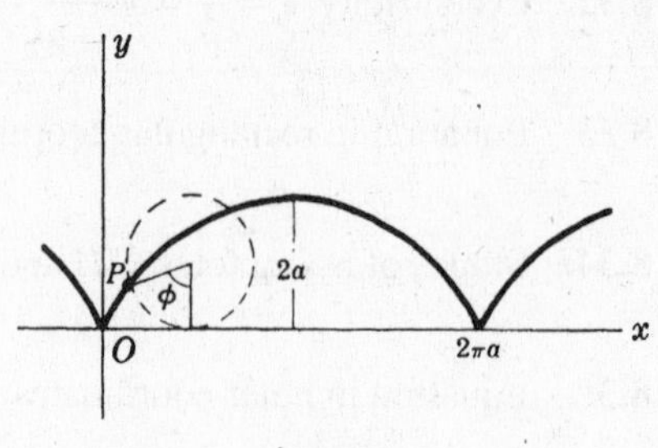

Fig. 9-2

9.6. Area of one arch $= 3\pi a^2$

9.7. Arc length of one arch $= 8a$

This is a curve described by a point P on a circle of radius a rolling along x axis.

Hypocycloid with Four Cusps

9.8. Equation in rectangular coordinates:

$$x^{2/3} + y^{2/3} = a^{2/3}$$

9.9. Equations in parametric form:

$$\begin{cases} x = a\cos^3\theta \\ y = a\sin^3\theta \end{cases}$$

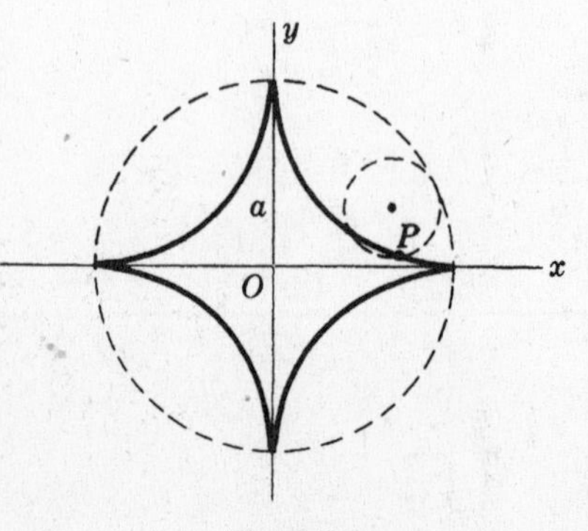

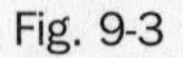
Fig. 9-3

9.10. Area bounded by curve $= \frac{3}{8}\pi a^2$

9.11. Arc length of entire curve $= 6a$

This is a curve described by a point P on a circle of radius $a/4$ as it rolls on the inside of a circle of radius a.

Cardioid

9.12. Equation: $r = 2a(1 + \cos\theta)$

9.13. Area bounded by curve $= 6\pi a^2$

9.14. Arc length of curve $= 16a$

This is the curve described by a point P of a circle of radius a as it rolls on the outside of a fixed circle of radius a. The curve is also a special case of the limacon of Pascal (see 9.32).

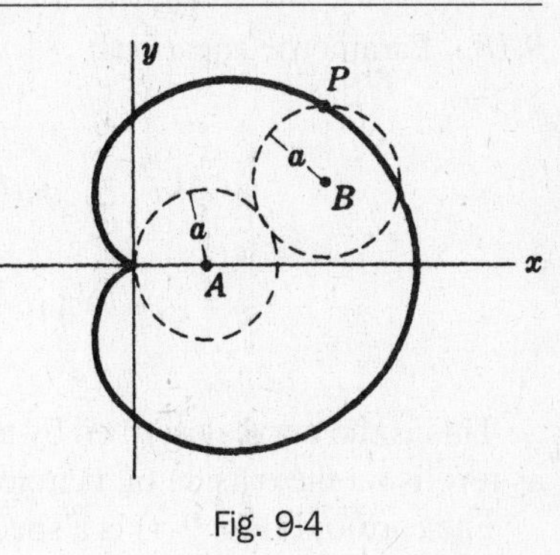

Fig. 9-4

Catenary

9.15. Equation: $y = \frac{a}{2}(e^{x/a} + e^{-x/a}) = a \cosh\frac{x}{a}$

This is the curve in which a heavy uniform chain would hang if suspended vertically from fixed points A and B.

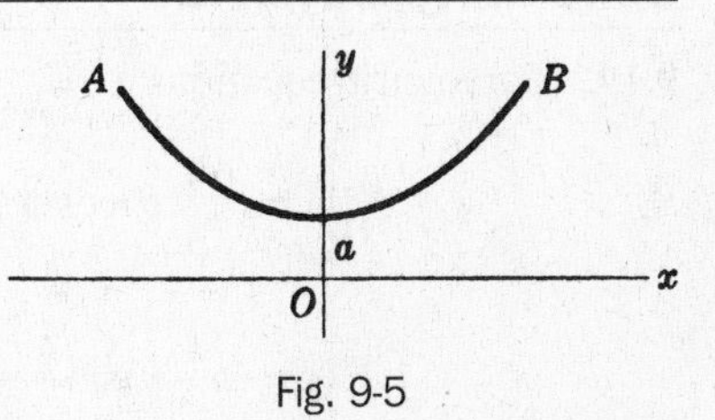

Fig. 9-5

Three-Leaved Rose

9.16. Equation: $r = a \cos 3\theta$

The equation $r = a \sin 3\theta$ is a similar curve obtained by rotating the curve of Fig. 9-6 counterclockwise through 30° or $\pi/6$ radians.

In general, $r = a \cos n\theta$ or $r = a \sin n\theta$ has n leaves if n is odd.

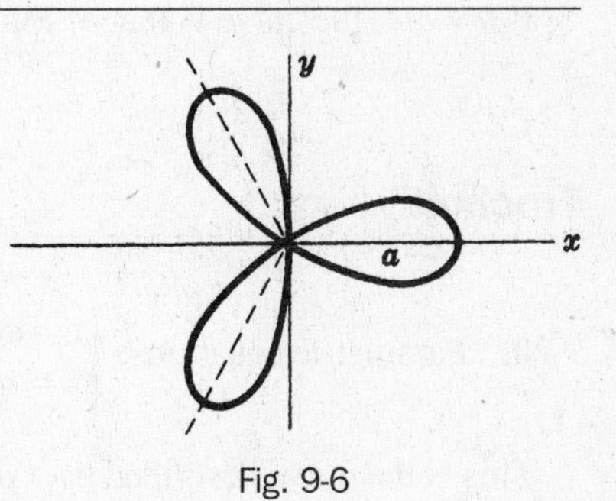

Fig. 9-6

Four-Leaved Rose

9.17. Equation: $r = a \cos 2\theta$

The equation $r = a \sin 2\theta$ is a similar curve obtained by rotating the curve of Fig. 9-7 counterclockwise through 45° or $\pi/4$ radians.

In general, $r = a \cos n\theta$ or $r = a \sin n\theta$ has $2n$ leaves if n is even.

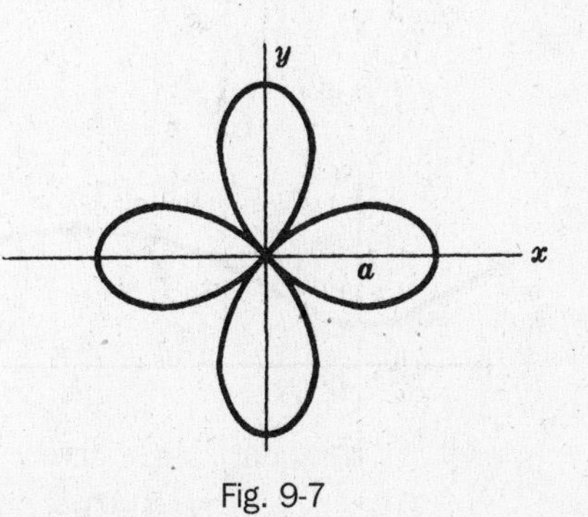

Fig. 9-7

Epicycloid

9.18. Parametric equations:

$$\begin{cases} x = (a+b)\cos\theta - b\cos\left(\dfrac{a+b}{b}\right)\theta \\ y = (a+b)\sin\theta - b\sin\left(\dfrac{a+b}{b}\right)\theta \end{cases}$$

This is the curve described by a point P on a circle of radius b as it rolls on the outside of a circle of radius a.

The cardioid (Fig. 9-4) is a special case of an epicycloid.

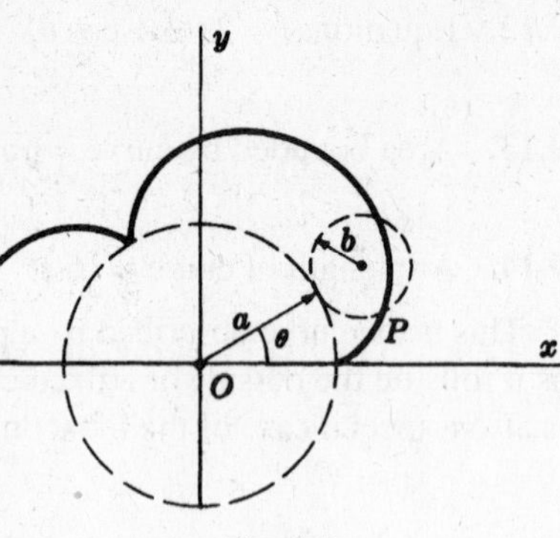

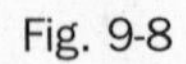
Fig. 9-8

General Hypocycloid

9.19. Parametric equations:

$$\begin{cases} x = (a-b)\cos\phi + b\cos\left(\dfrac{a-b}{b}\right)\phi \\ y = (a-b)\sin\phi - b\sin\left(\dfrac{a-b}{b}\right)\phi \end{cases}$$

This is the curve described by a point P on a circle of radius b as it rolls on the inside of a circle of radius a.

If $b = a/4$, the curve is that of Fig. 9-3.

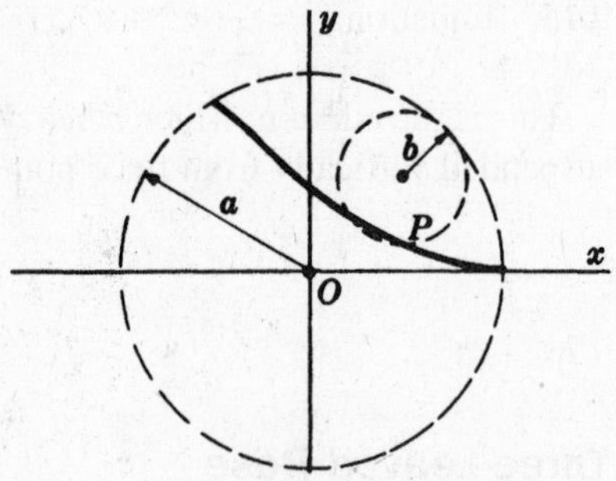

Fig. 9-9

Trochoid

9.20. Parametric equations: $\begin{cases} x = a\phi - b\sin\phi \\ y = a - b\cos\phi \end{cases}$

This is the curve described by a point P at distance b from the center of a circle of radius a as the circle rolls on the x axis.

If $b < a$, the curve is as shown in Fig. 9-10 and is called a *curtate cycloid.*

If $b > a$, the curve is as shown in Fig. 9-11 and is called a *prolate cycloid.*

If $b = a$, the curve is the cycloid of Fig. 9-2.

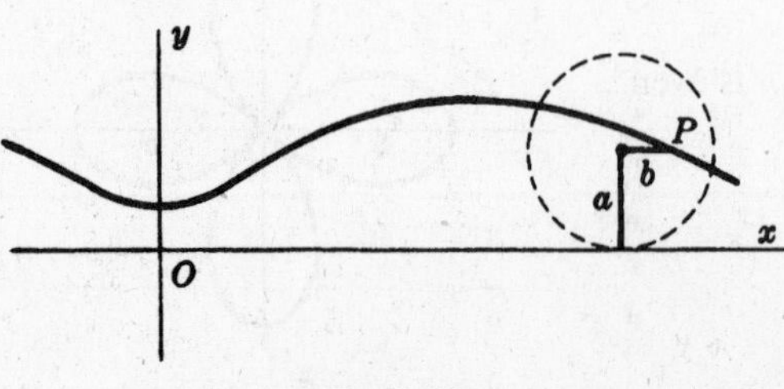

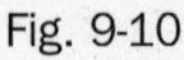
Fig. 9-10

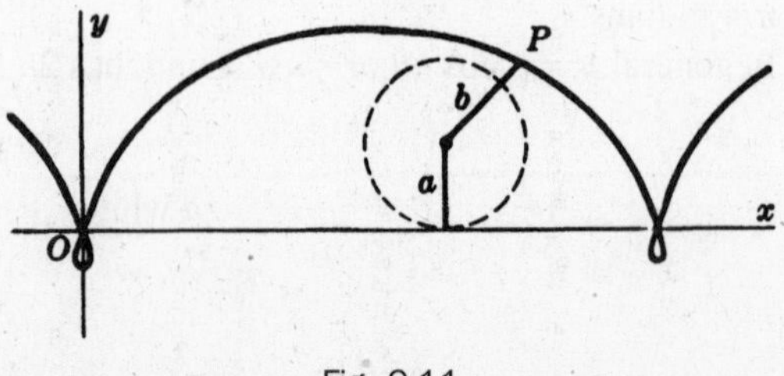

Fig. 9-11

Tractrix

9.21. Parametric equations: $\begin{cases} x = a(\ln \cot \frac{1}{2}\phi - \cos\phi) \\ y = a\sin\phi \end{cases}$

This is the curve described by endpoint P of a taut string PQ of length a as the other end Q is moved along the x axis.

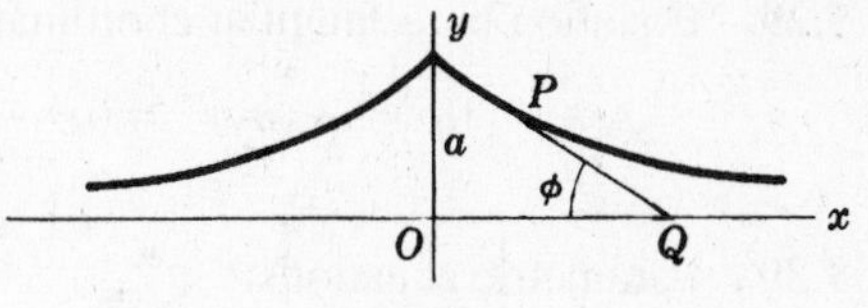

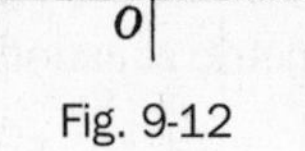

Fig. 9-12

Witch of Agnesi

9.22. Equation in rectangular coordinates: $y = \dfrac{8a^3}{x^2 + 4a^2}$

9.23. Parametric equations: $\begin{cases} x = 2a\cot\theta \\ y = a(1 - \cos 2\theta) \end{cases}$

In Fig. 9-13 the variable line QA intersects $y = 2a$ and the circle of radius a with center $(0, a)$ at A and B, respectively. Any point P on the "witch" is located by constructing lines parallel to the x and y axes through B and A, respectively, and determining the point P of intersection.

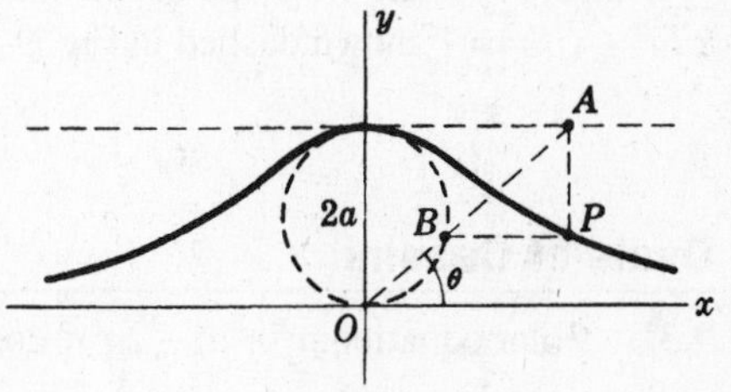

Fig. 9-13

Folium of Descartes

9.24. Equation in rectangular coordinates:

$$x^3 + y^3 = 3axy$$

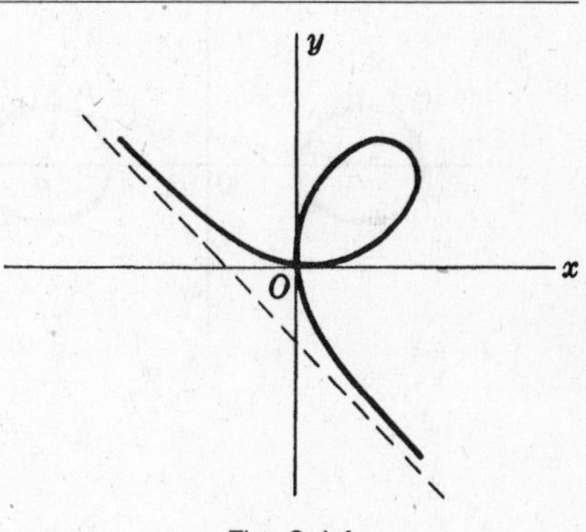

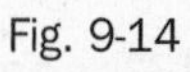

Fig. 9-14

9.25. Parametric equations:

$$\begin{cases} x = \dfrac{3at}{1+t^3} \\ y = \dfrac{3at^2}{1+t^3} \end{cases}$$

9.26. Area of loop $= \dfrac{3}{2}a^2$

9.27. Equation of asymptote: $x + y + a = 0$

Involute of a Circle

9.28. Parametric equations:

$$\begin{cases} x = a(\cos\phi + \phi\sin\phi) \\ y = a(\sin\phi - \phi\cos\phi) \end{cases}$$

This is the curve described by the endpoint P of a string as it unwinds from a circle of radius a while held taut.

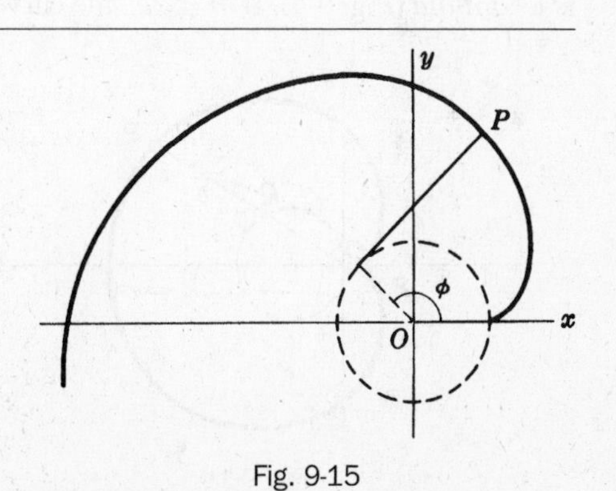

Fig. 9-15

Evolute of an Ellipse

9.29. Equation in rectangular coordinates:

$$(ax)^{2/3} + (by)^{2/3} = (a^2 - b^2)^{2/3}$$

9.30. Parametric equations:

$$\begin{cases} ax = (a^2 - b^2)\cos^3\theta \\ by = (a^2 - b^2)\sin^3\theta \end{cases}$$

This curve is the envelope of the normals to the ellipse $x^2/a^2 + y^2/b^2 = 1$ shown dashed in Fig. 9-16.

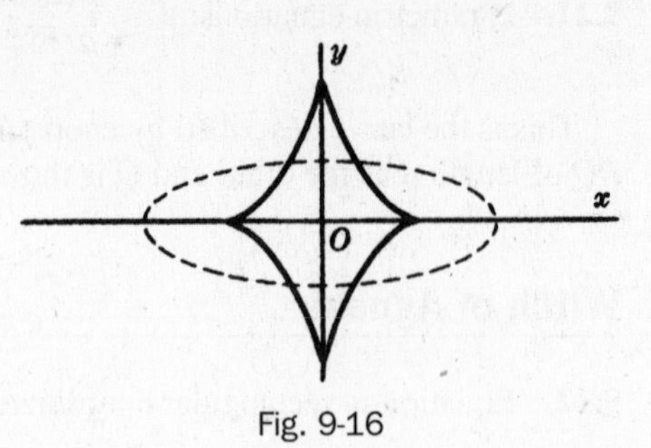

Fig. 9-16

Ovals of Cassini

9.31. Polar equation: $r^4 + a^4 - 2a^2r^2 \cos 2\theta = b^4$

This is the curve described by a point P such that the product of its distance from two fixed points (distance $2a$ apart) is a constant b^2.

The curve is as in Fig. 9-17 or Fig. 9-18 according as $b < a$ or $b > a$, respectively.

If $b = a$, the curve is a *lemniscate* (Fig. 9-1).

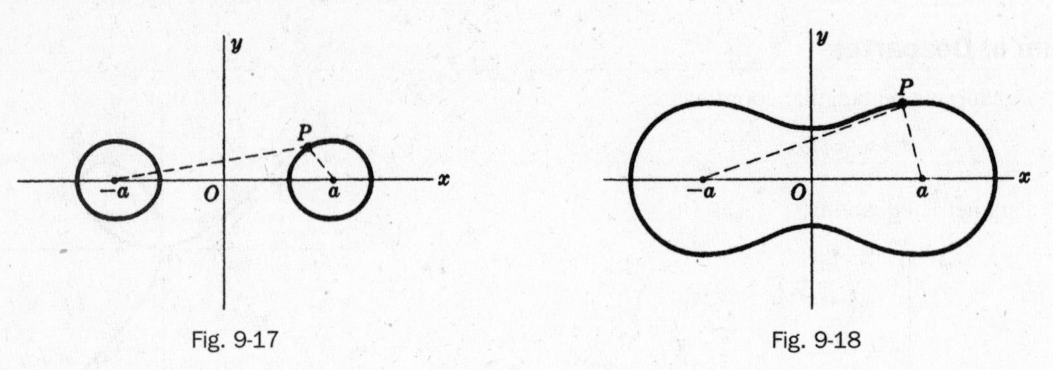

Fig. 9-17

Fig. 9-18

Limacon of Pascal

9.32. Polar equation: $r = b + a \cos\theta$

Let OQ be a line joining origin O to any point Q on a circle of diameter a passing through O. Then the curve is the locus of all points P such that $PQ = b$.

The curve is as in Fig. 9-19 or Fig. 9-20 according as $2a > b > a$ or $b < a$, respectively. If $b = a$, the curve is a *cardioid* (Fig. 9-4). If $b \geqq 2a$, the curve is convex.

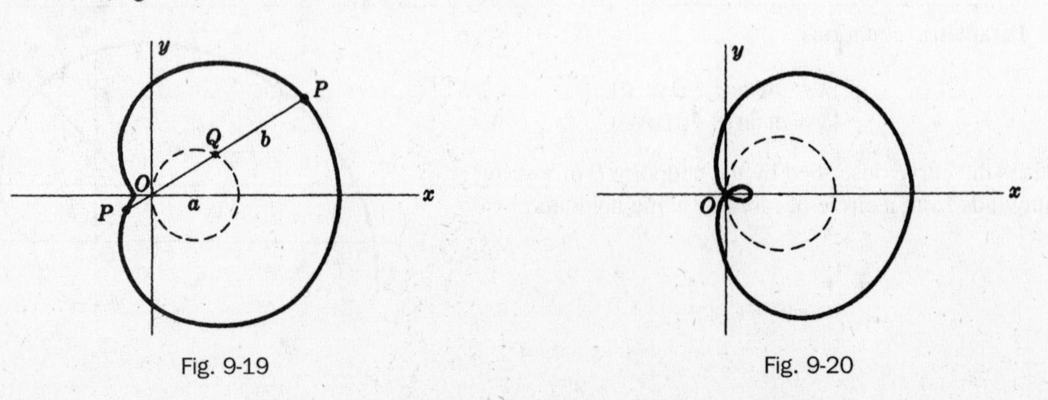

Fig. 9-19

Fig. 9-20

Cissoid of Diocles

9.33. Equation in rectangular coordinates:

$$y^2 = \frac{x^2}{2a - x}$$

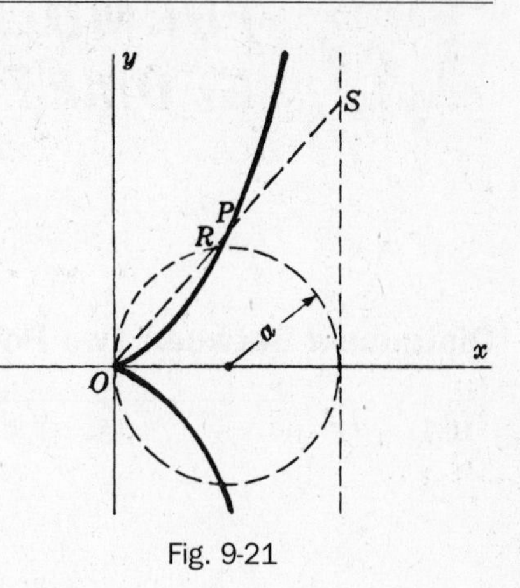

Fig. 9-21

9.34. Parametric equations:

$$\begin{cases} x = 2a\sin^2\theta \\ y = \dfrac{2a\sin^3\theta}{\cos\theta} \end{cases}$$

This is the curve described by a point P such that the distance OP = distance RS. It is used in the problem of *duplication of a cube,* i.e., finding the side of a cube which has twice the volume of a given cube.

Spiral of Archimedes

9.35. Polar equation: $r = a\theta$

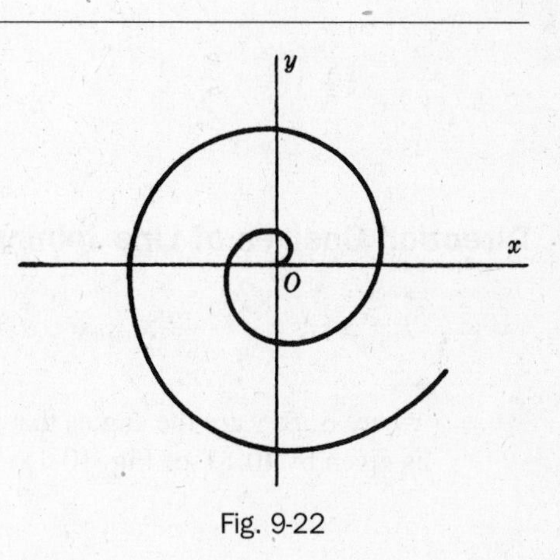

Fig. 9-22

10 FORMULAS from SOLID ANALYTIC GEOMETRY

Distance d Between Two Points $P_1(x_1, y_1, z_1)$ and $P_2(x_2, y_2, z_2)$

10.1. $d = \sqrt{(x_2 - x_1)^2 + (y_2 - y_1)^2 + (z_2 - z_1)^2}$

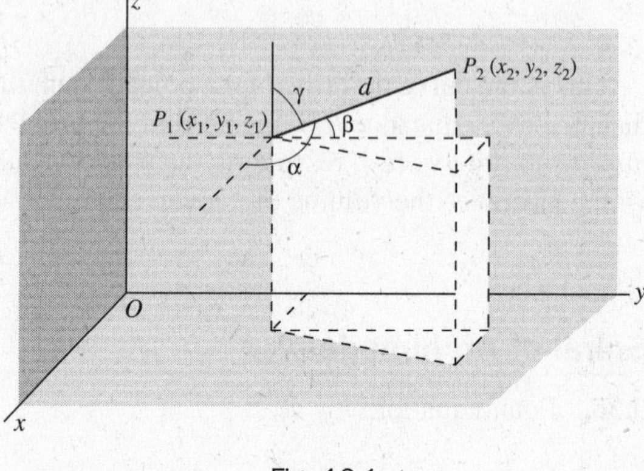

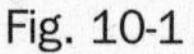

Fig. 10-1

Direction Cosines of Line Joining Points $P_1(x_1, y_1, z_1)$ and $P_2(x_2, y_2, z_2)$

10.2. $l = \cos\alpha = \dfrac{x_2 - x_1}{d}, \quad m = \cos\beta = \dfrac{y_2 - y_1}{d}, \quad n = \cos\gamma = \dfrac{z_2 - z_1}{d}$

where α, β, γ are the angles that line P_1P_2 makes with the positive x, y, z axes, respectively, and d is given by 10.1 (see Fig. 10-1).

Relationship Between Direction Cosines

10.3. $\cos^2\alpha + \cos^2\beta + \cos^2\gamma = 1 \quad \text{or} \quad l^2 + m^2 + n^2 = 1$

Direction Numbers

Numbers L, M, N, which are proportional to the direction cosines l, m, n, are called *direction numbers*. The relationship between them is given by

10.4. $l = \dfrac{L}{\sqrt{L^2 + M^2 + N^2}}, \quad m = \dfrac{M}{\sqrt{L^2 + M^2 + N^2}}, \quad n = \dfrac{N}{\sqrt{L^2 + M^2 + N^2}}$

Equations of Line Joining $P_1(x_1, y_1, z_1)$ and $P_2(x_2, y_2, z_2)$ in Standard Form

10.5. $\dfrac{x - x_1}{x_2 - x_1} = \dfrac{y - y_1}{y_2 - y_1} = \dfrac{z - z_1}{z_2 - z_1} \quad \text{or} \quad \dfrac{x - x_1}{l} = \dfrac{y - y_1}{m} = \dfrac{z - z_1}{n}$

These are also valid if l, m, n are replaced by L, M, N, respectively.

Equations of Line Joining $P_1(x_1, y_1, z_1)$ and $P_2(x_2, y_2, z_2)$ in Parametric Form

10.6. $x = x_1 + lt,\ \ y = y_1 + mt,\ \ z = z_1 + nt$

These are also valid if l, m, n are replaced by L, M, N, respectively.

Angle ϕ Between Two Lines with Direction Cosines l_1, m_1, n_1 and l_2, m_2, n_2

10.7. $\cos\phi = l_1 l_2 + m_1 m_2 + n_1 n_2$

General Equation of a Plane

10.8. $Ax + By + Cz + D = 0$ $\qquad$ (A, B, C, D are constants)

Equation of Plane Passing Through Points (x_1, y_1, z_1), (x_2, y_2, z_2), (x_3, y_3, z_3)

10.9.
$$\begin{vmatrix} x - x_1 & y - y_1 & z - z_1 \\ x_2 - x_1 & y_2 - y_1 & z_2 - z_1 \\ x_3 - x_1 & y_3 - y_1 & z_3 - z_1 \end{vmatrix} = 0$$

or

10.10.
$$\begin{vmatrix} y_2 - y_1 & z_2 - z_1 \\ y_3 - y_1 & z_3 - z_1 \end{vmatrix}(x - x_1) + \begin{vmatrix} z_2 - z_1 & x_2 - x_1 \\ z_3 - z_1 & x_3 - x_1 \end{vmatrix}(y - y_1) + \begin{vmatrix} x_2 - x_1 & y_2 - y_1 \\ x_3 - x_1 & y_3 - y_1 \end{vmatrix}(z - z_1) = 0$$

Equation of Plane in Intercept Form

10.11. $\dfrac{x}{a} + \dfrac{y}{b} + \dfrac{z}{c} = 1$

where a, b, c are the intercepts on the x, y, z axes, respectively.

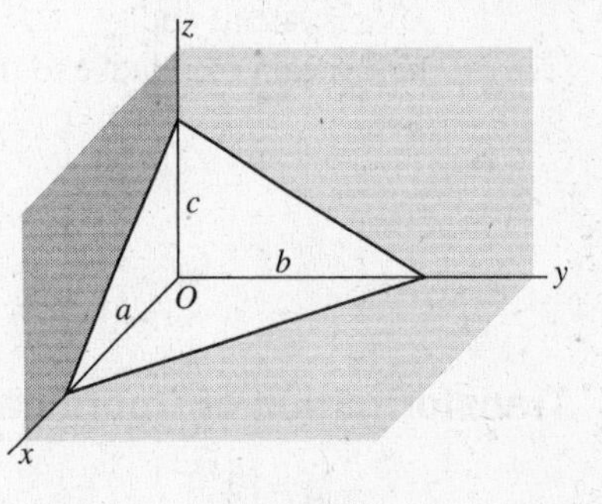

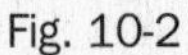
Fig. 10-2

Equations of Line Through (x_0, y_0, z_0) and Perpendicular to Plane $Ax + By + Cz + D = 0$

10.12. $\dfrac{x - x_0}{A} = \dfrac{y - y_0}{B} = \dfrac{z - z_0}{C}$ $\quad$ or $\quad x = x_0 + At,\ \ y = y_0 + Bt,\ \ z = z_0 + Ct$

Note that the direction numbers for a line perpendicular to the plane $Ax + By + Cz + D = 0$ are A, B, C.

Distance from Point (x_0, y_0, z_0) to Plane $Ax + By + Cz + D = 0$

10.13. $$\frac{Ax_0 + By_0 + Cz_0 + D}{\pm\sqrt{A^2 + B^2 + C^2}}$$

where the sign is chosen so that the distance is nonnegative.

Normal Form for Equation of Plane

10.14. $x\cos\alpha + y\cos\beta + z\cos\gamma = p$

where p = perpendicular distance from O to plane at P and α, β, γ are angles between OP and positive x, y, z axes.

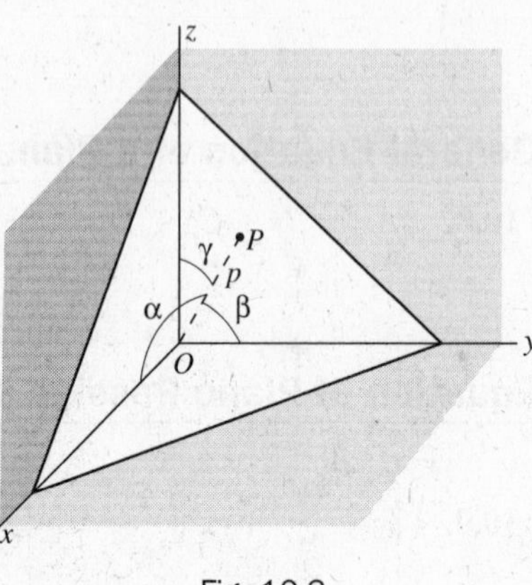

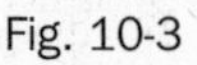
Fig. 10-3

Transformation of Coordinates Involving Pure Translation

10.15. $$\begin{cases} x = x' + x_0 \\ y = y' + y_0 \\ z = z' + z_0 \end{cases} \quad \text{or} \quad \begin{cases} x' = x - x_0 \\ y' = y - y_0 \\ z' = z - z_0 \end{cases}$$

where (x, y, z) are old coordinates (i.e., coordinates relative to xyz system), (x', y', z') are new coordinates (relative to $x'y'z'$ system) and (x_0, y_0, z_0) are the coordinates of the new origin O' relative to the old xyz coordinate system.

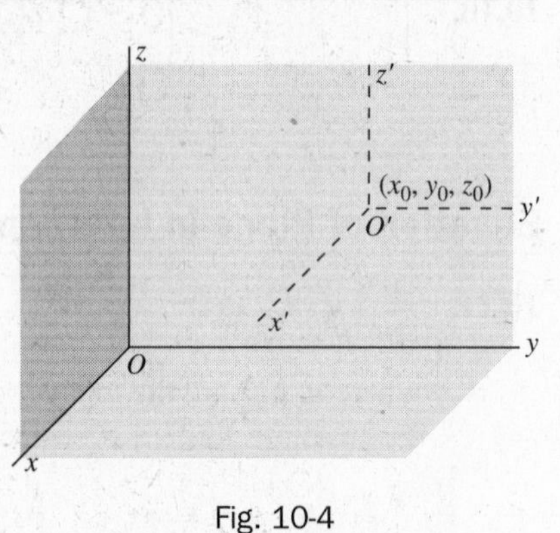

Fig. 10-4

Transformation of Coordinates Involving Pure Rotation

10.16. $$\begin{cases} x = l_1x' + l_2y' + l_3z' \\ y = m_1x' + m_2y' + m_3z' \\ z = n_1x' + n_2y' + n_3z' \end{cases}$$

or $$\begin{cases} x' = l_1x + m_1y + n_1z \\ y' = l_2x + m_2y + n_2z \\ z' = l_3x + m_3y + n_3z \end{cases}$$

where the origins of the xyz and $x'y'z'$ systems are the same and l_1, m_1, n_1; l_2, m_2, n_2; l_3, m_3, n_3 are the direction cosines of the x', y', z' axes relative to the x, y, z axes, respectively.

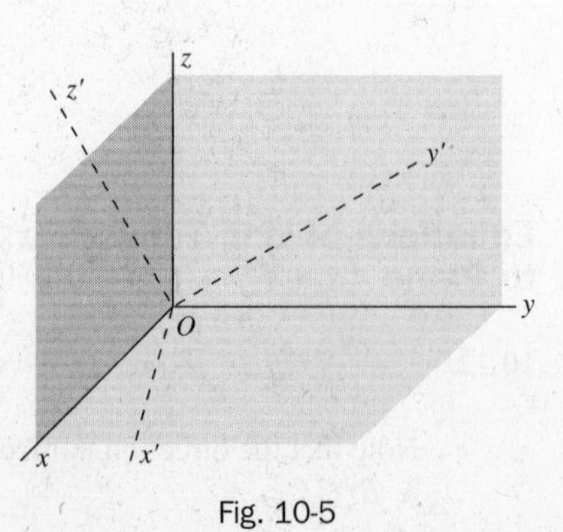

Fig. 10-5

Transformation of Coordinates Involving Translation and Rotation

10.17.
$$\begin{cases} x = l_1x' + l_2y' + l_3z' + x_0 \\ y = m_1x' + m_2y' + m_3z' + y_0 \\ z = n_1x' + n_2y' + n_3z' + z_0 \end{cases}$$

or
$$\begin{cases} x' = l_1(x - x_0) + m_1(y - y_0) + n_1(z - z_0) \\ y' = l_2(x - x_0) + m_2(y - y_0) + n_2(z - z_0) \\ z' = l_3(x - x_0) + m_3(y - y_0) + n_3(z - z_0) \end{cases}$$

where the origin O' of the $x'y'z'$ system has coordinates (x_0, y_0, z_0) relative to the xyz system and

$$l_1, m_1, n_1; \ l_2, m_2, n_2; \ l_3, m_3, n_3$$

are the direction cosines of the x', y', z' axes relative to the x, y, z axes, respectively.

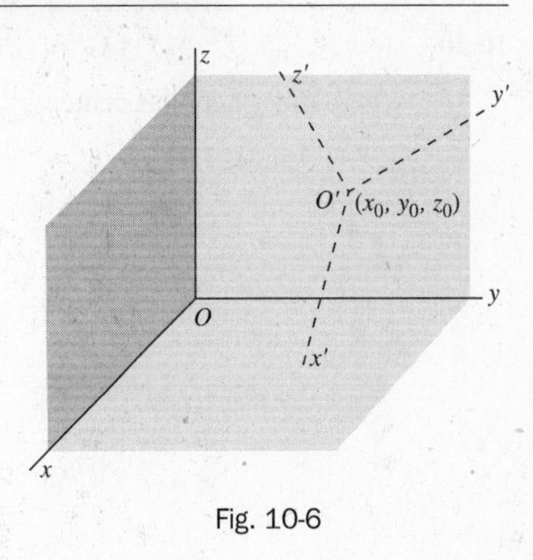

Fig. 10-6

Cylindrical Coordinates (r, θ, z)

A point P can be located by cylindrical coordinates (r, θ, z) (see Fig. 10-7) as well as rectangular coordinates (x, y, z).

The transformation between these coordinates is

10.18.
$$\begin{cases} x = r\cos\theta \\ y = r\sin\theta \\ z = z \end{cases} \quad \text{or} \quad \begin{cases} r = \sqrt{x^2 + y^2} \\ \theta = \tan^{-1}(y/x) \\ z = z \end{cases}$$

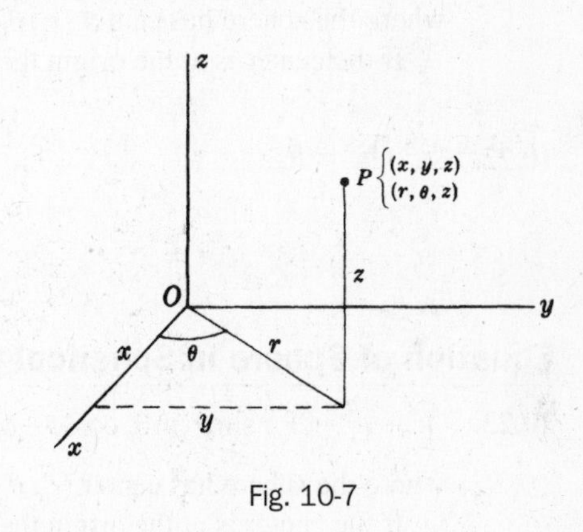

Fig. 10-7

Spherical Coordinates (r, θ, ϕ)

A point P can be located by spherical coordinates (r, θ, ϕ) (see Fig. 10-8) as well as rectangular coordinates (x, y, z).

The transformation between those coordinates is

10.19.
$$\begin{cases} x = r\sin\theta\cos\phi \\ y = r\sin\theta\sin\phi \\ z = r\cos\theta \end{cases}$$

or
$$\begin{cases} r = \sqrt{x^2 + y^2 + z^2} \\ \phi = \tan^{-1}(y/x) \\ \theta = \cos^{-1}(z/\sqrt{x^2 + y^2 + z^2}) \end{cases}$$

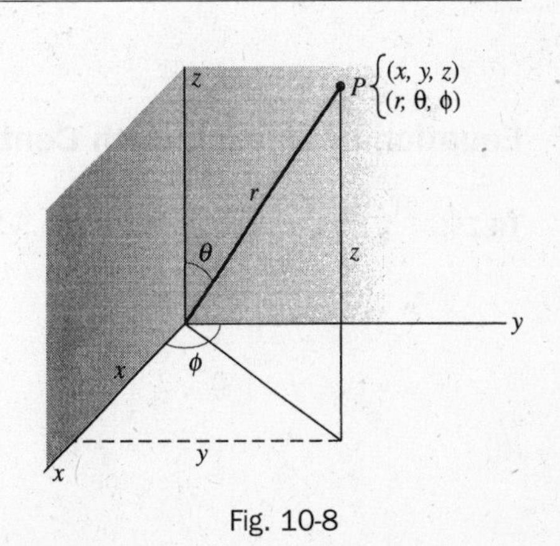

Fig. 10-8

Equation of Sphere in Rectangular Coordinates

10.20. $(x - x_0)^2 + (y - y_0)^2 + (z - z_0)^2 = R^2$

where the sphere has center (x_0, y_0, z_0) and radius R.

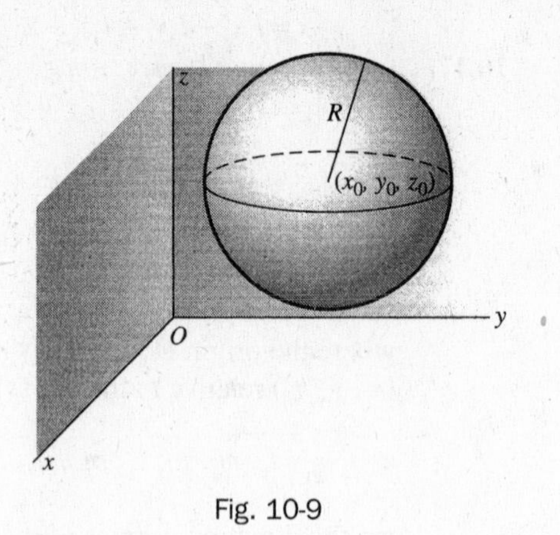

Fig. 10-9

Equation of Sphere in Cylindrical Coordinates

10.21. $r^2 - 2r_0 r \cos(\theta - \theta_0) + r_0^2 + (z - z_0)^2 = R^2$

where the sphere has center (r_0, θ_0, z_0) in cylindrical coordinates and radius R.
If the center is at the origin the equation is

10.22. $r^2 + z^2 = R^2$

Equation of Sphere in Spherical Coordinates

10.23. $r^2 + r_0^2 - 2r_0 r \sin\theta \sin\theta_0 \cos(\phi - \phi_0) = R^2$

where the sphere has center (r_0, θ_0, ϕ_0) in spherical coordinates and radius R.
If the center is at the origin the equation is

10.24. $r = R$

Equation of Ellipsoid with Center (x_0, y_0, z_0) and Semi-axes a, b, c

10.25. $$\frac{(x - x_0)^2}{a^2} + \frac{(y - y_0)^2}{b^2} + \frac{(z - z_0)^2}{c^2} = 1$$

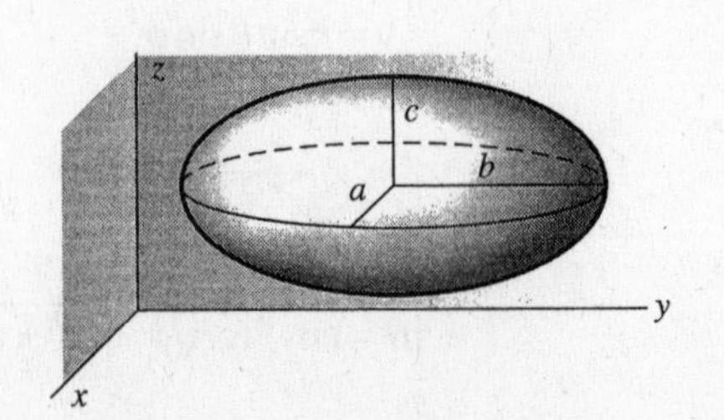

Fig. 10-10

Elliptic Cylinder with Axis as z Axis

10.26. $\dfrac{x^2}{a^2}+\dfrac{y^2}{b^2}=1$

where a, b are semi-axes of elliptic cross-section.
If $b=a$ it becomes a circular cylinder of radius a.

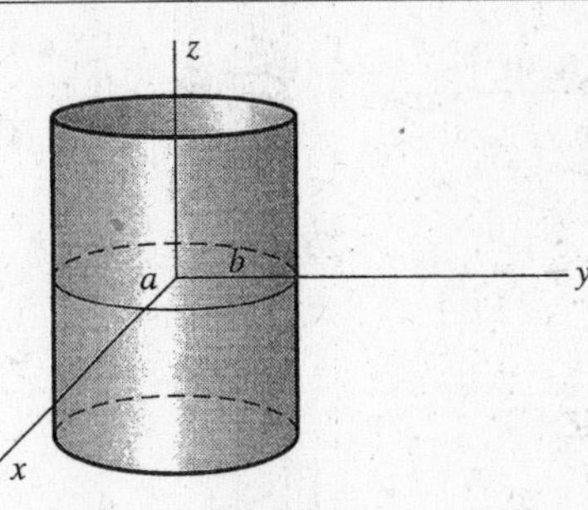

Fig. 10-11

Elliptic Cone with Axis as z Axis

10.27. $\dfrac{x^2}{a^2}+\dfrac{y^2}{b^2}=\dfrac{z^2}{c^2}$

Fig. 10-12

Hyperboloid of One Sheet

10.28. $\dfrac{x^2}{a^2}+\dfrac{y^2}{b^2}-\dfrac{z^2}{c^2}=1$

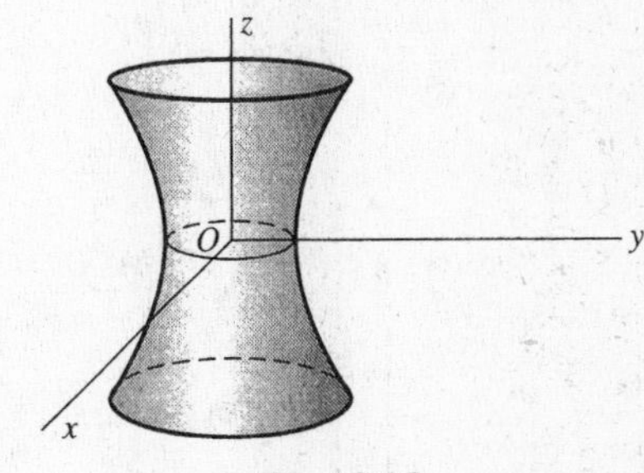

Fig. 10-13

Hyperboloid of Two Sheets

10.29. $\dfrac{x^2}{a^2}-\dfrac{y^2}{b^2}-\dfrac{z^2}{c^2}=1$

Note orientation of axes in Fig. 10-14.

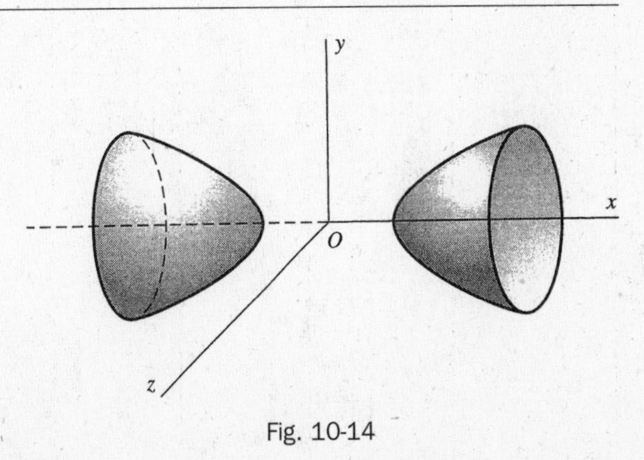

Fig. 10-14

Elliptic Paraboloid

10.30. $\dfrac{x^2}{a^2}+\dfrac{y^2}{b^2}=\dfrac{z}{c}$

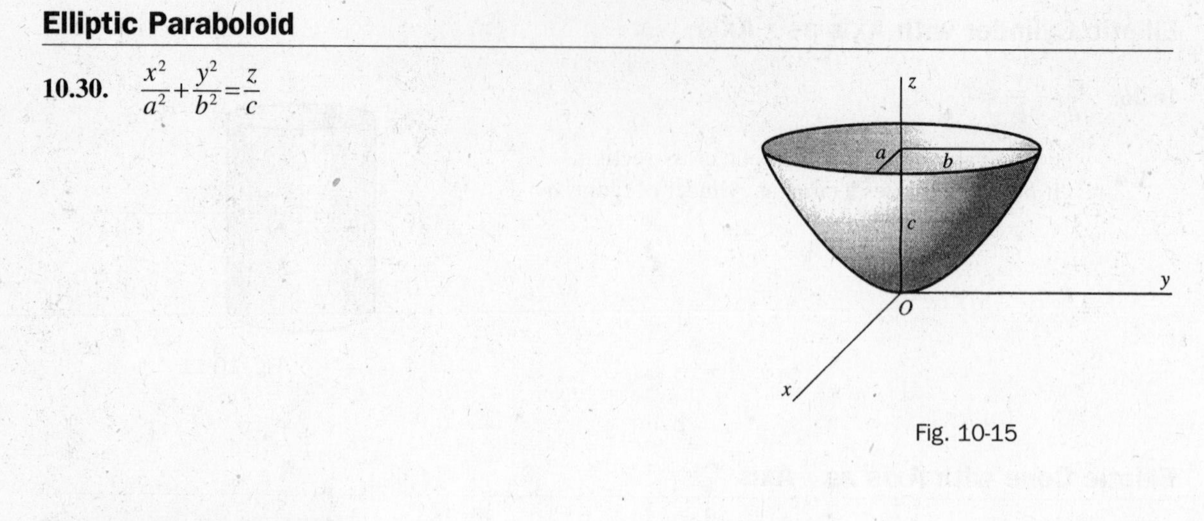

Fig. 10-15

Hyperbolic Paraboloid

10.31. $\dfrac{x^2}{a^2}-\dfrac{y^2}{b^2}=\dfrac{z}{c}$

Note orientation of axes in Fig. 10-16.

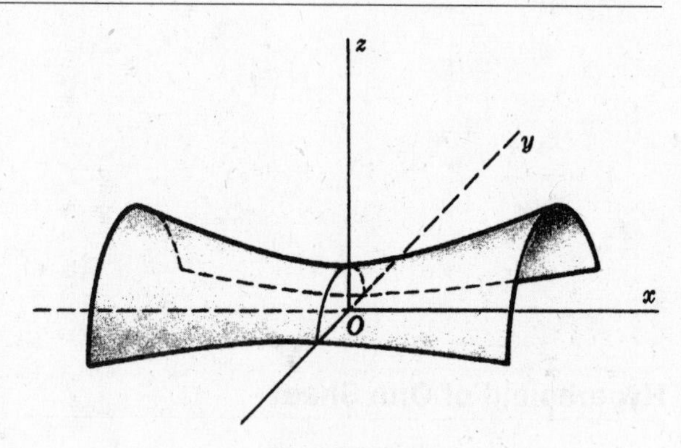

Fig. 10-16

11 SPECIAL MOMENTS of INERTIA

The table below shows the moments of inertia of various rigid bodies of mass M. In all cases it is assumed the body has uniform (i.e., constant) density.

TYPE OF RIGID BODY	MOMENT OF INERTIA
11.1. Thin rod of length a	
(a) about axis perpendicular to the rod through the center of mass	$\frac{1}{12}Ma^2$
(b) about axis perpendicular to the rod through one end	$\frac{1}{3}Ma^2$
11.2. Rectangular parallelepiped with sides a, b, c	
(a) about axis parallel to c and through center of face ab	$\frac{1}{12}M(a^2+b^2)$
(b) about axis through center of face bc and parallel to c	$\frac{1}{12}M(4a^2+b^2)$
11.3. Thin rectangular plate with sides a, b	
(a) about axis perpendicular to the plate through center	$\frac{1}{12}M(a^2+b^2)$
(b) about axis parallel to side b through center	$\frac{1}{12}Ma^2$
11.4. Circular cylinder of radius a and height h	
(a) about axis of cylinder	$\frac{1}{2}Ma^2$
(b) about axis through center of mass and perpendicular to cylindrical axis	$\frac{1}{12}M(h^2+3a^2)$
(c) about axis coinciding with diameter at one end	$\frac{1}{12}M(4h^2+3a^2)$
11.5. Hollow circular cylinder of outer radius a, inner radius b and height h	
(a) about axis of cylinder	$\frac{1}{2}M(a^2+b^2)$
(b) about axis through center of mass and perpendicular to cylindrical axis	$\frac{1}{12}M(3a^2+3b^2+h^2)$
(c) about axis coinciding with diameter at one end	$\frac{1}{12}M(3a^2+3b^2+4h^2)$
11.6. Circular plate of radius a	
(a) about axis perpendicular to plate through center	$\frac{1}{2}Ma^2$
(b) about axis coinciding with a diameter	$\frac{1}{4}Ma^2$

11.7.	Hollow circular plate or ring with outer radius a and inner radius b	
(a)	about axis perpendicular to plane of plate through center	$\frac{1}{2}M(a^2+b^2)$
(b)	about axis coinciding with a diameter	$\frac{1}{4}M(a^2+b^2)$
11.8.	Thin circular ring of radius a	
(a)	about axis perpendicular to plane of ring through center	Ma^2
(b)	about axis coinciding with diameter	$\frac{1}{2}Ma^2$
11.9.	Sphere of radius a	
(a)	about axis coinciding with a diameter	$\frac{2}{5}Ma^2$
(b)	about axis tangent to the surface	$\frac{7}{5}Ma^2$
11.10.	Hollow sphere of outer radius a and inner radius b	
(a)	about axis coinciding with a diameter	$\frac{2}{5}M(a^5-b^5)/(a^3-b^3)$
(b)	about axis tangent to the surface	$\frac{2}{5}M(a^5-b^5)/(a^3-b^3)+Ma^2$
11.11.	Hollow spherical shell of radius a	
(a)	about axis coinciding with a diameter	$\frac{2}{3}Ma^2$
(b)	about axis tangent to the surface	$\frac{5}{3}Ma^2$
11.12.	Ellipsoid with semi-axes a, b, c	
(a)	about axis coinciding with semi-axis c	$\frac{1}{5}M(a^2+b^2)$
(b)	about axis tangent to surface, parallel to semi-axis c and at distance a from center	$\frac{1}{5}M(6a^2+b^2)$
11.13.	Circular cone of radius a and height h	
(a)	about axis of cone	$\frac{3}{10}Ma^2$
(b)	about axis through vertex and perpendicular to axis	$\frac{3}{20}M(a^2+4h^2)$
(c)	about axis through center of mass and perpendicular to axis	$\frac{3}{80}M(4a^2+h^2)$
11.14.	Torus with outer radius a and inner radius b	
(a)	about axis through center of mass and perpendicular to the plane of torus	$\frac{1}{4}M(7a^2-6ab+3b^2)$
(b)	about axis through center of mass and in the plane of torus	$\frac{1}{4}M(9a^2-10ab+5b^2)$

Section III: Elementary Transcendental Functions

12 TRIGONOMETRIC FUNCTIONS

Definition of Trigonometric Functions for a Right Triangle

Triangle ABC has a right angle (90°) at C and sides of length a, b, c. The trigonometric functions of angle A are defined as follows:

12.1. $\quad$ *sine* of $A = \sin A = \dfrac{a}{c} = \dfrac{\text{opposite}}{\text{hypotenuse}}$

12.2. $\quad$ *cosine* of $A = \cos A = \dfrac{b}{c} = \dfrac{\text{adjacent}}{\text{hypotenuse}}$

12.3. $\quad$ *tangent* of $A = \tan A = \dfrac{a}{b} = \dfrac{\text{opposite}}{\text{adjacent}}$

12.4. $\quad$ *cotangent* of $A = \cot A = \dfrac{b}{a} = \dfrac{\text{adjacent}}{\text{opposite}}$

12.5. $\quad$ *secant* of $A = \sec A = \dfrac{c}{b} = \dfrac{\text{hypotenuse}}{\text{adjacent}}$

12.6. $\quad$ *cosecant* of $A = \csc A = \dfrac{c}{a} = \dfrac{\text{hypotenuse}}{\text{opposite}}$

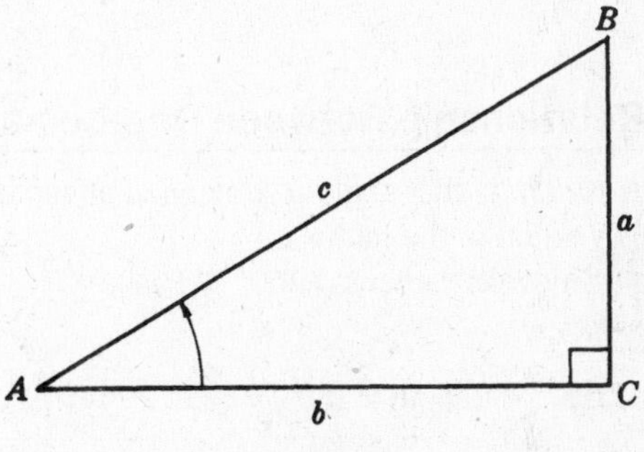

Fig. 12-1

Extensions to Angles Which May be Greater Than 90°

Consider an xy coordinate system (see Figs. 12-2 and 12-3). A point P in the xy plane has coordinates (x, y) where x is considered as positive along OX and negative along OX' while y is positive along OY and negative along OY'. The distance from origin O to point P is positive and denoted by $r = \sqrt{x^2 + y^2}$. The angle A described *counterclockwise* from OX is considered *positive*. If it is described *clockwise* from OX it is considered *negative*. We call $X'OX$ and $Y'OY$ the x and y axis, respectively.

The various quadrants are denoted by I, II, III, and IV called the first, second, third, and fourth quadrants, respectively. In Fig. 12-2, for example, angle A is in the second quadrant while in Fig. 12-3 angle A is in the third quadrant.

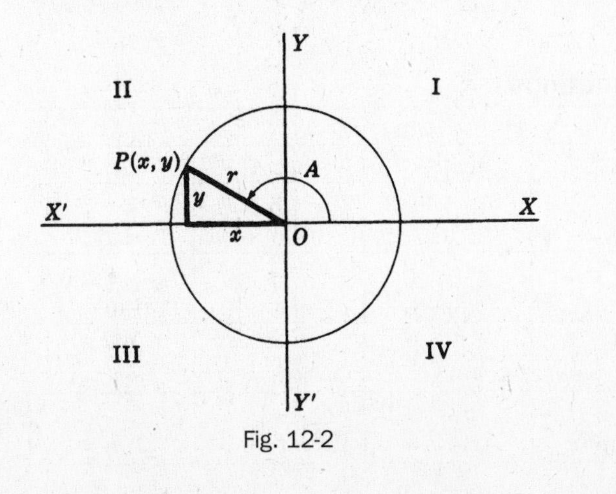

Fig. 12-2

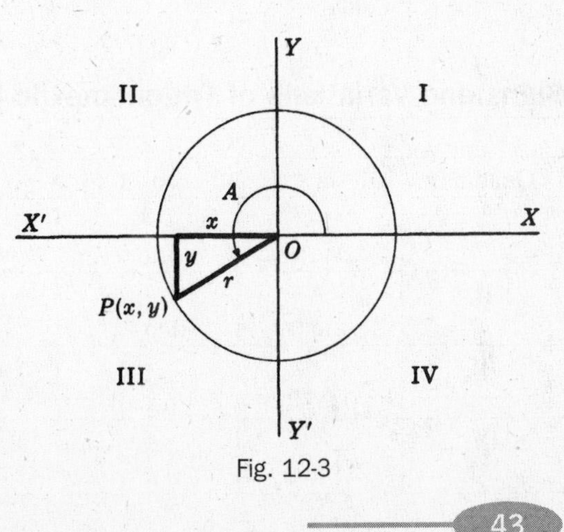

Fig. 12-3

For an angle A in any quadrant, the trigonometric functions of A are defined as follows.

12.7. $\sin A = y/r$

12.8. $\cos A = x/r$

12.9. $\tan A = y/x$

12.10. $\cot A = x/y$

12.11. $\sec A = r/x$

12.12. $\csc A = r/y$

Relationship Between Degrees and Radians

A *radian* is that angle θ subtended at center O of a circle by an arc MN equal to the radius r.

Since 2π radians = 360° we have

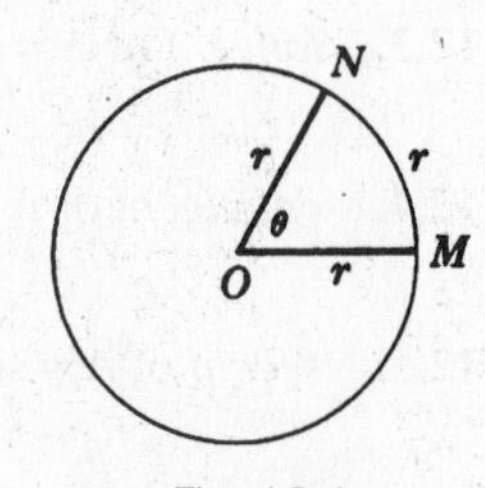

Fig. 12-4

12.13. $1 \text{ radian} = 180°/\pi = 57.29577\ 95130\ 8232 \ldots°$

12.14. $1° = \pi/180 \text{ radians} = 0.01745\ 32925\ 19943\ 29576\ 92 \ldots \text{ radians}$

Relationships Among Trigonometric Functions

12.15. $\tan A = \dfrac{\sin A}{\cos A}$

12.16. $\cot A = \dfrac{1}{\tan A} = \dfrac{\cos A}{\sin A}$

12.17. $\sec A = \dfrac{1}{\cos A}$

12.18. $\csc A = \dfrac{1}{\sin A}$

12.19. $\sin^2 A + \cos^2 A = 1$

12.20. $\sec^2 A - \tan^2 A = 1$

12.21. $\csc^2 A - \cot^2 A = 1$

Signs and Variations of Trigonometric Functions

Quadrant	$\sin A$	$\cos A$	$\tan A$	$\cot A$	$\sec A$	$\csc A$
I	+ 0 to 1	+ 1 to 0	+ 0 to ∞	+ ∞ to 0	+ 1 to ∞	+ ∞ to 1
II	+ 1 to 0	− 0 to −1	− −∞ to 0	− 0 to −∞	− −∞ to −1	+ 1 to ∞
III	− 0 to −1	− −1 to 0	+ 0 to ∞	+ ∞ to 0	− −1 to −∞	− −∞ to −1
IV	− −1 to 0	+ 0 to 1	− −∞ to 0	− 0 to −∞	+ ∞ to 1	− −1 to −∞

Exact Values for Trigonometric Functions of Various Angles

Angle A in degrees	Angle A in radians	$\sin A$	$\cos A$	$\tan A$	$\cot A$	$\sec A$	$\csc A$
0°	0	0	1	0	∞	1	∞
15°	$\pi/12$	$\frac{1}{4}(\sqrt{6}-\sqrt{2})$	$\frac{1}{4}(\sqrt{6}+\sqrt{2})$	$2-\sqrt{3}$	$2+\sqrt{3}$	$\sqrt{6}-\sqrt{2}$	$\sqrt{6}+\sqrt{2}$
30°	$\pi/6$	$\frac{1}{2}$	$\frac{1}{2}\sqrt{3}$	$\frac{1}{3}\sqrt{3}$	$\sqrt{3}$	$\frac{2}{3}\sqrt{3}$	2
45°	$\pi/4$	$\frac{1}{2}\sqrt{2}$	$\frac{1}{2}\sqrt{2}$	1	1	$\sqrt{2}$	$\sqrt{2}$
60°	$\pi/3$	$\frac{1}{2}\sqrt{3}$	$\frac{1}{2}$	$\sqrt{3}$	$\frac{1}{3}\sqrt{3}$	2	$\frac{2}{3}\sqrt{3}$
75°	$5\pi/12$	$\frac{1}{4}(\sqrt{6}+\sqrt{2})$	$\frac{1}{4}(\sqrt{6}-\sqrt{2})$	$2+\sqrt{3}$	$2-\sqrt{3}$	$\sqrt{6}+\sqrt{2}$	$\sqrt{6}-\sqrt{2}$
90°	$\pi/2$	1	0	$\pm\infty$	0	$\pm\infty$	1
105°	$7\pi/12$	$\frac{1}{4}(\sqrt{6}+\sqrt{2})$	$-\frac{1}{4}(\sqrt{6}-\sqrt{2})$	$-(2+\sqrt{3})$	$-(2-\sqrt{3})$	$-(\sqrt{6}+\sqrt{2})$	$\sqrt{6}-\sqrt{2}$
120°	$2\pi/3$	$\frac{1}{2}\sqrt{3}$	$-\frac{1}{2}$	$-\sqrt{3}$	$-\frac{1}{3}\sqrt{3}$	-2	$\frac{2}{3}\sqrt{3}$
135°	$3\pi/4$	$\frac{1}{2}\sqrt{2}$	$-\frac{1}{2}\sqrt{2}$	-1	-1	$-\sqrt{2}$	$\sqrt{2}$
150°	$5\pi/6$	$\frac{1}{2}$	$-\frac{1}{2}\sqrt{3}$	$-\frac{1}{3}\sqrt{3}$	$-\sqrt{3}$	$-\frac{2}{3}\sqrt{3}$	2
165°	$11\pi/12$	$\frac{1}{4}(\sqrt{6}-\sqrt{2})$	$-\frac{1}{4}(\sqrt{6}+\sqrt{2})$	$-(2-\sqrt{3})$	$-(2+\sqrt{3})$	$-(\sqrt{6}-\sqrt{2})$	$\sqrt{6}+\sqrt{2}$
180°	π	0	-1	0	$\mp\infty$	-1	$\pm\infty$
195°	$13\pi/12$	$-\frac{1}{4}(\sqrt{6}-\sqrt{2})$	$-\frac{1}{4}(\sqrt{6}+\sqrt{2})$	$2-\sqrt{3}$	$2+\sqrt{3}$	$-(\sqrt{6}-\sqrt{2})$	$-(\sqrt{6}+\sqrt{2})$
210°	$7\pi/6$	$-\frac{1}{2}$	$-\frac{1}{2}\sqrt{3}$	$\frac{1}{3}\sqrt{3}$	$\sqrt{3}$	$-\frac{2}{3}\sqrt{3}$	-2
225°	$5\pi/4$	$-\frac{1}{2}\sqrt{2}$	$-\frac{1}{2}\sqrt{2}$	1	1	$-\sqrt{2}$	$-\sqrt{2}$
240°	$4\pi/3$	$-\frac{1}{2}\sqrt{3}$	$-\frac{1}{2}$	$\sqrt{3}$	$\frac{1}{3}\sqrt{3}$	-2	$-\frac{2}{3}\sqrt{3}$
255°	$17\pi/12$	$-\frac{1}{4}(\sqrt{6}+\sqrt{2})$	$-\frac{1}{4}(\sqrt{6}-\sqrt{2})$	$2+\sqrt{3}$	$2-\sqrt{3}$	$-(\sqrt{6}+\sqrt{2})$	$-(\sqrt{6}-\sqrt{2})$
270°	$3\pi/2$	-1	0	$\pm\infty$	0	$\mp\infty$	-1
285°	$19\pi/12$	$-\frac{1}{4}(\sqrt{6}+\sqrt{2})$	$\frac{1}{4}(\sqrt{6}-\sqrt{2})$	$-(2+\sqrt{3})$	$-(2-\sqrt{3})$	$\sqrt{6}+\sqrt{2}$	$-(\sqrt{6}-\sqrt{2})$
300°	$5\pi/3$	$-\frac{1}{2}\sqrt{3}$	$\frac{1}{2}$	$-\sqrt{3}$	$-\frac{1}{3}\sqrt{3}$	2	$-\frac{2}{3}\sqrt{3}$
315°	$7\pi/4$	$-\frac{1}{2}\sqrt{2}$	$\frac{1}{2}\sqrt{2}$	-1	-1	$\sqrt{2}$	$-\sqrt{2}$
330°	$11\pi/6$	$-\frac{1}{2}$	$\frac{1}{2}\sqrt{3}$	$-\frac{1}{3}\sqrt{3}$	$-\sqrt{3}$	$\frac{2}{3}\sqrt{3}$	-2
345°	$23\pi/12$	$-\frac{1}{4}(\sqrt{6}-\sqrt{2})$	$\frac{1}{4}(\sqrt{6}+\sqrt{2})$	$-(2-\sqrt{3})$	$-(2+\sqrt{3})$	$\sqrt{6}-\sqrt{2}$	$-(\sqrt{6}+\sqrt{2})$
360°	2π	0	1	0	$\mp\infty$	1	$\mp\infty$

For other angles see Tables 2, 3, and 4.

Graphs of Trigonometric Functions

In each graph x is in radians.

12.22. $y = \sin x$

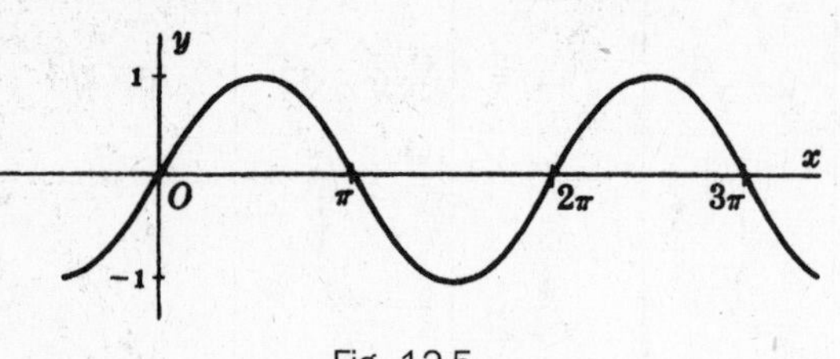

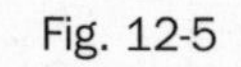
Fig. 12-5

12.23. $y = \cos x$

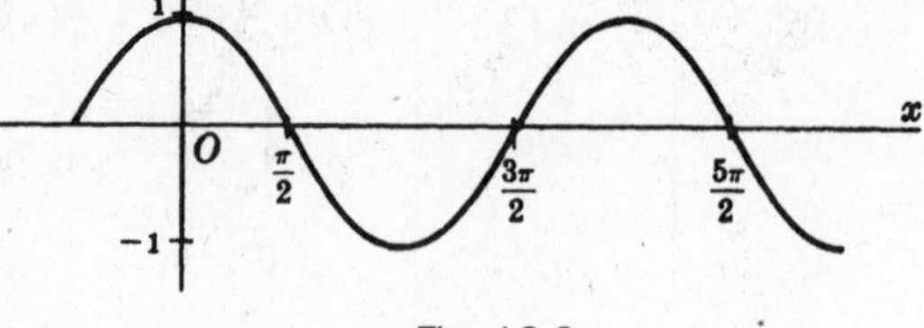

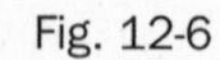
Fig. 12-6

12.24. $y = \tan x$

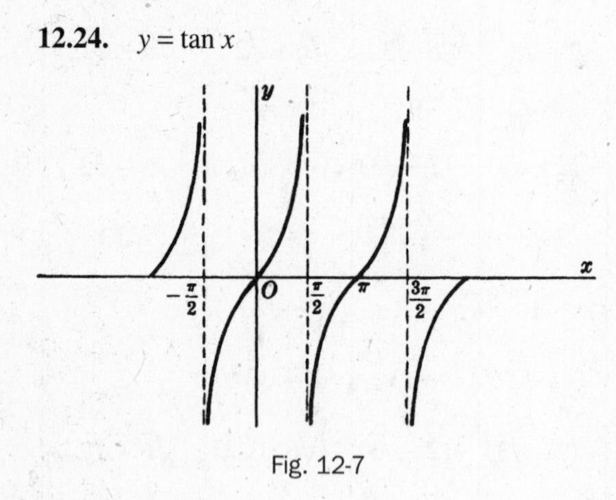

Fig. 12-7

12.25. $y = \cot x$

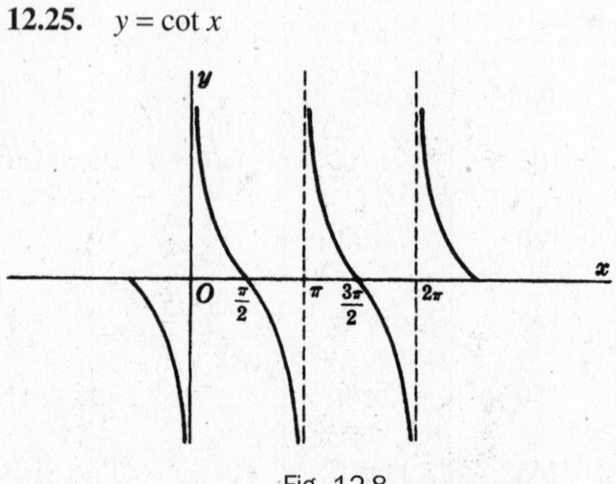

Fig. 12-8

12.26. $y = \sec x$

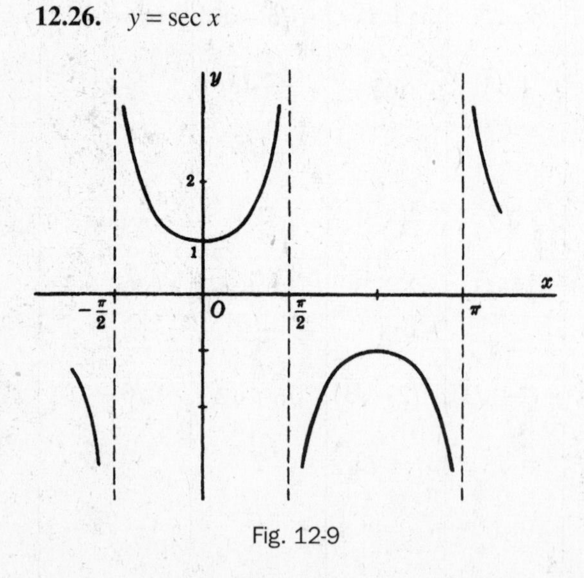

Fig. 12-9

12.27. $y = \csc x$

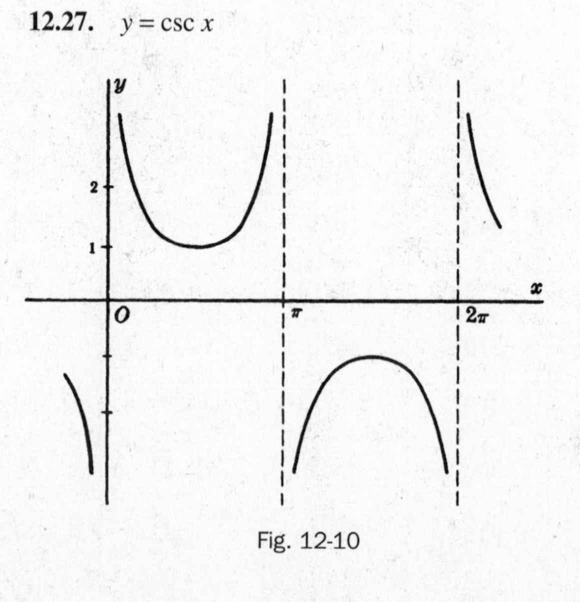

Fig. 12-10

Functions of Negative Angles

12.28. $\sin(-A) = -\sin A$

12.29. $\cos(-A) = \cos A$

12.30. $\tan(-A) = -\tan A$

12.31. $\csc(-A) = -\csc A$

12.32. $\sec(-A) = \sec A$

12.33. $\cot(-A) = -\cot A$

Addition Formulas

12.34. $\sin(A \pm B) = \sin A \cos B \pm \cos A \sin B$

12.35. $\cos(A \pm B) = \cos A \cos B \mp \sin A \sin B$

12.36. $\tan(A \pm B) = \dfrac{\tan A \pm \tan B}{1 \mp \tan A \tan B}$

12.37. $\cot(A \pm B) = \dfrac{\cot A \cot B \mp 1}{\cot B \pm \cot A}$

Functions of Angles in All Quadrants in Terms of Those in Quadrant I

	$-A$	$90° \pm A$ $\frac{\pi}{2} \pm A$	$180° \pm A$ $\pi \pm A$	$270° \pm A$ $\frac{3\pi}{2} \pm A$	$k(360°) \pm A$ $2k\pi \pm A$ k = integer
sin	$-\sin A$	$\cos A$	$\sin A$	$-\cos A$	$\pm \sin A$
cos	$\cos A$	$\mp \sin A$	$-\cos A$	$\mp \sin A$	$\cos A$
tan	$-\tan A$	$\mp \cot A$	$\pm \tan A$	$\mp \cot A$	$\pm \tan A$
csc	$-\csc A$	$\sec A$	$\mp \csc A$	$-\sec A$	$\pm \csc A$
sec	$\sec A$	$\mp \csc A$	$-\sec A$	$\pm \csc A$	$\sec A$
cot	$-\cot A$	$\mp \tan A$	$\pm \cot A$	$\mp \tan A$	$\pm \cot A$

Relationships Among Functions of Angles in Quadrant I

	$\sin A = u$	$\cos A = u$	$\tan A = u$	$\cot A = u$	$\sec A = u$	$\csc A = u$
$\sin A$	u	$\sqrt{1-u^2}$	$u/\sqrt{1+u^2}$	$1/\sqrt{1+u^2}$	$\sqrt{u^2-1}/u$	$1/u$
$\cos A$	$\sqrt{1-u^2}$	u	$1/\sqrt{1+u^2}$	$u/\sqrt{1+u^2}$	$1/u$	$\sqrt{u^2-1}/u$
$\tan A$	$u/\sqrt{1-u^2}$	$\sqrt{1-u^2}/u$	u	$1/u$	$\sqrt{u^2-1}$	$1/\sqrt{u^2-1}$
$\cot A$	$\sqrt{1-u^2}/u$	$u/\sqrt{1-u^2}$	$1/u$	u	$1/\sqrt{u^2-1}$	$\sqrt{u^2-1}$
$\sec A$	$1/\sqrt{1-u^2}$	$1/u$	$\sqrt{1+u^2}$	$\sqrt{1+u^2}/u$	u	$u/\sqrt{u^2-1}$
$\csc A$	$1/u$	$1/\sqrt{1-u^2}$	$\sqrt{1+u^2}/u$	$\sqrt{1+u^2}$	$u/\sqrt{u^2-1}$	u

For extensions to other quadrants use appropriate signs as given in the preceding table.

Double Angle Formulas

12.38. $\sin 2A = 2\sin A\cos A$

12.39. $\cos 2A = \cos^2 A - \sin^2 A = 1 - 2\sin^2 A = 2\cos^2 A - 1$

12.40. $\tan 2A = \dfrac{2\tan A}{1-\tan^2 A}$

Half Angle Formulas

12.41. $\sin\dfrac{A}{2} = \pm\sqrt{\dfrac{1-\cos A}{2}}\quad\begin{bmatrix}+\text{ if } A/2 \text{ is in quadrant I or II}\\ -\text{ if } A/2 \text{ is in quadrant III or IV}\end{bmatrix}$

12.42. $\cos\dfrac{A}{2} = \pm\sqrt{\dfrac{1+\cos A}{2}}\quad\begin{bmatrix}+\text{ if } A/2 \text{ is in quadrant I or IV}\\ -\text{ if } A/2 \text{ is in quadrant II or III}\end{bmatrix}$

12.43. $\tan\dfrac{A}{2} = \pm\sqrt{\dfrac{1-\cos A}{1+\cos A}}\quad\begin{bmatrix}+\text{ if } A/2 \text{ is in quadrant I or III}\\ -\text{ if } A/2 \text{ is in quadrant II or IV}\end{bmatrix}$

$$= \frac{\sin A}{1+\cos A} = \frac{1-\cos A}{\sin A} = \csc A - \cot A$$

Multiple Angle Formulas

12.44. $\sin 3A = 3\sin A - 4\sin^3 A$

12.45. $\cos 3A = 4\cos^3 A - 3\cos A$

12.46. $\tan 3A = \dfrac{3\tan A - \tan^3 A}{1-3\tan^2 A}$

12.47. $\sin 4A = 4\sin A\cos A - 8\sin^3 A\cos A$

12.48. $\cos 4A = 8\cos^4 A - 8\cos^2 A + 1$

12.49. $\tan 4A = \dfrac{4\tan A - 4\tan^3 A}{1-6\tan^2 A + \tan^4 A}$

12.50. $\sin 5A = 5\sin A - 20\sin^3 A + 16\sin^5 A$

12.51. $\cos 5A = 16\cos^5 A - 20\cos^3 A + 5\cos A$

12.52. $\tan 5A = \dfrac{\tan^5 A - 10\tan^3 A + 5\tan A}{1-10\tan^2 A + 5\tan^4 A}$

See also formulas 12.68 and 12.69.

Powers of Trignometric Functions

12.53. $\sin^2 A = \frac{1}{2} - \frac{1}{2}\cos 2A$

12.54. $\cos^2 A = \frac{1}{2} + \frac{1}{2}\cos 2A$

12.55. $\sin^3 A = \frac{3}{4}\sin A - \frac{1}{4}\sin 3A$

12.56. $\cos^3 A = \frac{3}{4}\cos A + \frac{1}{4}\cos 3A$

12.57. $\sin^4 A = \frac{3}{8} - \frac{1}{2}\cos 2A + \frac{1}{8}\cos 4A$

12.58. $\cos^4 A = \frac{3}{8} + \frac{1}{2}\cos 2A + \frac{1}{8}\cos 4A$

12.59. $\sin^5 A = \frac{5}{8}\sin A - \frac{5}{16}\sin 3A + \frac{1}{16}\sin 5A$

12.60. $\cos^5 A = \frac{5}{8}\cos A + \frac{5}{16}\cos 3A + \frac{1}{16}\cos 5A$

See also formulas 12.70 through 12.73.

Sum, Difference, and Product of Trignometric Functions

12.61. $\sin A + \sin B = 2\sin\frac{1}{2}(A+B)\cos\frac{1}{2}(A-B)$

12.62. $\sin A - \sin B = 2\cos\frac{1}{2}(A+B)\sin\frac{1}{2}(A-B)$

12.63. $\cos A + \cos B = 2\cos\frac{1}{2}(A+B)\cos\frac{1}{2}(A-B)$

12.64. $\cos A - \cos B = 2\sin\frac{1}{2}(A+B)\sin\frac{1}{2}(B-A)$

12.65. $\sin A \sin B = \frac{1}{2}\{\cos(A-B) - \cos(A-B)\}$

12.66. $\cos A \cos B = \frac{1}{2}\{\cos(A-B) + \cos(A+B)\}$

12.67. $\sin A \cos B = \frac{1}{2}\{\sin(A-B) + \sin(A+B)\}$

General Formulas

12.68. $$\sin nA = \sin A\left\{(2\cos A)^{n-1} - \binom{n-2}{1}(2\cos A)^{n-3} + \binom{n-3}{2}(2\cos A)^{n-5} - \cdots\right\}$$

12.69. $$\cos nA = \frac{1}{2}\left\{(2\cos A)^n - \frac{n}{1}(2\cos A)^{n-2} + \frac{n}{2}\binom{n-3}{1}(2\cos A)^{n-4} - \frac{n}{3}\binom{n-4}{2}(2\cos A)^{n-6} + \cdots\right\}$$

12.70. $$\sin^{2n-1} A = \frac{(-1)^{n-1}}{2^{2n-2}}\left\{\sin(2n-1)A - \binom{2n-1}{1}\sin(2n-3)A + \cdots (-1)^{n-1}\binom{2n-1}{n-1}\sin A\right\}$$

12.71. $$\cos^{2n-1} A = \frac{1}{2^{2n-2}}\left\{\cos(2n-1)A + \binom{2n-1}{1}\cos(2n-3)A + \cdots + \binom{2n-1}{n-1}\cos A\right\}$$

12.72. $$\sin^{2n} A = \frac{1}{2^{2n}}\binom{2n}{n} + \frac{(-1)^n}{2^{2n-1}}\left\{\cos 2nA - \binom{2n}{1}\cos(2n-2)A + \cdots (-1)^{n-1}\binom{2n}{n-1}\cos 2A\right\}$$

12.73. $$\cos^{2n} A = \frac{1}{2^{2n}}\binom{2n}{n} + \frac{1}{2^{2n-1}}\left\{\cos 2nA + \binom{2n}{1}\cos(2n-2)A + \cdots + \binom{2n}{n-1}\cos 2A\right\}$$

Inverse Trigonometric Functions

If $x = \sin y$, then $y = \sin^{-1}x$, i.e. *the angle whose sine is* x or *inverse sine of* x is a many-valued function of x which is a collection of single-valued functions called *branches*. Similarly, the other inverse trigonometric functions are multiple-valued.

For many purposes a particular branch is required. This is called the *principal branch* and the values for this branch are called *principal values.*

Principal Values for Inverse Trigonometric Functions

Principal values for $x \geqq 0$	Principal values for $x < 0$
$0 \leqq \sin^{-1} x \leqq \pi/2$	$-\pi/2 \leqq \sin^{-1} x < 0$
$0 \leqq \cos^{-1} x \leqq \pi/2$	$\pi/2 < \cos^{-1} x \leqq \pi$
$0 \leqq \tan^{-1} x < \pi/2$	$-\pi/2 < \tan^{-1} x < 0$
$0 < \cot^{-1} x \leqq \pi/2$	$\pi/2 < \cot^{-1} x < \pi$
$0 \leqq \sec^{-1} x < \pi/2$	$\pi/2 < \sec^{-1} x \leqq \pi$
$0 < \csc^{-1} x \leqq \pi/2$	$-\pi/2 \leqq \csc^{-1} x < 0$

Relations Between Inverse Trigonometric Functions

In all cases it is assumed that principal values are used.

12.74. $\sin^{-1} x + \cos^{-1} x = \pi/2$

12.75. $\tan^{-1} x + \cot^{-1} x = \pi / 2$

12.76. $\sec^{-1} x + \csc^{-1} x = \pi/2$

12.77. $\csc^{-1} x = \sin^{-1}(1/x)$

12.78. $\sec^{-1} x = \cos^{-1}(1/x)$

12.79. $\cot^{-1} x = \tan^{-1}(1/x)$

12.80. $\sin^{-1}(-x) = -\sin^{-1} x$

12.81. $\cos^{-1}(-x) = \pi - \cos^{-1} x$

12.82. $\tan^{-1}(-x) = -\tan^{-1} x$

12.83. $\cot^{-1}(-x) = \pi - \cot^{-1} x$

12.84. $\sec^{-1}(-x) = \pi - \sec^{-1} x$

12.85. $\csc^{-1}(-x) = -\csc^{-1} x$

Graphs of Inverse Trigonometric Functions

In each graph y is in radians. Solid portions of curves correspond to principal values.

12.86. $y = \sin^{-1} x$

12.87. $y = \cos^{-1} x$

12.88. $y = \tan^{-1} x$

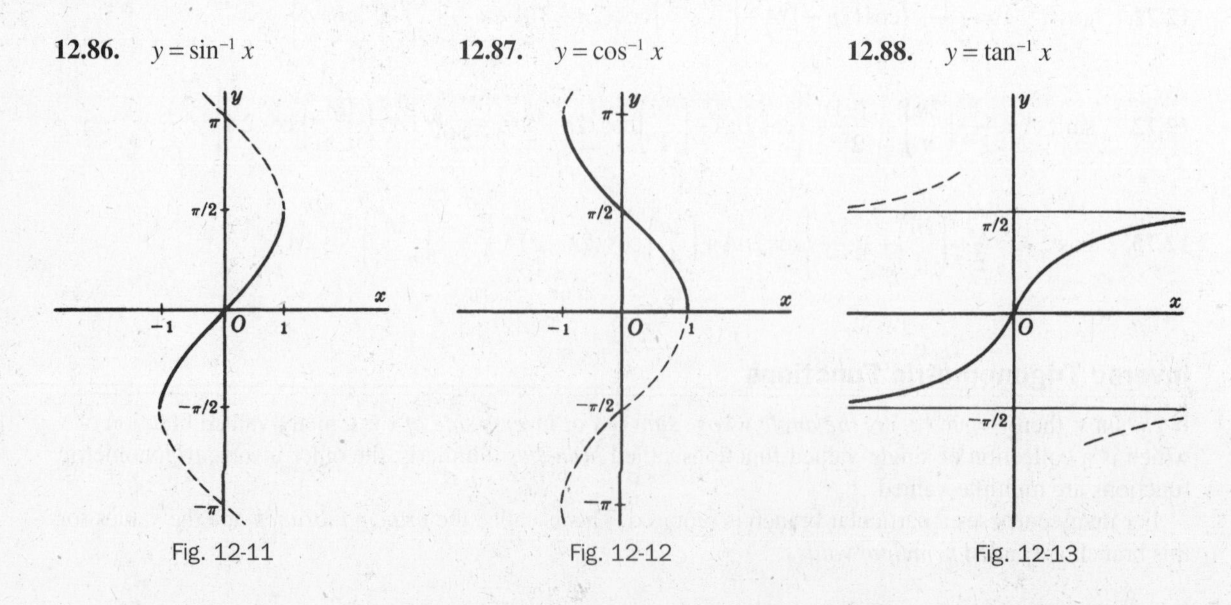

Fig. 12-11 Fig. 12-12 Fig. 12-13

12.89. $y = \cot^{-1} x$

12.90. $y = \sec^{-1} x$

12.91. $y = \csc^{-1} x$

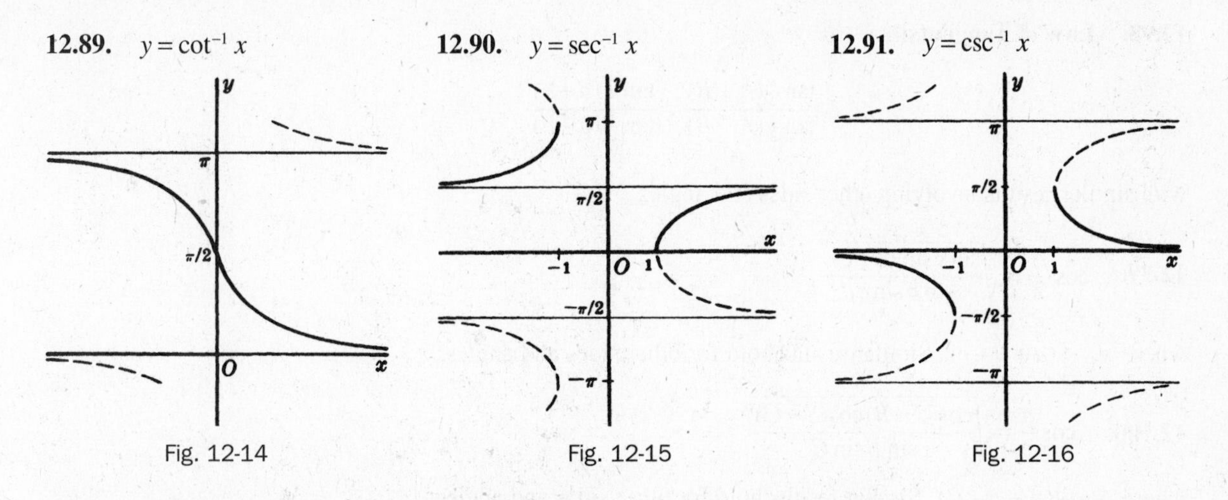

Fig. 12-14 Fig. 12-15 Fig. 12-16

Relationships Between Sides and Angles of a Plane Triangle

The following results hold for any plane triangle ABC with sides a, b, c and angles A, B, C.

12.92. Law of Sines:

$$\frac{a}{\sin A} = \frac{b}{\sin B} = \frac{c}{\sin C}$$

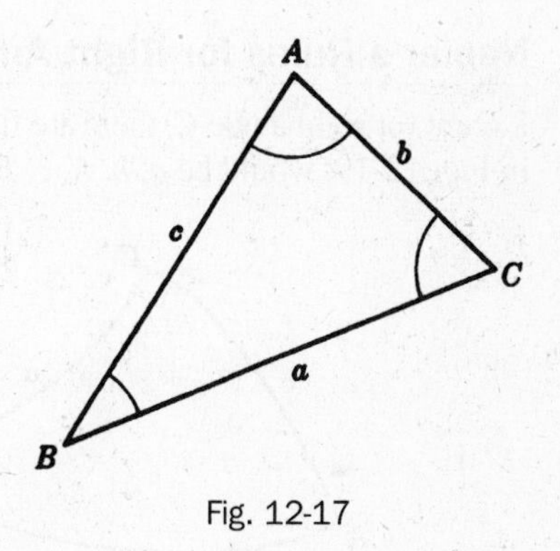

Fig. 12-17

12.93. Law of Cosines:

$$c^2 = a^2 + b^2 - 2ab \cos C$$

with similar relations involving the other sides and angles.

12.94. Law of Tangents:

$$\frac{a+b}{a-b} = \frac{\tan \frac{1}{2}(A+B)}{\tan \frac{1}{2}(A-B)}$$

with similar relations involving the other sides and angles.

12.95. $\sin A = \dfrac{2}{bc}\sqrt{s(s-a)(s-b)(s-c)}$

where $s = \frac{1}{2}(a+b+c)$ is the semiperimeter of the triangle. Similar relations involving angles B and C can be obtained.

See also formula 7.5.

Relationships Between Sides and Angles of a Spherical Triangle

Spherical triangle ABC is on the surface of a sphere as shown in Fig. 12-18. Sides a, b, c (which are arcs of great circles) are measured by their angles subtended at center O of the sphere. A, B, C are the angles opposite sides a, b, c, respectively. Then the following results hold.

12.96. Law of Sines:

$$\frac{\sin a}{\sin A} = \frac{\sin b}{\sin B} = \frac{\sin c}{\sin C}$$

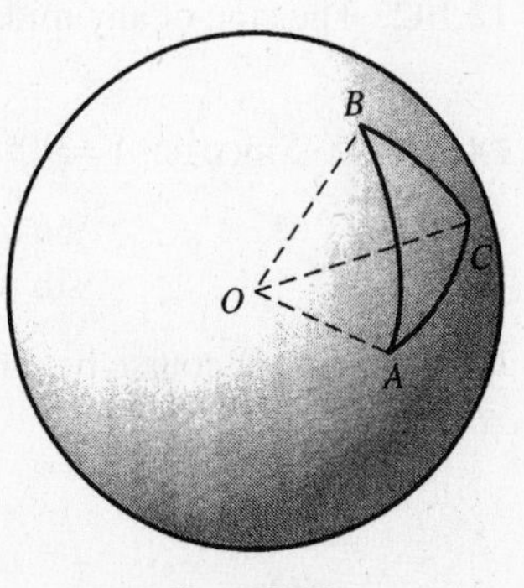

Fig. 12-18

12.97. Law of Cosines:

$$\cos a = \cos b \cos c + \sin b \sin c \cos A$$
$$\cos A = -\cos B \cos C + \sin B \sin C \cos a$$

with similar results involving other sides and angles.

12.98. Law of Tangents:

$$\frac{\tan\frac{1}{2}(A+B)}{\tan\frac{1}{2}(A-B)} = \frac{\tan\frac{1}{2}(a+b)}{\tan\frac{1}{2}(a-b)}$$

with similar results involving other sides and angles.

12.99. $\cos\frac{A}{2} = \sqrt{\frac{\sin s \sin(s-c)}{\sin b \sin c}}$

where $s = \frac{1}{2}(a+b+c)$. Similar results hold for other sides and angles.

12.100. $\cos\frac{a}{2} = \sqrt{\frac{\cos(S-B)\cos(S-C)}{\sin B \sin C}}$

where $S = \frac{1}{2}(A+B+C)$. Similar results hold for other sides and angles.

See also formula 7.44.

Napier's Rules for Right Angled Spherical Triangles

Except for right angle C, there are five parts of spherical triangle ABC which, if arranged in the order as given in Fig. 12-19, would be a, b, A, c, B.

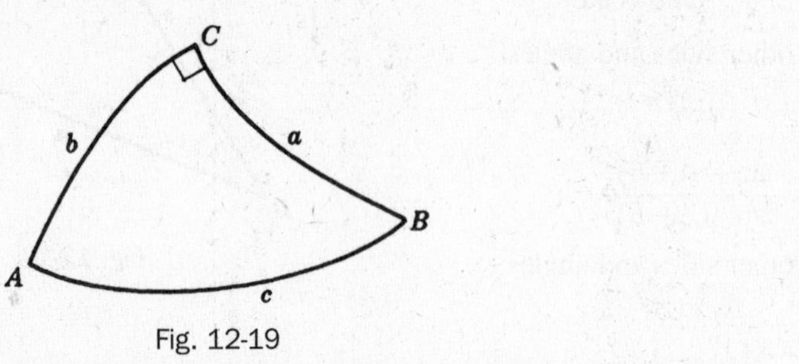

Fig. 12-19

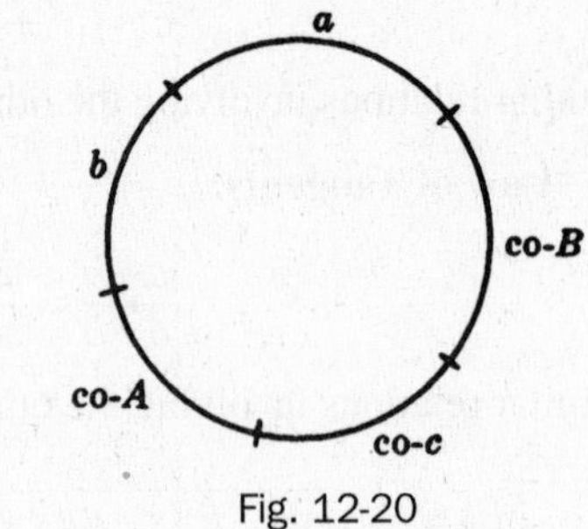

Fig. 12-20

Suppose these quantities are arranged in a circle as in Fig. 12-20 where we attach the prefix "co" (indicating *complement*) to hypotenuse c and angles A and B.

Any one of the parts of this circle is called a *middle part*, the two neighboring parts are called *adjacent parts,* and the two remaining parts are called *opposite parts*. Then Napier's rules are

12.101. The sine of any middle part equals the product of the tangents of the adjacent parts.

12.102. The sine of any middle part equals the product of the cosines of the opposite parts.

EXAMPLE: Since co-$A = 90° - A$, co-$B = 90° - B$, we have

$$\sin a = \tan b\,(\text{co-}B) \quad \text{or} \quad \sin a = \tan b \cot B$$
$$\sin(\text{co-}A) = \cos a \cos(\text{co-}B) \quad \text{or} \quad \cos A = \cos a \sin B$$

These can of course be obtained also from the results of 12.97.

13 EXPONENTIAL and LOGARITHMIC FUNCTIONS

Laws of Exponents

In the following p, q are real numbers, a, b are positive numbers, and m, n are positive integers.

13.1. $a^p \cdot a^q = a^{p+q}$ **13.2.** $a^p/a^q = a^{p-q}$ **13.3.** $(a^p)^q = a^{pq}$

13.4. $a^0 = 1, \; a \neq 0$ **13.5.** $a^{-p} = 1/a^p$ **13.6.** $(ab)^p = a^p b^p$

13.7. $\sqrt[n]{a} = a^{1/n}$ **13.8.** $\sqrt[n]{a^m} = a^{m/n}$ **13.9.** $\sqrt[n]{a/b} = \sqrt[n]{a}/\sqrt[n]{b}$

In a^p, p is called the *exponent*, a is the *base*, and a^p is called the pth *power of a*. The function $y = a^x$ is called an *exponential function*.

Logarithms and Antilogarithms

If $a^p = N$ where $a \neq 0$ or 1, then $p = \log_a N$ is called the *logarithm* of N to the base a. The number $N = a^p$ is called the *antilogarithm* of p to the base a, written $\text{antilog}_a\, p$.

Example: Since $3^2 = 9$ we have $\log_3 9 = 2$. $\text{antilog}_3\, 2 = 9$.

The function $y = \log_a x$ is called a *logarithmic function*.

Laws of Logarithms

13.10. $\log_a MN = \log_a M + \log_a N$

13.11. $\log_a \dfrac{M}{N} = \log_a M - \log_a N$

13.12. $\log_a M^p = p \log_a M$

Common Logarithms and Antilogarithms

Common logarithms and antilogarithms (also called *Briggsian*) are those in which the base $a = 10$. The common logarithm of N is denoted by $\log_{10} N$ or briefly $\log N$. For numerical values of common logarithms, see Table 1.

Natural Logarithms and Antilogarithms

Natural logarithms and antilogarithms (also called *Napierian*) are those in which the base $a = e = 2.71828\ 18 \ldots$ [see page 3]. The natural logarithm of N is denoted by $\log_e N$ or In N. For numerical values of natural logarithms see Table 7. For values of natural antilogarithms (i.e., a table giving e^x for values of x) see Table 8.

Change of Base of Logarithms

The relationship between logarithms of a number N to different bases a and b is given by

13.13. $\log_a N = \dfrac{\log_b N}{\log_b a}$

In particular,

13.14. $\log_e N = \ln N = 2.30258\ 50929\ 94 \ldots \log_{10} N$

13.15. $\log_{10} N = \log N = 0.43429\ 44819\ 03 \ldots \log_e N$

Relationship Between Exponential and Trigonometric Functions

13.16. $e^{i\theta} = \cos\theta + i\sin\theta, \qquad e^{-i\theta} = \cos\theta - i\sin\theta$

These are called *Euler's identities.* Here i is the imaginary unit [see page 10].

13.17. $\sin\theta = \dfrac{e^{i\theta} - e^{-i\theta}}{2i}$

13.18. $\cos\theta = \dfrac{e^{i\theta} + e^{-i\theta}}{2}$

13.19. $\tan\theta = \dfrac{e^{i\theta} - e^{-i\theta}}{i(e^{i\theta} + e^{-i\theta})} = -i\left(\dfrac{e^{i\theta} - e^{-i\theta}}{e^{i\theta} + e^{-i\theta}}\right)$

13.20. $\cot\theta = i\left(\dfrac{e^{i\theta} + e^{-i\theta}}{e^{i\theta} - e^{-i\theta}}\right)$

13.21. $\sec\theta = \dfrac{2}{e^{i\theta} + e^{-i\theta}}$

13.22. $\csc\theta = \dfrac{2i}{e^{i\theta} - e^{-i\theta}}$

Periodicity of Exponential Functions

13.23. $e^{i(\theta + 2k\pi)} = e^{i\theta} \qquad k = \text{integer}$

From this it is seen that e^x has period $2\pi i$.

Polar Form of Complex Numbers Expressed as an Exponential

The polar form (see 4.7) of a complex number $z = x + iy$ can be written in terms of exponentials as follows:

13.24. $z = x + iy = r(\cos\theta + i\sin\theta) = re^{i\theta}$

Operations with Complex Numbers in Polar Form

Formulas 4.8 to 4.11 are equivalent to the following:

13.25. $(r_1 e^{i\theta_1})(r_2 e^{i\theta_2}) = r_1 r_2 e^{i(\theta_1+\theta_2)}$

13.26. $\dfrac{r_1 e^{i\theta_1}}{r_2 e^{i\theta_2}} = \dfrac{r_1}{r_2} e^{i(\theta_1-\theta_2)}$

13.27. $(re^{i\theta})^p = r^p e^{ip\theta}$ (De Moivre's theorem)

13.28. $(re^{i\theta})^{1/n} = [re^{i(\theta+2k\pi)}]^{1/n} = r^{1/n} e^{i(\theta+2k\pi)/n}$

Logarithm of a Complex Number

13.29. $\ln(re^{i\theta}) = \ln r + i\theta + 2k\pi i \quad k = \text{integer}$

14 HYPERBOLIC FUNCTIONS

Definition of Hyperbolic Functions

14.1. *Hyperbolic sine* of x $= \sinh x = \dfrac{e^x - e^{-x}}{2}$

14.2. *Hyperbolic cosine* of x $= \cosh x = \dfrac{e^x + e^{-x}}{2}$

14.3. *Hyperbolic tangent* of x $= \tanh x = \dfrac{e^x - e^{-x}}{e^x + e^{-x}}$

14.4. *Hyperbolic cotangent* of x $= \coth x = \dfrac{e^x + e^{-x}}{e^x - e^{-x}}$

14.5. *Hyperbolic secant* of x $= \operatorname{sech} x = \dfrac{2}{e^x + e^{-x}}$

14.6. *Hyperbolic cosecant* of x $= \operatorname{csch} x = \dfrac{2}{e^x - e^{-x}}$

Relationships Among Hyperbolic Functions

14.7. $\tanh x = \dfrac{\sinh x}{\cosh x}$

14.8. $\coth x = \dfrac{1}{\tanh x} = \dfrac{\cosh x}{\sinh x}$

14.9. $\operatorname{sech} x = \dfrac{1}{\cosh x}$

14.10. $\operatorname{csch} x = \dfrac{1}{\sinh x}$

14.11. $\cosh^2 x - \sinh^2 x = 1$

14.12. $\operatorname{sech}^2 x + \tanh^2 x = 1$

14.13. $\coth^2 x - \operatorname{csch}^2 x = 1$

Functions of Negative Arguments

14.14. $\sinh(-x) = -\sinh x$ **14.15.** $\cosh(-x) = \cosh x$ **14.16.** $\tanh(-x) = -\tanh x$

14.17. $\operatorname{csch}(-x) = -\operatorname{csch} x$ **14.18.** $\operatorname{sech}(-x) = \operatorname{sech} x$ **14.19.** $\coth(-x) = -\coth x$

Addition Formulas

14.20. $\sinh(x \pm y) = \sinh x \cosh y \pm \cosh x \sinh y$

14.21. $\cosh(x \pm y) = \cosh x \cosh y \pm \sinh x \sinh y$

14.22. $\tanh(x \pm y) = \dfrac{\tanh x \pm \tanh y}{1 \pm \tanh x \tanh y}$

14.23. $\coth(x \pm y) = \dfrac{\coth x \coth y \pm 1}{\coth y \pm \coth x}$

Double Angle Formulas

14.24. $\sinh 2x = 2 \sinh x \cosh x$

14.25. $\cosh 2x = \cosh^2 x + \sinh^2 x = 2\cosh^2 x - 1 = 1 + 2\sinh^2 x$

14.26. $\tanh 2x = \dfrac{2 \tanh x}{1 + \tanh^2 x}$

Half Angle Formulas

14.27. $\sinh \dfrac{x}{2} = \pm \sqrt{\dfrac{\cosh x - 1}{2}}$ $[+ \text{ if } x > 0, - \text{ if } x < 0]$

14.28. $\cosh \dfrac{x}{2} = \sqrt{\dfrac{\cosh x + 1}{2}}$

14.29. $\tanh \dfrac{x}{2} = \pm \sqrt{\dfrac{\cosh x - 1}{\cosh x + 1}}$ $[+ \text{ if } x > 0, - \text{ if } x < 0]$

$$= \frac{\sinh x}{\cosh x + 1} = \frac{\cosh x - 1}{\sinh x}$$

Multiple Angle Formulas

14.30. $\sinh 3x = 3 \sinh x + 4 \sinh^3 x$

14.31. $\cosh 3x = 4 \cosh^3 x - 3 \cosh x$

14.32. $\tanh 3x = \dfrac{3 \tanh x + \tanh^3 x}{1 + 3 \tanh^2 x}$

14.33. $\sinh 4x = 8 \sinh^3 x \cosh x + 4 \sinh x \cosh x$

14.34. $\cosh 4x = 8 \cosh^4 x - 8 \cosh^2 x + 1$

14.35. $\tanh 4x = \dfrac{4 \tanh x + 4 \tanh^3 x}{1 + 6 \tanh^2 x + \tanh^4 x}$

Powers of Hyperbolic Functions

14.36. $\sinh^2 x = \frac{1}{2}\cosh 2x - \frac{1}{2}$

14.37. $\cosh^2 x = \frac{1}{2}\cosh 2x + \frac{1}{2}$

14.38. $\sinh^3 x = \frac{1}{4}\sinh 3x - \frac{3}{4}\sinh x$

14.39. $\cosh^3 x = \frac{1}{4}\cosh 3x + \frac{3}{4}\cosh x$

14.40. $\sinh^4 x = \frac{3}{8} - \frac{1}{2}\cosh 2x + \frac{1}{8}\cosh 4x$

14.41. $\cosh^4 x = \frac{3}{8} + \frac{1}{2}\cosh 2x + \frac{1}{8}\cosh 4x$

Sum, Difference, and Product of Hyperbolic Functions

14.42. $\sinh x + \sinh y = 2\sinh \frac{1}{2}(x+y)\cosh \frac{1}{2}(x-y)$

14.43. $\sinh x - \sinh y = 2\cosh \frac{1}{2}(x+y)\sinh \frac{1}{2}(x-y)$

14.44. $\cosh x + \cosh y = 2\cosh \frac{1}{2}(x+y)\cosh \frac{1}{2}(x-y)$

14.45. $\cosh x - \cosh y = 2\sinh \frac{1}{2}(x+y)\sinh \frac{1}{2}(x-y)$

14.46. $\sinh x \sinh y = \frac{1}{2}\{\cosh(x+y) - \cosh(x-y)\}$

14.47. $\cosh x \cosh y = \frac{1}{2}\{\cosh(x+y) + \cosh(x-y)\}$

14.48. $\sinh x \cosh y = \frac{1}{2}\{\sinh(x+y) + \sinh(x-y)\}$

Expression of Hyperbolic Functions in Terms of Others

In the following we assume $x > 0$. If $x < 0$, use the appropriate sign as indicated by formulas 14.14 to 14.19.

	$\sinh x = u$	$\cosh x = u$	$\tanh x = u$	$\coth x = u$	$\operatorname{sech} x = u$	$\operatorname{csch} x = u$
$\sinh x$	u	$\sqrt{u^2-1}$	$u/\sqrt{1-u^2}$	$1/\sqrt{u^2-1}$	$\sqrt{1-u^2}/u$	$1/u$
$\cosh x$	$\sqrt{1+u^2}$	u	$1/\sqrt{1-u^2}$	$u/\sqrt{u^2-1}$	$1/u$	$\sqrt{1+u^2}/u$
$\tanh x$	$u/\sqrt{1+u^2}$	$\sqrt{u^2-1}/u$	u	$1/u$	$\sqrt{1-u^2}$	$1/\sqrt{1+u^2}$
$\coth x$	$\sqrt{u^2+1}/u$	$u/\sqrt{u^2-1}$	$1/u$	u	$1/\sqrt{1-u^2}$	$\sqrt{1+u^2}$
$\operatorname{sech} x$	$1/\sqrt{1+u^2}$	$1/u$	$\sqrt{1-u^2}$	$\sqrt{u^2-1}/u$	u	$u/\sqrt{1+u^2}$
$\operatorname{csch} x$	$1/u$	$1/\sqrt{u^2-1}$	$\sqrt{1-u^2}/u$	$\sqrt{u^2-1}$	$u/\sqrt{1-u^2}$	u

Graphs of Hyperbolic Functions

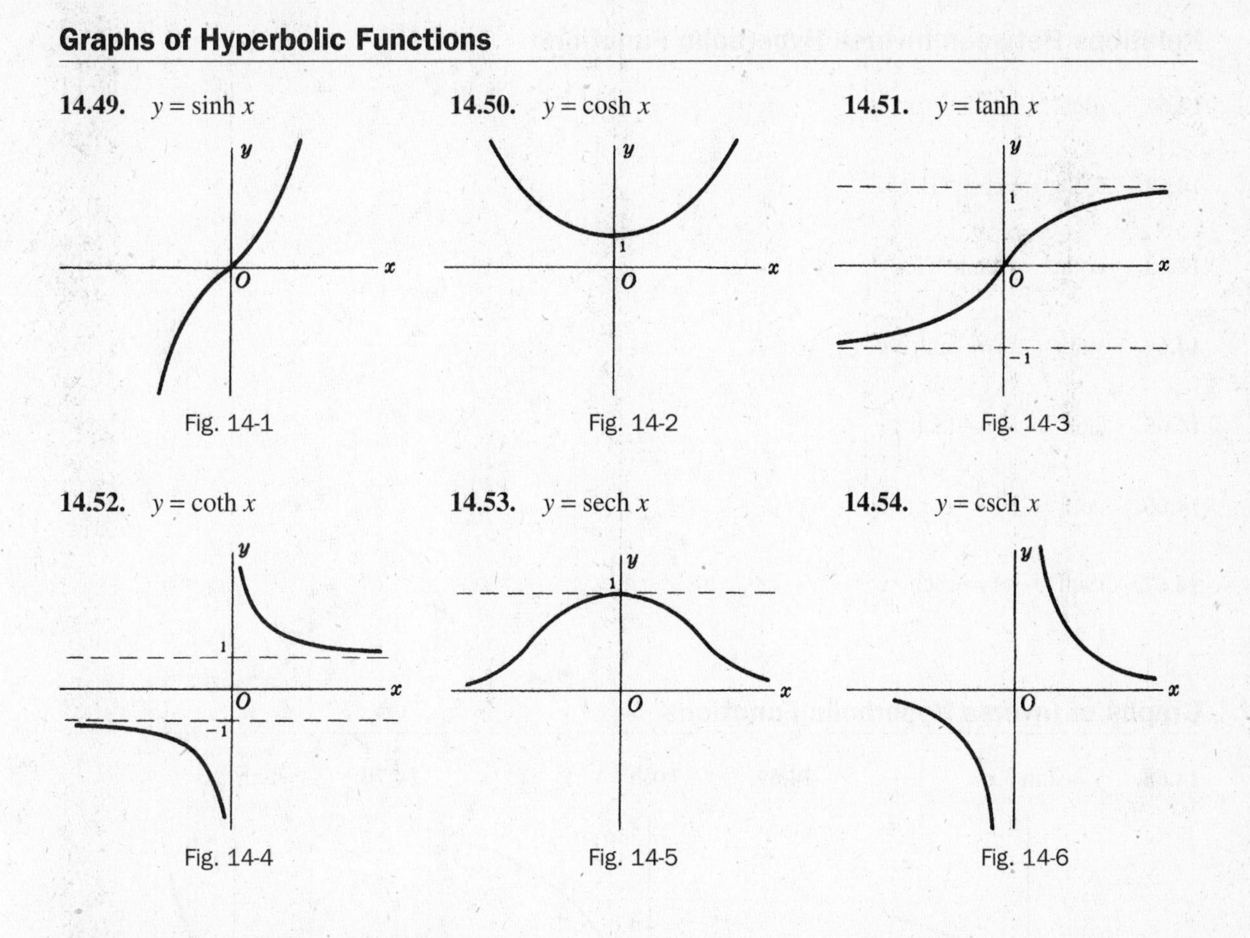

Fig. 14-1 Fig. 14-2 Fig. 14-3

Fig. 14-4 Fig. 14-5 Fig. 14-6

Inverse Hyperbolic Functions

If $x = \sinh y$, then $y = \sinh^{-1} x$ is called the *inverse hyperbolic sine* of x. Similarly we define the other inverse hyperbolic functions. The inverse hyperbolic functions are multiple-valued and as in the case of inverse trigonometric functions [see page 49] we restrict ourselves to principal values for which they can be considered as single-valued.

The following list shows the principal values (unless otherwise indicated) of the inverse hyperbolic functions expressed in terms of logarithmic functions which are taken as real valued.

14.55. $\sinh^{-1} x = \ln(x + \sqrt{x^2 + 1})$ $\quad -\infty < x < \infty$

14.56. $\cosh^{-1} x = \ln(x + \sqrt{x^2 - 1})$ $\quad x \geqq 1 \quad (\cosh^{-1} x > 0$ is principal value)

14.57. $\tanh^{-1} x = \dfrac{1}{2}\ln\left(\dfrac{1+x}{1-x}\right)$ $\quad -1 < x < 1$

14.58. $\coth^{-1} x = \dfrac{1}{2}\ln\left(\dfrac{x+1}{x-1}\right)$ $\quad x > 1$ or $x < -1$

14.59. $\operatorname{sech}^{-1} x = \ln\left(\dfrac{1}{x} + \sqrt{\dfrac{1}{x^2} - 1}\right)$ $\quad 0 < x \leqq 1 \quad (\operatorname{sech}^{-1} x > 0$ is principal value)

14.60. $\operatorname{csch}^{-1} x = \ln\left(\dfrac{1}{x} + \sqrt{\dfrac{1}{x^2} + 1}\right)$ $\quad x \neq 0$

Relations Between Inverse Hyperbolic Functions

14.61. $\operatorname{csch}^{-1} x = \sinh^{-1}(1/x)$

14.62. $\operatorname{sech}^{-1} x = \cosh^{-1}(1/x)$

14.63. $\coth^{-1} x = \tanh^{-1}(1/x)$

14.64. $\sinh^{-1}(-x) = -\sinh^{-1} x$

14.65. $\tanh^{-1}(-x) = -\tanh^{-1} x$

14.66. $\coth^{-1}(-x) = -\coth^{-1} x$

14.67. $\operatorname{csch}^{-1}(-x) = -\operatorname{csch}^{-1} x$

Graphs of Inverse Hyperbolic Functions

14.68. $y = \sinh^{-1} x$

14.69. $y = \cosh^{-1} x$

14.70. $y = \tanh^{-1} x$

14.71. $y = \coth^{-1} x$

14.72. $y = \operatorname{sech}^{-1} x$

14.73. $y = \operatorname{csch}^{-1} x$

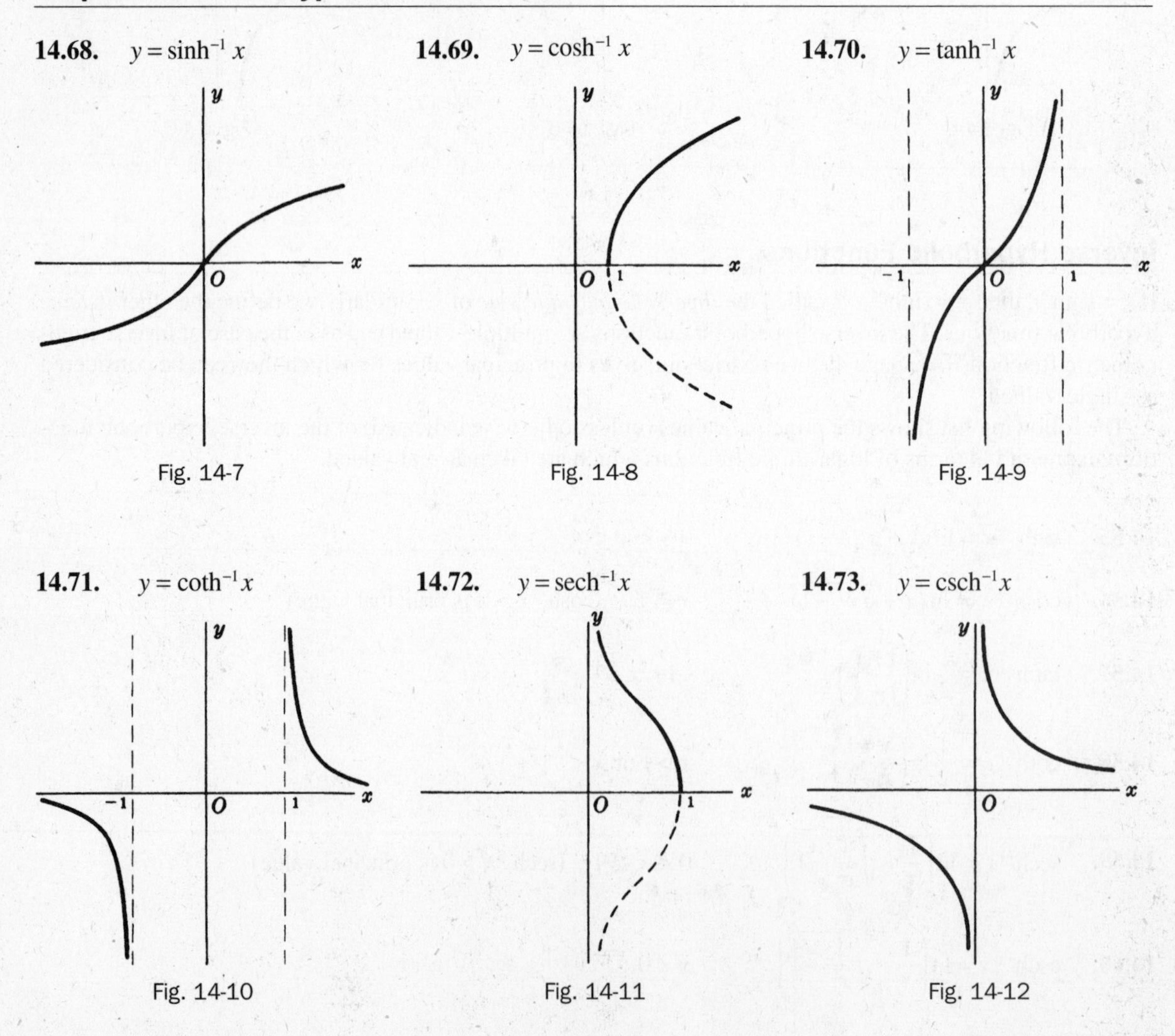

Fig. 14-7 Fig. 14-8 Fig. 14-9 Fig. 14-10 Fig. 14-11 Fig. 14-12

Relationship Between Hyperbolic and Trigonometric Functions

14.74. $\sin(ix) = i\sinh x$

14.75. $\cos(ix) = \cosh x$

14.76. $\tan(ix) = i\tanh x$

14.77. $\csc(ix) = -i\,\text{csch}\,x$

14.78. $\sec(ix) = \text{sech}\,x$

14.79. $\cot(ix) = -i\coth x$

14.80. $\sinh(ix) = i\sin x$

14.81. $\cosh(ix) = \cos x$

14.82. $\tanh(ix) = i\tan x$

14.83. $\text{csch}\,(ix) = -i\csc x$

14.84. $\text{sech}\,(ix) = \sec x$

14.85. $\coth(ix) = -i\cot x$

Periodicity of Hyperbolic Functions

In the following k is any integer.

14.86. $\sinh(x + 2k\pi i) = \sinh x$

14.87. $\cosh(x + 2k\pi i) = \cosh x$

14.88. $\tanh(x + k\pi i) = \tanh x$

14.89. $\text{csch}\,(x + 2k\pi i) = \text{csch}\,x$

14.90. $\text{sech}\,(x + 2k\pi i) = \text{sech}\,x$

14.91. $\coth(x + k\pi i) = \coth x$

Relationship Between Inverse Hyperbolic and Inverse Trigonometric Functions

14.92. $\sin^{-1}(ix) = i\sin^{-1} x$

14.93. $\sinh^{-1}(ix) = i\sin^{-1} x$

14.94. $\cos^{-1} x = \pm i\cosh^{-1} x$

14.95. $\cosh^{-1} x = \pm\, i\cos^{-1} x$

14.96. $\tan^{-1}(ix) = i\tanh^{-1} x$

14.97. $\tanh^{-1}(ix) = i\tan^{-1} x$

14.98. $\cot^{-1}(ix) = i\coth^{-1} x$

14.99. $\coth^{-1}(ix) = -i\cot^{-1} x$

14.100. $\sec^{-1} x = \pm i\,\text{sech}^{-1} x$

14.101. $\text{sech}^{-1} x = \pm\, i\sec^{-1} x$

14.102. $\csc^{-1}(ix) = -i\,\text{csch}^{-1} x$

14.103. $\text{csch}^{-1}(ix) = -i\csc^{-1} x$

Section IV: Calculus

15 DERIVATIVES

Definition of a Derivative

Suppose $y = f(x)$. The derivative of y or $f(x)$ is defined as

15.1. $$\frac{dy}{dx} = \lim_{h\to 0}\frac{f(x+h)-f(x)}{h} = \lim_{\Delta x\to 0}\frac{f(x+\Delta x)-f(x)}{\Delta x}$$

where $h = \Delta x$. The derivative is also denoted by y', df/dx or $f'(x)$. The process of taking a derivative is called *differentiation.*

General Rules of Differentiation

In the following, u, v, w are functions of x; a, b, c, n are constants (restricted if indicated); $e = 2.71828 \ldots$ is the natural base of logarithms; $\ln u$ is the natural logarithm of u (i.e., the logarithm to the base e) where it is assumed that $u > 0$ and all angles are in radians.

15.2. $$\frac{d}{dx}(c) = 0$$

15.3. $$\frac{d}{dx}(cx) = c$$

15.4. $$\frac{d}{dx}(cx^n) = ncx^{n-1}$$

15.5. $$\frac{d}{dx}(u \pm v \pm w \pm \cdots) = \frac{du}{dx} \pm \frac{dv}{dx} \pm \frac{dw}{dx} \pm \cdots$$

15.6. $$\frac{d}{dx}(cu) = c\frac{du}{dx}$$

15.7. $$\frac{d}{dx}(uv) = u\frac{dv}{dx} + v\frac{du}{dx}$$

15.8. $$\frac{d}{dx}(uvw) = uv\frac{dw}{dx} + uw\frac{dv}{dx} + vw\frac{du}{dx}$$

15.9. $$\frac{d}{dx}\left(\frac{u}{v}\right) = \frac{v(du/dx) - u(dv/dx)}{v^2}$$

15.10. $$\frac{d}{dx}(u^n) = nu^{n-1}\frac{du}{dx}$$

15.11. $$\frac{dy}{dx} = \frac{dy}{du}\frac{du}{dx}$$ (Chain rule)

15.12. $\dfrac{du}{dx} = \dfrac{1}{dx/du}$

15.13. $\dfrac{dy}{dx} = \dfrac{dy/du}{dx/du}$

Derivatives of Trigonometric and Inverse Trigonometric Functions

15.14. $\dfrac{d}{dx}\sin u = \cos u \dfrac{du}{dx}$

15.15. $\dfrac{d}{dx}\cos u = -\sin u \dfrac{du}{dx}$

15.16. $\dfrac{d}{dx}\tan u = \sec^2 u \dfrac{du}{dx}$

15.17. $\dfrac{d}{dx}\cot u = -\csc^2 u \dfrac{du}{dx}$

15.18. $\dfrac{d}{dx}\sec u = \sec u \tan u \dfrac{du}{dx}$

15.19. $\dfrac{d}{dx}\csc u = -\csc u \cot u \dfrac{du}{dx}$

15.20. $\dfrac{d}{dx}\sin^{-1} u = \dfrac{1}{\sqrt{1-u^2}}\dfrac{du}{dx} \quad \left[-\dfrac{\pi}{2} < \sin^{-1} u < \dfrac{\pi}{2}\right]$

15.21. $\dfrac{d}{dx}\cos^{-1} u = \dfrac{-1}{\sqrt{1-u^2}}\dfrac{du}{dx} \quad [0 < \cos^{-1} u < \pi]$

15.22. $\dfrac{d}{dx}\tan^{-1} u = \dfrac{1}{1+u^2}\dfrac{du}{dx} \quad \left[-\dfrac{\pi}{2} < \tan^{-1} u < \dfrac{\pi}{2}\right]$

15.23. $\dfrac{d}{dx}\cot^{-1} u = \dfrac{-1}{1+u^2}\dfrac{du}{dx} \quad [0 < \cot^{-1} u < \pi]$

15.24. $\dfrac{d}{dx}\sec^{-1} u = \dfrac{1}{|u|\sqrt{u^2-1}}\dfrac{du}{dx} = \dfrac{\pm 1}{u\sqrt{u^2-1}}\dfrac{du}{dx} \quad \begin{bmatrix} + \text{ if } 0 < \sec^{-1} u < \pi/2 \\ - \text{ if } \pi/2 < \sec^{-1} u < \pi \end{bmatrix}$

15.25. $\dfrac{d}{dx}\csc^{-1} u = \dfrac{-1}{|u|\sqrt{u^2-1}}\dfrac{du}{dx} = \dfrac{\mp 1}{u\sqrt{u^2-1}}\dfrac{du}{dx} \quad \begin{bmatrix} - \text{ if } 0 < \csc^{-1} u < \pi/2 \\ + \text{ if } -\pi/2 < \csc^{-1} u < 0 \end{bmatrix}$

Derivatives of Exponential and Logarithmic Functions

15.26. $\dfrac{d}{dx}\log_a u = \dfrac{\log_a e}{u}\dfrac{du}{dx} \quad a \neq 0, 1$

15.27. $\dfrac{d}{dx}\ln u = \dfrac{d}{dx}\log_e u = \dfrac{1}{u}\dfrac{du}{dx}$

15.28. $\dfrac{d}{dx}a^u = a^u \ln a \dfrac{du}{dx}$

15.29. $\frac{d}{dx}e^u = e^u \frac{du}{dx}$

15.30. $\frac{d}{dx}u^v = \frac{d}{dx}e^{v\ln u} = e^{v\ln u}\frac{d}{dx}[v\ln u] = vu^{v-1}\frac{du}{dx} + u^v \ln u\frac{dv}{dx}$

Derivatives of Hyperbolic and Inverse Hyperbolic Functions

15.31. $\frac{d}{dx}\sinh u = \cosh u\frac{du}{dx}$

15.32. $\frac{d}{dx}\cosh u = \sinh u\frac{du}{dx}$

15.33. $\frac{d}{dx}\tanh u = \operatorname{sech}^2 u\frac{du}{dx}$

15.34. $\frac{d}{dx}\coth u = -\operatorname{csch}^2 u\frac{du}{dx}$

15.35. $\frac{d}{dx}\operatorname{sech} u = -\operatorname{sech} u \tanh u\frac{du}{dx}$

15.36. $\frac{d}{dx}\operatorname{csch} u = -\operatorname{csch} u \coth u\frac{du}{dx}$

15.37. $\frac{d}{dx}\sinh^{-1} u = \frac{1}{\sqrt{u^2+1}}\frac{du}{dx}$

15.38. $\frac{d}{dx}\cosh^{-1} u = \frac{\pm 1}{\sqrt{u^2-1}}\frac{du}{dx}$ $\left[\begin{array}{l} + \text{ if } \cosh^{-1} u > 0,\ u > 1 \\ - \text{ if } \cosh^{-1} u < 0,\ u > 1 \end{array}\right]$

15.39. $\frac{d}{dx}\tanh^{-1} u = \frac{1}{1-u^2}\frac{du}{dx}$ $[-1 < u < 1]$

15.40. $\frac{d}{dx}\coth^{-1} u = \frac{1}{1-u^2}\frac{du}{dx}$ $[u > 1 \text{ or } u < -1]$

15.41. $\frac{d}{dx}\operatorname{sech}^{-1} u = \frac{\mp 1}{u\sqrt{1-u^2}}\frac{du}{dx}$ $\left[\begin{array}{l} - \text{ if } \operatorname{sech}^{-1} u > 0,\ 0 < u < 1 \\ + \text{ if } \operatorname{sech}^{-1} u < 0,\ 0 < u < 1 \end{array}\right]$

15.42. $\frac{d}{dx}\operatorname{csch}^{-1} u = \frac{-1}{|u|\sqrt{1+u^2}}\frac{du}{dx} = \frac{\mp 1}{u\sqrt{1+u^2}}\frac{du}{dx}$ $[- \text{ if } u > 0,\ + \text{ if } u < 0]$

Higher Derivatives

The second, third, and higher derivatives are defined as follows.

15.43. Second derivative $= \frac{d}{dx}\left(\frac{dy}{dx}\right) = \frac{d^2y}{dx^2} = f''(x) = y''$

15.44. Third derivative $= \frac{d}{dx}\left(\frac{d^2y}{dx^2}\right) = \frac{d^3y}{dx^3} = f'''(x) = y'''$

15.45. nth derivative $= \frac{d}{dx}\left(\frac{d^{n-1}y}{dx^{n-1}}\right) = \frac{d^ny}{dx^n} = f^{(n)}(x) = y^{(n)}$

Leibniz's Rule for Higher Derivatives of Products

Let D^p stand for the operator $\frac{d^p}{dx^p}$ so that $D^p u = \frac{d^p u}{dx^p} =$ the pth derivative of u. Then

15.46. $$D^n(uv) = uD^n v + \binom{n}{1}(Du)(D^{n-1}v) + \binom{n}{2}(D^2u)(D^{n-2}v) + \cdots + vD^n u$$

where $\binom{n}{1}, \binom{n}{2}, \ldots$ are the binomial coefficients (see 3.5).

As special cases we have

15.47. $$\frac{d^2}{dx^2}(uv) = u\frac{d^2v}{dx^2} + 2\frac{du}{dx}\frac{dv}{dx} + v\frac{d^2u}{dx^2}$$

15.48. $$\frac{d^3}{dx^3}(uv) = u\frac{d^3v}{dx^3} + 3\frac{du}{dx}\frac{d^2v}{dx^2} + 3\frac{d^2u}{dx^2}\frac{dv}{dx} + v\frac{d^3u}{dx^3}$$

Differentials

Let $y = f(x)$ and $\Delta y = f(x + \Delta x) - f(x)$. Then

15.49. $$\frac{\Delta y}{\Delta x} = \frac{f(x + \Delta x) - f(x)}{\Delta x} = f'(x) + \epsilon = \frac{dy}{dx} + \epsilon$$

where $\epsilon \to 0$ as $\Delta x \to 0$. Thus,

15.50. $$\Delta y = f'(x)\Delta x + \epsilon \Delta x$$

If we call $\Delta x = dx$ the differential of x, then we define the differential of y to be

15.51. $$dy = f'(x)\,dx$$

Rules for Differentials

The rules for differentials are exactly analogous to those for derivatives. As examples we observe that

15.52. $$d(u \pm v \pm w \pm \cdots) = du \pm dv \pm dw \pm \cdots$$

15.53. $$d(uv) = u\,dv + v\,du$$

15.54. $$d\left(\frac{u}{v}\right) = \frac{v\,du - u\,dv}{v^2}$$

15.55. $$d(u^n) = nu^{n-1}du$$

15.56. $$d(\sin u) = \cos u\,du$$

15.57. $$d(\cos u) = -\sin u\,du$$

Partial Derivatives

Let $z = f(x, y)$ be a function of the two variables x and y. Then we define the *partial derivative* of z or $f(x, y)$ with respect to x, keeping y constant, to be

15.58. $$\frac{\partial f}{\partial x} = \lim_{\Delta x \to 0} \frac{f(x + \Delta x, y) - f(x, y)}{\Delta x}$$

This partial derivative is also denoted by $\partial z/\partial x$, f_x, or z_x.

Similarly the partial derivative of $z = f(x, y)$ with respect to y, keeping x constant, is defined to be

15.59. $$\frac{\partial f}{\partial y} = \lim_{\Delta y \to 0} \frac{f(x, y + \Delta y) - f(x, y)}{\Delta y}$$

This partial derivative is also denoted by $\partial z/\partial y$, f_y, or z_y.

Partial derivatives of higher order can be defined as follows:

15.60. $$\frac{\partial^2 f}{\partial x^2} = \frac{\partial}{\partial x}\left(\frac{\partial f}{\partial x}\right), \quad \frac{\partial^2 f}{\partial y^2} = \frac{\partial}{\partial y}\left(\frac{\partial f}{\partial y}\right)$$

15.61. $$\frac{\partial^2 f}{\partial x \partial y} = \frac{\partial}{\partial x}\left(\frac{\partial f}{\partial y}\right), \quad \frac{\partial^2 f}{\partial y \partial x} = \frac{\partial}{\partial y}\left(\frac{\partial f}{\partial x}\right)$$

The results in 15.61 will be equal if the function and its partial derivatives are continuous; that is, in such cases, the order of differentiation makes no difference.

Extensions to functions of more than two variables are exactly analogous.

Multivariable Differentials

The differential of $z = f(x, y)$ is defined as

15.62. $$dz = df = \frac{\partial f}{\partial x} dx + \frac{\partial f}{\partial y} dy$$

where $dx = \Delta x$ and $dy = \Delta y$. Note that dz is a function of four variables, namely x, y, dx, dy, and is linear in the variables dx and dy.

Extensions to functions of more than two variables are exactly analogous.

EXAMPLE: Let $z = x^2 + 5xy + 2y^3$. Then

$$z_x = 2x + 5y \quad \text{and} \quad z_y = 5x + 6y^2$$

and hence

$$dz = (2x + 5y)\, dx + (5x + 6y^2)\, dy$$

Suppose we want to find dz for $dx = 2$, $dy = 3$ and at the point $P(4, 1)$, i.e., when $x = 4$ and $y = 1$. Substitution yields

$$dz = (8 + 5)2 + (20 + 6)3 = 26 + 78 = 104$$

16 INDEFINITE INTEGRALS

Definition of an Indefinite Integral

If $\dfrac{dy}{dx} = f(x)$, then y is the function whose derivative is $f(x)$ and is called the *anti-derivative* of $f(x)$ or the *indefinite integral* of $f(x)$, denoted by $\int f(x)\,dx$. Similarly if $y = \int f(u)\,du$, then $\dfrac{dy}{du} = f(u)$. Since the derivative of a constant is zero, all indefinite integrals differ by an arbitrary constant.

For the definition of a definite integral, see 18.1. The process of finding an integral is called *integration*.

General Rules of Integration

In the following, u, v, w are functions of x; a, b, p, q, n any constants, restricted if indicated; $e = 2.71828\ldots$ is the natural base of logarithms; $\ln u$ denotes the natural logarithm of u where it is assumed that $u > 0$ (in general, to extend formulas to cases where $u < 0$ as well, replace $\ln u$ by $\ln |u|$); all angles are in radians; all constants of integration are omitted but implied.

16.1. $\int a\,dx = ax$

16.2. $\int af(x)\,dx = a\int f(x)\,dx$

16.3. $\int (u \pm v \pm w \pm \cdots)\,dx = \int u\,dx \pm \int v\,dx \pm \int w\,dx \pm \cdots$

16.4. $\int u\,dv = uv - \int v\,du$ (Integration by parts)

For generalized integration by parts, see 16.48.

16.5. $\int f(ax)\,dx = \dfrac{1}{a}\int f(u)\,du$

16.6. $\int F\{f(x)\}\,dx = \int F(u)\dfrac{dx}{du}\,du = \int \dfrac{F(u)}{f'(x)}\,du$ where $u = f(x)$

16.7. $\int u^n\,du = \dfrac{u^{n+1}}{n+1}$, $n \neq -1$ (For $n = -1$, see 16.8)

16.8. $\int \dfrac{du}{u} = \ln u$ if $u > 0$ or $\ln(-u)$ if $u < 0$

$= \ln |u|$

16.9. $\int e^u\,du = e^u$

16.10. $\int a^u\,du = \int e^{u\ln a}\,du = \dfrac{e^{u\ln a}}{\ln a} = \dfrac{a^u}{\ln a}$, $a > 0$, $a \neq 1$

16.11. $\int \sin u\,du = -\cos u$

16.12. $\int \cos u\,du = \sin u$

16.13. $\int \tan u\,du = \ln \sec u = -\ln \cos u$

16.14. $\int \cot u\,du = \ln \sin u$

16.15. $\int \sec u\,du = \ln(\sec u + \tan u) = \ln \tan\left(\frac{u}{2} + \frac{\pi}{4}\right)$

16.16. $\int \csc u\,du = \ln(\csc u - \cot u) = \ln \tan \frac{u}{2}$

16.17. $\int \sec^2 u\,du = \tan u$

16.18. $\int \csc^2 u\,du = -\cot u$

16.19. $\int \tan^2 u\,du = \tan u - u$

16.20. $\int \cot^2 u\,du = -\cot u - u$

16.21. $\int \sin^2 u\,du = \frac{u}{2} - \frac{\sin 2u}{4} = \frac{1}{2}(u - \sin u \cos u)$

16.22. $\int \cos^2 u\,du = \frac{u}{2} + \frac{\sin 2u}{4} = \frac{1}{2}(u + \sin u \cos u)$

16.23. $\int \sec u \tan u\,du = \sec u$

16.24. $\int \csc u \cot u\,du = -\csc u$

16.25. $\int \sinh u\,du = \cosh u$

16.26. $\int \cosh u\,du = \sinh u$

16.27. $\int \tanh u\,du = \ln \cosh u$

16.28. $\int \coth u\,du = \ln \sinh u$

16.29. $\int \operatorname{sech} u\,du = \sin^{-1}(\tanh u) \text{ or } 2\tan^{-1} e^u$

16.30. $\int \operatorname{csch} u\,du = \ln \tanh \frac{u}{2} \text{ or } -\coth^{-1} e^u$

16.31. $\int \operatorname{sech}^2 u\,du = \tanh u$

16.32. $\int \operatorname{csch}^2 u\, du = -\coth u$

16.33. $\int \tanh^2 u\, du = u - \tanh u$

16.34. $\int \coth^2 u\, du = u - \coth u$

16.35. $\int \sinh^2 u\, du = \dfrac{\sinh 2u}{4} - \dfrac{u}{2} = \dfrac{1}{2}(\sinh u \cosh u - u)$

16.36. $\int \cosh^2 u\, du = \dfrac{\sinh 2u}{4} + \dfrac{u}{2} = \dfrac{1}{2}(\sinh u \cosh u + u)$

16.37. $\int \operatorname{sech} u \tanh u\, du = -\operatorname{sech} u$

16.38. $\int \operatorname{csch} u \coth u\, du = -\operatorname{csch} u$

16.39. $\int \dfrac{du}{u^2 + a^2} = \dfrac{1}{a} \tan^{-1} \dfrac{u}{a}$

16.40. $\int \dfrac{du}{u^2 - a^2} = \dfrac{1}{2a} \ln\left(\dfrac{u-a}{u+a}\right) = -\dfrac{1}{a} \coth^{-1} \dfrac{u}{a} \quad u^2 > a^2$

16.41. $\int \dfrac{du}{a^2 - u^2} = \dfrac{1}{2a} \ln\left(\dfrac{a+u}{a-u}\right) = \dfrac{1}{a} \tanh^{-1} \dfrac{u}{a} \quad u^2 < a^2$

16.42. $\int \dfrac{du}{\sqrt{a^2 - u^2}} = \sin^{-1} \dfrac{u}{a}$

16.43. $\int \dfrac{du}{\sqrt{u^2 + a^2}} = \ln(u + \sqrt{u^2 + a^2}) \quad \text{or} \quad \sinh^{-1} \dfrac{u}{a}$

16.44. $\int \dfrac{du}{\sqrt{u^2 - a^2}} = \ln(u + \sqrt{u^2 - a^2})$

16.45. $\int \dfrac{du}{u\sqrt{u^2 - a^2}} = \dfrac{1}{a} \sec^{-1} \left|\dfrac{u}{a}\right|$

16.46. $\int \dfrac{du}{u\sqrt{u^2 + a^2}} = -\dfrac{1}{a} \ln\left(\dfrac{a + \sqrt{u^2 + a^2}}{u}\right)$

16.47. $\int \dfrac{du}{u\sqrt{a^2 - u^2}} = -\dfrac{1}{a} \ln\left(\dfrac{a + \sqrt{a^2 - u^2}}{u}\right)$

16.48. $\int f^{(n)} g\, dx = f^{(n-1)} g - f^{(n-2)} g' + f^{(n-3)} g'' - \cdots (-1)^n \int f g^{(n)} dx$

This is called *generalized integration* by parts.

Important Transformations

Often in practice an integral can be simplified by using an appropriate transformation or substitution together with Formula 16.6. The following list gives some transformations and their effects.

16.49. $\int F(ax+b)dx = \frac{1}{a}\int F(u)\,du$ where $u = ax+b$

16.50. $\int F(\sqrt{ax+b})\,dx = \frac{2}{a}\int u\,F(u)\,du$ where $u = \sqrt{ax+b}$

16.51. $\int F(\sqrt[n]{ax+b})\,dx = \frac{n}{a}\int u^{n-1}F(u)\,du$ where $u = \sqrt[n]{ax+b}$

16.52. $\int F(\sqrt{a^2-x^2})\,dx = a\int F(a\cos u)\cos u\,du$ where $x = a\sin u$

16.53. $\int F(\sqrt{x^2+a^2})\,dx = a\int F(a\sec u)\sec^2 u\,du$ where $x = a\tan u$

16.54. $\int F(\sqrt{x^2-a^2})\,dx = a\int F(a\tan u)\sec u\tan u\,du$ where $x = a\sec u$

16.55. $\int F(e^{ax})\,dx = \frac{1}{a}\int \frac{F(u)}{u}\,du$ where $u = e^{ax}$

16.56. $\int F(\ln x)\,dx = \int F(u)e^u\,du$ where $u = \ln x$

16.57. $\int F\left(\sin^{-1}\frac{x}{a}\right)dx = a\int F(u)\cos u\,du$ where $u = \sin^{-1}\frac{x}{a}$

Similar results apply for other inverse trigonometric functions.

16.58. $\int F(\sin x, \cos x)\,dx = 2\int F\left(\frac{2u}{1+u^2}, \frac{1-u^2}{1+u^2}\right)\frac{du}{1+u^2}$ where $u = \tan\frac{x}{2}$

17 TABLES of SPECIAL INDEFINITE INTEGRALS

Here we provide tables of special indefinite integrals. As stated in the remarks on page 67, here a, b, p, q, n are constants, restricted if indicated; $e = 2.71828\ldots$ is the natural base of logarithms; $\ln u$ denotes the natural logarithm of u, where it is assumed that $u > 0$ (in general, to extend formulas to cases where $u < 0$ as well, replace $\ln u$ by $\ln |u|$); all angles are in radians; and all constants of integration are omitted but implied. It is assumed in all cases that division by zero is excluded.

Our integrals are divided into types which involve the following algebraic expressions and functions:

(1) $ax + b$
(2) $\sqrt{ax+b}$
(3) $ax + b$ and $px + q$
(4) $\sqrt{ax+b}$ and $px + q$
(5) $\sqrt{ax+b}$ and $\sqrt{px+q}$
(6) $x^2 + a^2$
(7) $x^2 - a^2$, with $x^2 > a^2$
(8) $a^2 - x^2$, with $x^2 < a^2$
(9) $\sqrt{x^2+a^2}$
(10) $\sqrt{x^2-a^2}$
(11) $\sqrt{a^2-x^2}$
(12) $ax^2 + bx + c$
(13) $\sqrt{ax^2+bx+c}$
(14) $x^3 + a^3$
(15) $x^4 \pm a^4$
(16) $x^n \pm a^n$
(17) $\sin ax$
(18) $\cos ax$
(19) $\sin ax$ and $\cos ax$
(20) $\tan ax$
(21) $\cot ax$
(22) $\sec ax$
(23) $\csc ax$
(24) inverse trigonometric functions
(25) e^{ax}
(26) $\ln x$
(27) $\sinh ax$
(28) $\cosh ax$
(29) $\sinh ax$ and $\cosh ax$
(30) $\tanh ax$
(31) $\coth ax$
(32) $\operatorname{sech} ax$
(33) $\operatorname{csch} ax$
(34) inverse hyperbolic functions

Some integrals contain the Bernouilli numbers B_n and the Euler numbers E_n defined in Chapter 23.

(1) Integrals Involving $ax + b$

17.1.1. $$\int \frac{dx}{ax+b} = \frac{1}{a}\ln(ax+b)$$

17.1.2. $$\int \frac{x\,dx}{ax+b} = \frac{x}{a} - \frac{b}{a^2}\ln(ax+b)$$

17.1.3. $$\int \frac{x^2 dx}{ax+b} = \frac{(ax+b)^2}{2a^3} - \frac{2b(ax+b)}{a^3} + \frac{b^2}{a^3}\ln(ax+b)$$

17.1.4. $$\int \frac{dx}{x(ax+b)} = \frac{1}{b}\ln\left(\frac{x}{ax+b}\right)$$

17.1.5. $$\int \frac{dx}{x^2(ax+b)} = -\frac{1}{bx} + \frac{a}{b^2}\ln\left(\frac{ax+b}{x}\right)$$

17.1.6. $$\int \frac{dx}{(ax+b)^2} = \frac{-1}{a(ax+b)}$$

17.1.7. $$\int \frac{x\,dx}{(ax+b)^2} = \frac{b}{a^2(ax+b)} + \frac{1}{a^2}\ln(ax+b)$$

17.1.8. $$\int \frac{x^2 dx}{(ax+b)^2} = \frac{ax+b}{a^3} - \frac{b^2}{a^3(ax+b)} - \frac{2b}{a^3}\ln(ax+b)$$

17.1.9. $$\int \frac{dx}{x(ax+b)^2} = \frac{1}{b(ax+b)} + \frac{1}{b^2}\ln\left(\frac{x}{ax+b}\right)$$

17.1.10. $\int \frac{dx}{x^2(ax+b)^2} = \frac{-a}{b^2(ax+b)} - \frac{1}{b^2x} + \frac{2a}{b^3}\ln\left(\frac{ax+b}{x}\right)$

17.1.11. $\int \frac{dx}{(ax+b)^3} = \frac{-1}{2(ax+b)^2}$

17.1.12. $\int \frac{x\,dx}{(ax+b)^3} = \frac{-1}{a^2(ax+b)} + \frac{b}{2a^2(ax+b)^2}$

17.1.13. $\int \frac{x^2dx}{(ax+b)^3} = \frac{2b}{a^3(ax+b)} - \frac{b^2}{2a^3(ax+b)^2} + \frac{1}{a^3}\ln(ax+b)$

17.1.14. $\int (ax+b)^n\,dx = \frac{(ax+b)^{n+1}}{(n+1)a}$. If $n=-1$, see 17.1.1.

17.1.15. $\int x(ax+b)^n dx = \frac{(ax+b)^{n+2}}{(n+2)a^2} - \frac{b(ax+b)^{n+1}}{(n+1)a^2}, \quad n \neq -1, -2$

If $n = -1, -2$, see 17.1.2 and 17.1.7.

17.1.16. $\int x^2(ax+b)^n dx = \frac{(ax+b)^{n+3}}{(n+3)a^3} - \frac{2b(ax+b)^{n+2}}{(n+2)a^3} + \frac{b^2(ax+b)^{n+1}}{(n+1)a^3}$

If $n = -1, -2, -3$, see 17.1.3, 17.1.8, and 17.1.13.

17.1.17. $$\int x^m(ax+b)^n\,dx = \begin{cases} \frac{x^{m+1}(ax+b)^n}{m+n+1} + \frac{nb}{m+n+1}\int x^m(ax+b)^{n-1}dx \\ \frac{x^m(ax+b)^{n+1}}{(m+n+1)a} - \frac{mb}{(m+n+1)a}\int x^{m-1}(ax+b)^n dx \\ \frac{-x^{m+1}(ax+b)^{n+1}}{(n+1)b} + \frac{m+n+2}{(n+1)b}\int x^m(ax+b)^{n+1}dx \end{cases}$$

(2) Integrals Involving $\sqrt{ax+b}$

17.2.1. $\int \frac{dx}{\sqrt{ax+b}} = \frac{2\sqrt{ax+b}}{a}$

17.2.2. $\int \frac{x\,dx}{\sqrt{ax+b}} = \frac{2(ax-2b)}{3a^2}\sqrt{ax+b}$

17.2.3. $\int \frac{x^2dx}{\sqrt{ax+b}} = \frac{2(3a^2x^2 - 4abx + 8b^2)}{15a^3}\sqrt{ax+b}$

17.2.4. $$\int \frac{dx}{x\sqrt{ax+b}} = \begin{cases} \frac{1}{\sqrt{b}}\ln\left(\frac{\sqrt{ax+b}-\sqrt{b}}{\sqrt{ax+b}+\sqrt{b}}\right) \\ \frac{2}{\sqrt{-b}}\tan^{-1}\sqrt{\frac{ax+b}{-b}} \end{cases}$$

17.2.5. $\int \frac{dx}{x^2\sqrt{ax+b}} = -\frac{\sqrt{ax+b}}{bx} - \frac{a}{2b}\int \frac{dx}{x\sqrt{ax+b}}$ (see 17.2.12.)

17.2.6. $\int \sqrt{ax+b}\,dx = \frac{2\sqrt{(ax+b)^3}}{3a}$

17.2.7. $\int x\sqrt{ax+b}\,dx = \frac{2(3ax-2b)}{15a^2}\sqrt{(ax+b)^3}$

17.2.8. $\int x^2\sqrt{ax+b}\,dx = \dfrac{2(15a^2x^2 - 12abx + 8b^2)}{105a^3}\sqrt{(ax+b)^3}$

17.2.9. $\int \dfrac{\sqrt{ax+b}}{x}\,dx = 2\sqrt{ax+b} + b\int \dfrac{dx}{x\sqrt{ax+b}}$ (See 17.2.12.)

17.2.10. $\int \dfrac{\sqrt{ax+b}}{x^2}\,dx = -\dfrac{\sqrt{ax+b}}{x} + \dfrac{a}{2}\int \dfrac{dx}{x\sqrt{ax+b}}$ (See 17.2.12.)

17.2.11. $\int \dfrac{x^m}{\sqrt{ax+b}}\,dx = \dfrac{2x^m\sqrt{ax+b}}{(2m+1)a} - \dfrac{2mb}{(2m+1)a}\int \dfrac{x^{m-1}}{\sqrt{ax+b}}\,dx$

17.2.12. $\int \dfrac{dx}{x^m\sqrt{ax+b}} = -\dfrac{\sqrt{ax+b}}{(m-1)bx^{m-1}} - \dfrac{(2m-3)a}{(2m-2)b}\int \dfrac{dx}{x^{m-1}\sqrt{ax+b}}$

17.2.13. $\int x^m\sqrt{ax+b}\,dx = \dfrac{2x^m}{(2m+3)a}(ax+b)^{3/2} - \dfrac{2mb}{(2m+3)a}\int x^{m-1}\sqrt{ax+b}\,dx$

17.2.14. $\int \dfrac{\sqrt{ax+b}}{x^m}\,dx = -\dfrac{\sqrt{ax+b}}{(m-1)x^{m-1}} + \dfrac{a}{2(m-1)}\int \dfrac{dx}{x^{m-1}\sqrt{ax+b}}$

17.2.15. $\int \dfrac{\sqrt{ax+b}}{x^m}\,dx = \dfrac{-(ax+b)^{3/2}}{(m-1)bx^{m-1}} - \dfrac{(2m-5)a}{(2m-2)b}\int \dfrac{\sqrt{ax+b}}{x^{m-1}}\,dx$

17.2.16. $\int (ax+b)^{m/2}\,dx = \dfrac{2(ax+b)^{(m+2)/2}}{a^2(m+2)}$

17.2.17. $\int x(ax+b)^{m/2}dx = \dfrac{2(ax+b)^{(m+4)/2}}{a^2(m+4)} - \dfrac{2b(ax+b)^{(m+2)/2}}{a^2(m+2)}$

17.2.18. $\int x^2(ax+b)^{m/2}\,dx = \dfrac{2(ax+b)^{(m+6)/2}}{a^3(m+6)} - \dfrac{4b(ax+b)^{(m+4)/2}}{a^3(m+4)} + \dfrac{2b^2(ax+b)^{(m+2)/2}}{a^3(m+2)}$

17.2.19. $\int \dfrac{(ax+b)^{m/2}}{x}\,dx = \dfrac{2(ax+b)^{m/2}}{m} + b\int \dfrac{(ax+b)^{(m-2)/2}}{x}\,dx$

17.2.20. $\int \dfrac{(ax+b)^{m/2}}{x^2}\,dx = -\dfrac{(ax+b)^{(m+2)/2}}{bx} + \dfrac{ma}{2b}\int \dfrac{(ax+b)^{m/2}}{x}\,dx$

17.2.21. $\int \dfrac{dx}{x(ax+b)^{m/2}} = \dfrac{2}{(m-2)b(ax+b)^{(m-2)/2}} + \dfrac{1}{b}\int \dfrac{dx}{x(ax+b)^{(m-2)/2}}$

(3) Integrals Involving $ax + b$ and $px + q$

17.3.1. $\int \dfrac{dx}{(ax+b)(px+q)} = \dfrac{1}{bp-aq}\ln\left(\dfrac{px+q}{ax+b}\right)$

17.3.2. $\int \dfrac{x\,dx}{(ax+b)(px+q)} = \dfrac{1}{bp-aq}\left\{\dfrac{b}{a}\ln(ax+b) - \dfrac{q}{p}\ln(px+q)\right\}$

17.3.3. $$\int \frac{dx}{(ax+b)^2(px+q)} = \frac{1}{bp-aq}\left\{\frac{1}{ax+b} + \frac{p}{bp-aq}\ln\left(\frac{px+q}{ax+b}\right)\right\}$$

17.3.4. $$\int \frac{x\,dx}{(ax+b)^2(px+q)} = \frac{1}{bp-aq}\left\{\frac{q}{bp-aq}\ln\left(\frac{ax+b}{px+q}\right) - \frac{b}{a(ax+b)}\right\}$$

17.3.5. $$\int \frac{x^2dx}{(ax+b)^2(px+q)} = \frac{b^2}{(bp-aq)a^2(ax+b)} + \frac{1}{(bp-aq)^2}\left\{\frac{q^2}{p}\ln(px+q) + \frac{b(bp-2aq)}{a^2}\ln(ax+b)\right\}$$

17.3.6. $$\int \frac{dx}{(ax+b)^m(px+q)^n} = \frac{-1}{(n-1)(bp-aq)}\left\{\frac{1}{(ax+b)^{m-1}(px+q)^{n-1}} + a(m+n-2)\int \frac{dx}{(ax+b)^m(px+q)^{n-1}}\right\}$$

17.3.7. $$\int \frac{ax+b}{px+q}dx = \frac{ax}{p} + \frac{bp-aq}{p^2}\ln(px+q)$$

17.3.8. $$\int \frac{(ax+b)^m}{(px+q)^n}dx = \begin{cases} \dfrac{-1}{(n-1)(bp-aq)}\left\{\dfrac{(ax+b)^{m+1}}{(px+q)^{n-1}} + (n-m-2)a\displaystyle\int \frac{(ax+b)^m}{(px+q)^{n-1}}dx\right\} \\ \dfrac{-1}{(n-m-1)p}\left\{\dfrac{(ax+b)^m}{(px+q)^{n-1}} + m(bp-aq)\displaystyle\int \frac{(ax+b)^{m-1}}{(px+q)^n}dx\right\} \\ \dfrac{-1}{(n-1)p}\left\{\dfrac{(ax+b)^m}{(px+q)^{n-1}} - ma\displaystyle\int \frac{(ax+b)^{m-1}}{(px+q)^{n-1}}dx\right\} \end{cases}$$

(4) Integrals Involving $\sqrt{ax+b}$ and $px + q$

17.4.1. $$\int \frac{px+q}{\sqrt{ax+b}}dx = \frac{2(apx+3aq-2bp)}{3a^2}\sqrt{ax+b}$$

17.4.2. $$\int \frac{dx}{(px+q)\sqrt{ax+b}} = \begin{cases} \dfrac{1}{\sqrt{bp-aq}\sqrt{p}}\ln\left(\dfrac{\sqrt{p(ax+b)}-\sqrt{bp-aq}}{\sqrt{p(ax+b)}+\sqrt{bp-aq}}\right) \\ \dfrac{2}{\sqrt{aq-bp}\sqrt{p}}\tan^{-1}\sqrt{\dfrac{p(ax+b)}{aq-bp}} \end{cases}$$

17.4.3. $$\int \frac{\sqrt{ax+b}}{px+q}dx = \begin{cases} \dfrac{2\sqrt{ax+b}}{p} + \dfrac{\sqrt{bp-aq}}{p\sqrt{p}}\ln\left(\dfrac{\sqrt{p(ax+b)}-\sqrt{bp-aq}}{\sqrt{p(ax+b)}+\sqrt{bp-aq}}\right) \\ \dfrac{2\sqrt{ax+b}}{p} - \dfrac{2\sqrt{aq-bp}}{p\sqrt{p}}\tan^{-1}\sqrt{\dfrac{p(ax+b)}{aq-bp}} \end{cases}$$

17.4.4. $$\int (px+q)^n\sqrt{ax+b}\,dx = \frac{2(px+q)^{n+1}\sqrt{ax+b}}{(2n+3)p} + \frac{bp-aq}{(2n+3)p}\int \frac{(px+q)^n}{\sqrt{ax+b}}$$

17.4.5. $$\int \frac{dx}{(px+q)^n\sqrt{ax+b}} = \frac{\sqrt{ax+b}}{(n-1)(aq-bp)(px+q)^{n-1}} + \frac{(2n-3)a}{2(n-1)(aq-bp)}\int \frac{dx}{(px+q)^{n-1}\sqrt{ax+b}}$$

17.4.6. $$\int \frac{(px+q)^n}{\sqrt{ax+b}}dx = \frac{2(px+q)^n\sqrt{ax+b}}{(2n+1)a} + \frac{2n(aq-bp)}{(2n+1)a}\int \frac{(px+q)^{n-1}dx}{\sqrt{ax+b}}$$

17.4.7. $$\int \frac{\sqrt{ax+b}}{(px+q)^n}dx = \frac{-\sqrt{ax+b}}{(n-1)p(px+q)^{n-1}} + \frac{a}{2(n-1)p}\int \frac{dx}{(px+q)^{n-1}\sqrt{ax+b}}$$

(5) Integrals Involving $\sqrt{ax+b}$ and $\sqrt{px+q}$

17.5.1. $$\int \frac{dx}{\sqrt{(ax+b)(px+q)}} = \begin{cases} \dfrac{2}{\sqrt{ap}} \ln\left(\sqrt{a(px+q)} + \sqrt{p(ax+b)}\right) \\ \dfrac{2}{\sqrt{-ap}} \tan^{-1} \sqrt{\dfrac{-p(ax+b)}{a(px+q)}} \end{cases}$$

17.5.2. $$\int \frac{x\,dx}{\sqrt{(ax+b)(px+q)}} = \frac{\sqrt{(ax+b)(px+q)}}{ap} - \frac{bp+aq}{2ap} \int \frac{dx}{\sqrt{(ax+b)(px+q)}}$$

17.5.3. $$\int \sqrt{(ax+b)(px+q)}\,dx = \frac{2apx+bp+aq}{4ap}\sqrt{(ax+b)(px+q)} - \frac{(bp-aq)^2}{8ap} \int \frac{dx}{\sqrt{(ax+b)(px+q)}}$$

17.5.4. $$\int \sqrt{\frac{px+q}{ax+b}}\,dx = \frac{\sqrt{(ax+b)(px+q)}}{a} + \frac{aq-bp}{2a} \int \frac{dx}{\sqrt{(ax+b)(px+q)}}$$

17.5.5. $$\int \frac{dx}{(px+q)\sqrt{(ax+b)(px+q)}} = \frac{2\sqrt{ax+b}}{(aq-bp)\sqrt{px+q}}$$

(6) Integrals Involving $x^2 + a^2$

17.6.1. $$\int \frac{dx}{x^2+a^2} = \frac{1}{a} \tan^{-1} \frac{x}{a}$$

17.6.2. $$\int \frac{x\,dx}{x^2+a^2} = \frac{1}{2} \ln(x^2+a^2)$$

17.6.3. $$\int \frac{x^2dx}{x^2+a^2} = x - a \tan^{-1} \frac{x}{a}$$

17.6.4. $$\int \frac{x^3dx}{x^2+a^2} = \frac{x^2}{2} - \frac{a^2}{2} \ln(x^2+a^2)$$

17.6.5. $$\int \frac{dx}{x(x^2+a^2)} = \frac{1}{2a^2} \ln\left(\frac{x^2}{x^2+a^2}\right)$$

17.6.6. $$\int \frac{dx}{x^2(x^2+a^2)} = -\frac{1}{a^2x} - \frac{1}{a^3} \tan^{-1} \frac{x}{a}$$

17.6.7. $$\int \frac{dx}{x^3(x^2+a^2)} = -\frac{1}{2a^2x^2} - \frac{1}{2a^4} \ln\left(\frac{x^2}{x^2+a^2}\right)$$

17.6.8. $$\int \frac{dx}{(x^2+a^2)^2} = \frac{x}{2a^2(x^2+a^2)} + \frac{1}{2a^3} \tan^{-1} \frac{x}{a}$$

17.6.9. $$\int \frac{x\,dx}{(x^2+a^2)^2} = \frac{-1}{2(x^2+a^2)}$$

17.6.10. $$\int \frac{x^2dx}{(x^2+a^2)^2} = \frac{-x}{2(x^2+a^2)} + \frac{1}{2a} \tan^{-1} \frac{x}{a}$$

17.6.11. $\int \frac{x^3 dx}{(x^2+a^2)^2} = \frac{a^2}{2(x^2+a^2)} + \frac{1}{2}\ln(x^2+a^2)$

17.6.12. $\int \frac{dx}{x(x^2+a^2)^2} = \frac{1}{2a^2(x^2+a^2)} + \frac{1}{2a^4}\ln\left(\frac{x^2}{x^2+a^2}\right)$

17.6.13. $\int \frac{dx}{x^2(x^2+a^2)^2} = -\frac{1}{a^4 x} - \frac{x}{2a^4(x^2+a^2)} - \frac{3}{2a^5}\tan^{-1}\frac{x}{a}$

17.6.14. $\int \frac{dx}{x^3(x^2+a^2)^2} = -\frac{1}{2a^4x^2} - \frac{1}{2a^4(x^2+a^2)} - \frac{1}{a^6}\ln\left(\frac{x^2}{x^2+a^2}\right)$

17.6.15. $\int \frac{dx}{(x^2+a^2)^n} = \frac{x}{2(n-1)a^2(x^2+a^2)^{n-1}} + \frac{2n-3}{(2n-2)a^2}\int \frac{dx}{(x^2+a^2)^{n-1}}$

17.6.16. $\int \frac{x\,dx}{(x^2+a^2)^n} = \frac{-1}{2(n-1)(x^2+a^2)^{n-1}}$

17.6.17. $\int \frac{dx}{x(x^2+a^2)^n} = \frac{1}{2(n-1)a^2(x^2+a^2)^{n-1}} + \frac{1}{a^2}\int \frac{dx}{x(x^2+a^2)^{n-1}}$

17.6.18. $\int \frac{x^m dx}{(x^2+a^2)^n} = \int \frac{x^{m-2}dx}{(x^2+a^2)^{n-1}} - a^2 \int \frac{x^{m-2}dx}{(x^2+a^2)^n}$

17.6.19. $\int \frac{dx}{x^m(x^2+a^2)^n} = \frac{1}{a^2}\int \frac{dx}{x^m(x^2+a^2)^{n-1}} - \frac{1}{a^2}\int \frac{dx}{x^{m-2}(x^2+a^2)^n}$

(7) Integrals Involving $x^2 - a^2$, $x^2 > a^2$

17.7.1. $\int \frac{dx}{x^2-a^2} = \frac{1}{2a}\ln\left(\frac{x-a}{x+a}\right)$ or $-\frac{1}{a}\coth^{-1}\frac{x}{a}$

17.7.2. $\int \frac{x\,dx}{x^2-a^2} = \frac{1}{2}\ln(x^2-a^2)$

17.7.3. $\int \frac{x^2 dx}{x^2-a^2} = x + \frac{a}{2}\ln\left(\frac{x-a}{x+a}\right)$

17.7.4. $\int \frac{x^3 dx}{x^2-a^2} = \frac{x^2}{2} + \frac{a^2}{2}\ln(x^2-a^2)$

17.7.5. $\int \frac{dx}{x(x^2-a^2)} = \frac{1}{2a^2}\ln\left(\frac{x^2-a^2}{x^2}\right)$

17.7.6. $\int \frac{dx}{x^2(x^2-a^2)} = \frac{1}{a^2 x} + \frac{1}{2a^3}\ln\left(\frac{x-a}{x+a}\right)$

17.7.7. $\int \frac{dx}{x^3(x^2-a^2)} = \frac{1}{2a^2x^2} - \frac{1}{2a^4}\ln\left(\frac{x^2}{x^2-a^2}\right)$

17.7.8. $\int \frac{dx}{(x^2-a^2)^2} = \frac{-x}{2a^2(x^2-a^2)} - \frac{1}{4a^3}\ln\left(\frac{x-a}{x+a}\right)$

17.7.9. $\displaystyle\int \frac{x\,dx}{(x^2-a^2)^2} = \frac{-1}{2(x^2-a^2)}$

17.7.10. $\displaystyle\int \frac{x^2dx}{(x^2-a^2)^2} = \frac{-x}{2(x^2-a^2)} + \frac{1}{4a}\ln\left(\frac{x-a}{x+a}\right)$

17.7.11. $\displaystyle\int \frac{x^3dx}{(x^2-a^2)^2} = \frac{-a^2}{2(x^2-a^2)} + \frac{1}{2}\ln(x^2-a^2)$

17.7.12. $\displaystyle\int \frac{dx}{x(x^2-a^2)^2} = \frac{-1}{2a^2(x^2-a^2)} + \frac{1}{2a^4}\ln\left(\frac{x^2}{x^2-a^2}\right)$

17.7.13. $\displaystyle\int \frac{dx}{x^2(x^2-a^2)^2} = -\frac{1}{a^4x} - \frac{x}{2a^4(x^2-a^2)} - \frac{3}{4a^5}\ln\left(\frac{x-a}{x+a}\right)$

17.7.14. $\displaystyle\int \frac{dx}{x^3(x^2-a^2)^2} = -\frac{1}{2a^4x^2} - \frac{1}{2a^4(x^2-a^2)} + \frac{1}{a^6}\ln\left(\frac{x^2}{x^2-a^2}\right)$

17.7.15. $\displaystyle\int \frac{dx}{(x^2-a^2)^n} = \frac{-x}{2(n-1)a^2(x^2-a^2)^{n-1}} - \frac{2n-3}{(2n-2)a^2}\int \frac{dx}{(x^2-a^2)^{n-1}}$

17.7.16. $\displaystyle\int \frac{x\,dx}{(x^2-a^2)^n} = \frac{-1}{2(n-1)(x^2-a^2)^{n-1}}$

17.7.17. $\displaystyle\int \frac{dx}{x(x^2-a^2)^n} = \frac{-1}{2(n-1)a^2(x^2-a^2)^{n-1}} - \frac{1}{a^2}\int \frac{dx}{x(x^2-a^2)^{n-1}}$

17.7.18. $\displaystyle\int \frac{x^m dx}{(x^2-a^2)^n} = \int \frac{x^{m-2}dx}{(x^2-a^2)^{n-1}} + a^2\int \frac{x^{m-2}dx}{(x^2-a^2)^n}$

17.7.19. $\displaystyle\int \frac{dx}{x^m(x^2-a^2)^n} = \frac{1}{a^2}\int \frac{dx}{x^{m-2}(x^2-a^2)^n} - \frac{1}{a^2}\int \frac{dx}{x^m(x^2-a^2)^{n-1}}$

(8) Integrals Involving $x^2 - a^2$, $x^2 < a^2$

17.8.1. $\displaystyle\int \frac{dx}{a^2-x^2} = \frac{1}{2a}\ln\left(\frac{a+x}{a-x}\right)$ or $\displaystyle\frac{1}{a}\tanh^{-1}\frac{x}{a}$

17.8.2. $\displaystyle\int \frac{x\,dx}{a^2-x^2} = -\frac{1}{2}\ln(a^2-x^2)$

17.8.3. $\displaystyle\int \frac{x^2dx}{a^2-x^2} = -x + \frac{a}{2}\ln\left(\frac{a+x}{a-x}\right)$

17.8.4. $\displaystyle\int \frac{x^3dx}{a^2-x^2} = -\frac{x^2}{2} - \frac{a^2}{2}\ln(a^2-x^2)$

17.8.5. $\displaystyle\int \frac{dx}{x(a^2-x^2)} = \frac{1}{2a^2}\ln\left(\frac{x^2}{a^2-x^2}\right)$

17.8.6. $\displaystyle\int \frac{dx}{x^2(a^2-x^2)} = -\frac{1}{a^2x} + \frac{1}{2a^3}\ln\left(\frac{a+x}{a-x}\right)$

17.8.7. $\displaystyle\int \frac{dx}{x^3(a^2-x^2)} = -\frac{1}{2a^2x^2} + \frac{1}{2a^4}\ln\left(\frac{x^2}{a^2-x^2}\right)$

17.8.8. $\displaystyle\int \frac{dx}{(a^2-x^2)^2} = \frac{x}{2a^2(a^2-x^2)} + \frac{1}{4a^3}\ln\left(\frac{a+x}{a-x}\right)$

17.8.9. $\displaystyle\int \frac{x\,dx}{(a^2-x^2)^2} = \frac{1}{2(a^2-x^2)}$

17.8.10. $\displaystyle\int \frac{x^2dx}{(a^2-x^2)^2} = \frac{x}{2(a^2-x^2)} - \frac{1}{4a}\ln\left(\frac{a+x}{a-x}\right)$

17.8.11. $\displaystyle\int \frac{x^3dx}{(a^2-x^2)^2} = \frac{a^2}{2(a^2-x^2)} + \frac{1}{2}\ln(a^2-x^2)$

17.8.12. $\displaystyle\int \frac{dx}{x(a^2-x^2)^2} = \frac{1}{2a^2(a^2-x^2)} + \frac{1}{2a^4}\ln\left(\frac{x^2}{a^2-x^2}\right)$

17.8.13. $\displaystyle\int \frac{dx}{x^2(a^2-x^2)^2} = \frac{-1}{a^4x} + \frac{x}{2a^4(a^2-x^2)} + \frac{3}{4a^5}\ln\left(\frac{a+x}{a-x}\right)$

17.8.14. $\displaystyle\int \frac{dx}{x^3(a^2-x^2)^2} = \frac{-1}{2a^4x^2} + \frac{1}{2a^4(a^2-x^2)} + \frac{1}{a^6}\ln\left(\frac{x^2}{a^2-x^2}\right)$

17.8.15. $\displaystyle\int \frac{dx}{(a^2-x^2)^n} = \frac{x}{2(n-1)a^2(a^2-x^2)^{n-1}} + \frac{2n-3}{(2n-2)a^2}\int \frac{dx}{(a^2-x^2)^{n-1}}$

17.8.16. $\displaystyle\int \frac{x\,dx}{(a^2-x^2)^n} = \frac{1}{2(n-1)(a^2-x^2)^{n-1}}$

17.8.17. $\displaystyle\int \frac{dx}{x(a^2-x^2)^n} = \frac{1}{2(n-1)a^2(a^2-x^2)^{n-1}} + \frac{1}{a^2}\int \frac{dx}{x(a^2-x^2)^{n-1}}$

17.8.18. $\displaystyle\int \frac{x^m dx}{(a^2-x^2)^n} = a^2\int \frac{x^{m-2}dx}{(a^2-x^2)^n} - \int \frac{x^{m-2}dx}{(a^2-x^2)^{n-1}}$

17.8.19. $\displaystyle\int \frac{dx}{x^m(a^2-x^2)^n} = \frac{1}{a^2}\int \frac{dx}{x^m(a^2-x^2)^{n-1}} + \frac{1}{a^2}\int \frac{dx}{x^{m-2}(a^2-x^2)^n}$

(9) Integrals Involving $\sqrt{x^2+a^2}$

17.9.1. $\displaystyle\int \frac{dx}{\sqrt{x^2+a^2}} = \ln(x+\sqrt{x^2+a^2})$ or $\sinh^{-1}\dfrac{x}{a}$

17.9.2. $\displaystyle\int \frac{x\,dx}{\sqrt{x^2+a^2}} = \sqrt{x^2+a^2}$

17.9.3. $\displaystyle\int \frac{x^2dx}{\sqrt{x^2+a^2}} = \frac{x\sqrt{x^2+a^2}}{2} - \frac{a^2}{2}\ln(x+\sqrt{x^2+a^2})$

17.9.4. $\displaystyle\int \frac{x^3dx}{\sqrt{x^2+a^2}} = \frac{(x^2+a^2)^{3/2}}{3} - a^2\sqrt{x^2+a^2}$

17.9.5. $\int \frac{dx}{x\sqrt{x^2+a^2}} = -\frac{1}{a}\ln\left(\frac{a+\sqrt{x^2+a^2}}{x}\right)$

17.9.6. $\int \frac{dx}{x^2\sqrt{x^2+a^2}} = -\frac{\sqrt{x^2+a^2}}{a^2x}$

17.9.7. $\int \frac{dx}{x^3\sqrt{x^2+a^2}} = -\frac{\sqrt{x^2+a^2}}{2a^2x^2} + \frac{1}{2a^3}\ln\left(\frac{a+\sqrt{x^2+a^2}}{x}\right)$

17.9.8. $\int \sqrt{x^2+a^2}\,dx = \frac{x\sqrt{x^2+a^2}}{2} + \frac{a^2}{2}\ln(x+\sqrt{x^2+a^2})$

17.9.9. $\int x\sqrt{x^2+a^2}\,dx = \frac{(x^2+a^2)^{3/2}}{3}$

17.9.10. $\int x^2\sqrt{x^2+a^2}\,dx = \frac{x(x^2+a^2)^{3/2}}{4} - \frac{a^2x\sqrt{x^2+a^2}}{8} - \frac{a^4}{8}\ln(x+\sqrt{x^2+a^2})$

17.9.11. $\int x^3\sqrt{x^2+a^2}\,dx = \frac{(x^2+a^2)^{5/2}}{5} - \frac{a^2(x^2+a^2)^{3/2}}{3}$

17.9.12. $\int \frac{\sqrt{x^2+a^2}}{x}dx = \sqrt{x^2+a^2} - a\ln\left(\frac{a+\sqrt{x^2+a^2}}{x}\right)$

17.9.13. $\int \frac{\sqrt{x^2+a^2}}{x^2}dx = -\frac{\sqrt{x^2+a^2}}{x} + \ln(x+\sqrt{x^2+a^2})$

17.9.14. $\int \frac{\sqrt{x^2+a^2}}{x^3}dx = -\frac{\sqrt{x^2+a^2}}{2x^2} - \frac{1}{2a}\ln\left(\frac{a+\sqrt{x^2+a^2}}{x}\right)$

17.9.15. $\int \frac{dx}{(x^2+a^2)^{3/2}} = \frac{x}{a^2\sqrt{x^2+a^2}}$

17.9.16. $\int \frac{x\,dx}{(x^2+a^2)^{3/2}} = \frac{-1}{\sqrt{x^2+a^2}}$

17.9.17. $\int \frac{x^2dx}{(x^2+a^2)^{3/2}} = \frac{-x}{\sqrt{x^2+a^2}} + \ln(x+\sqrt{x^2+a^2})$

17.9.18. $\int \frac{x^3dx}{(x^2+a^2)^{3/2}} = \sqrt{x^2+a^2} + \frac{a^2}{\sqrt{x^2+a^2}}$

17.9.19. $\int \frac{dx}{x(x^2+a^2)^{3/2}} = \frac{1}{a^2\sqrt{x^2+a^2}} - \frac{1}{a^3}\ln\left(\frac{a+\sqrt{x^2+a^2}}{x}\right)$

17.9.20. $\int \frac{dx}{x^2(x^2+a^2)^{3/2}} = -\frac{\sqrt{x^2+a^2}}{a^4x} - \frac{x}{a^4\sqrt{x^2+a^2}}$

17.9.21. $\int \frac{dx}{x^3(x^2+a^2)^{3/2}} = \frac{-1}{2a^2x^2\sqrt{x^2+a^2}} - \frac{3}{2a^4\sqrt{x^2+a^2}} + \frac{3}{2a^5}\ln\left(\frac{a+\sqrt{x^2+a^2}}{x}\right)$

17.9.22. $\int (x^2+a^2)^{3/2}dx = \frac{x(x^2+a^2)^{3/2}}{4} + \frac{3a^2x\sqrt{x^2+a^2}}{8} + \frac{3}{8}a^4\ln(x+\sqrt{x^2+a^2})$

17.9.23. $\int x(x^2+a^2)^{3/2}dx = \frac{(x^2+a^2)^{5/2}}{5}$

17.9.24. $\int x^2(x^2+a^2)^{3/2}dx = \frac{x(x^2+a^2)^{5/2}}{6} - \frac{a^2x(x^2+a^2)^{3/2}}{24} - \frac{a^4x\sqrt{x^2+a^2}}{16} - \frac{a^6}{16}\ln(x+\sqrt{x^2+a^2})$

17.9.25. $\int x^3(x^2+a^2)^{3/2}dx = \frac{(x^2+a^2)^{7/2}}{7} - \frac{a^2(x^2+a^2)^{5/2}}{5}$

17.9.26. $\int \frac{(x^2+a^2)^{3/2}}{x}dx = \frac{(x^2+a^2)^{3/2}}{3} + a^2\sqrt{x^2+a^2} - a^3\ln\left(\frac{a+\sqrt{x^2+a^2}}{x}\right)$

17.9.27. $\int \frac{(x^2+a^2)^{3/2}}{x^2}dx = -\frac{(x^2+a^2)^{3/2}}{x} + \frac{3x\sqrt{x^2+a^2}}{2} + \frac{3}{2}a^2\ln(x+\sqrt{x^2+a^2})$

17.9.28. $\int \frac{(x^2+a^2)^{3/2}}{x^3}dx = -\frac{(x^2+a^2)^{3/2}}{2x^2} + \frac{3}{2}\sqrt{x^2+a^2} - \frac{3}{2}a\ln\left(\frac{a+\sqrt{x^2+a^2}}{x}\right)$

(10) Integrals Involving $\sqrt{x^2-a^2}$

17.10.1. $\int \frac{dx}{\sqrt{x^2-a^2}} = \ln(x+\sqrt{x^2-a^2}), \quad \int \frac{x\,dx}{\sqrt{x^2-a^2}} = \sqrt{x^2-a^2}$

17.10.2. $\int \frac{x^2dx}{\sqrt{x^2-a^2}} = \frac{x\sqrt{x^2-a^2}}{2} + \frac{a^2}{2}\ln(x+\sqrt{x^2-a^2})$

17.10.3. $\int \frac{x^3dx}{\sqrt{x^2-a^2}} = \frac{(x^2-a^2)^{3/2}}{3} + a^2\sqrt{x^2-a^2}$

17.10.4. $\int \frac{dx}{x\sqrt{x^2-a^2}} = \frac{1}{a}\sec^{-1}\left|\frac{x}{a}\right|$

17.10.5. $\int \frac{dx}{x^2\sqrt{x^2-a^2}} = \frac{\sqrt{x^2-a^2}}{a^2x}$

17.10.6. $\int \frac{dx}{x^3\sqrt{x^2-a^2}} = \frac{\sqrt{x^2-a^2}}{2a^2x^2} + \frac{1}{2a^3}\sec^{-1}\left|\frac{x}{a}\right|$

17.10.7. $\int \sqrt{x^2-a^2}\,dx = \frac{x\sqrt{x^2-a^2}}{2} - \frac{a^2}{2}\ln(x+\sqrt{x^2-a^2})$

17.10.8. $\int x\sqrt{x^2-a^2}\,dx = \frac{(x^2-a^2)^{3/2}}{3}$

17.10.9. $\int x^2\sqrt{x^2-a^2}\,dx = \frac{x(x^2-a^2)^{3/2}}{4} + \frac{a^2x\sqrt{x^2-a^2}}{8} - \frac{a^4}{8}\ln(x+\sqrt{x^2-a^2})$

17.10.10. $\int x^3\sqrt{x^2-a^2}\,dx = \frac{(x^2-a^2)^{5/2}}{5} + \frac{a^2(x^2-a^2)^{3/2}}{3}$

17.10.11. $\int \frac{\sqrt{x^2-a^2}}{x}dx = \sqrt{x^2-a^2} - a\sec^{-1}\left|\frac{x}{a}\right|$

17.10.12. $\int \frac{\sqrt{x^2-a^2}}{x^2}dx = -\frac{\sqrt{x^2-a^2}}{x} + \ln(x+\sqrt{x^2-a^2})$

17.10.13. $\int \frac{\sqrt{x^2-a^2}}{x^3}\,dx = -\frac{\sqrt{x^2-a^2}}{2x^2} + \frac{1}{2a}\sec^{-1}\left|\frac{x}{a}\right|$

17.10.14. $\int \frac{dx}{(x^2-a^2)^{3/2}} = -\frac{x}{a^2\sqrt{x^2-a^2}}$

17.10.15. $\int \frac{x\,dx}{(x^2-a^2)^{3/2}} = \frac{-1}{\sqrt{x^2-a^2}}$

17.10.16. $\int \frac{x^2dx}{(x^2-a^2)^{3/2}} = -\frac{x}{\sqrt{x^2-a^2}} + \ln(x+\sqrt{x^2-a^2})$

17.10.17. $\int \frac{x^3dx}{(x^2-a^2)^{3/2}} = \sqrt{x^2-a^2} - \frac{a^2}{\sqrt{x^2-a^2}}$

17.10.18. $\int \frac{dx}{x(x^2-a^2)^{3/2}} = \frac{-1}{a^2\sqrt{x^2-a^2}} - \frac{1}{a^3}\sec^{-1}\left|\frac{x}{a}\right|$

17.10.19. $\int \frac{dx}{x^2(x^2-a^2)^{3/2}} = -\frac{\sqrt{x^2-a^2}}{a^4x} - \frac{x}{a^4\sqrt{x^2-a^2}}$

17.10.20. $\int \frac{dx}{x^3(x^2-a^2)^{3/2}} = \frac{1}{2a^2x^2\sqrt{x^2-a^2}} - \frac{3}{2a^4\sqrt{x^2-a^2}} - \frac{3}{2a^5}\sec^{-1}\left|\frac{x}{a}\right|$

17.10.21. $\int (x^2-a^2)^{3/2}\,dx = \frac{x(x^2-a^2)^{3/2}}{4} - \frac{3a^2x\sqrt{x^2-a^2}}{8} + \frac{3}{8}a^4\ln(x+\sqrt{x^2-a^2})$

17.10.22. $\int x(x^2-a^2)^{3/2}dx = \frac{(x^2-a^2)^{5/2}}{5}$

17.10.23. $\int x^2(x^2-a^2)^{3/2}dx = \frac{x(x^2-a^2)^{5/2}}{6} + \frac{a^2x(x^2-a^2)^{3/2}}{24} - \frac{a^4x\sqrt{x^2-a^2}}{16} + \frac{a^6}{16}\ln(x+\sqrt{x^2-a^2})$

17.10.24. $\int x^3(x^2-a^2)^{3/2}dx = \frac{(x^2-a^2)^{7/2}}{7} + \frac{a^2(x^2-a^2)^{5/2}}{5}$

17.10.25. $\int \frac{(x^2-a^2)^{3/2}}{x}\,dx = \frac{(x^2-a^2)^{3/2}}{3} - a^2\sqrt{x^2-a^2} + a^3\sec^{-1}\left|\frac{x}{a}\right|$

17.10.26. $\int \frac{(x^2-a^2)^{3/2}}{x^2}\,dx = -\frac{(x^2-a^2)^{3/2}}{x} + \frac{3x\sqrt{x^2-a^2}}{2} - \frac{3}{2}a^2\ln(x+\sqrt{x^2-a^2})$

17.10.27. $\int \frac{(x^2-a^2)^{3/2}}{x^3}\,dx = -\frac{(x^2-a^2)^{3/2}}{2x^2} + \frac{3\sqrt{x^2-a^2}}{2} - \frac{3}{2}a\sec^{-1}\left|\frac{x}{a}\right|$

(11) Integrals Involving $\sqrt{a^2-x^2}$

17.11.1. $\int \frac{dx}{\sqrt{a^2-x^2}} = \sin^{-1}\frac{x}{a}$

17.11.2. $\int \frac{x\,dx}{\sqrt{a^2-x^2}} = -\sqrt{a^2-x^2}$

17.11.3. $\int \frac{x^2 dx}{\sqrt{a^2 - x^2}} = -\frac{x\sqrt{a^2 - x^2}}{2} + \frac{a^2}{2}\sin^{-1}\frac{x}{a}$

17.11.4. $\int \frac{x^3 dx}{\sqrt{a^2 - x^2}} = \frac{(a^2 - x^2)^{3/2}}{3} - a^2\sqrt{a^2 - x^2}$

17.11.5. $\int \frac{dx}{x\sqrt{a^2 - x^2}} = -\frac{1}{a}\ln\left(\frac{a + \sqrt{a^2 - x^2}}{x}\right)$

17.11.6. $\int \frac{dx}{x^2\sqrt{a^2 - x^2}} = -\frac{\sqrt{a^2 - x^2}}{a^2 x}$

17.11.7. $\int \frac{dx}{x^3\sqrt{a^2 - x^2}} = -\frac{\sqrt{a^2 - x^2}}{2a^2x^2} - \frac{1}{2a^3}\ln\left(\frac{a + \sqrt{a^2 - x^2}}{x}\right)$

17.11.8. $\int \sqrt{a^2 - x^2}\, dx = \frac{x\sqrt{a^2 - x^2}}{2} + \frac{a^2}{2}\sin^{-1}\frac{x}{a}$

17.11.9. $\int x\sqrt{a^2 - x^2}\, dx = -\frac{(a^2 - x^2)^{3/2}}{3}$

17.11.10. $\int x^2\sqrt{a^2 - x^2}\, dx = -\frac{x(a^2 - x^2)^{3/2}}{4} + \frac{a^2x\sqrt{a^2 - x^2}}{8} + \frac{a^4}{8}\sin^{-1}\frac{x}{a}$

17.11.11. $\int x^3\sqrt{a^2 - x^2}\, dx = \frac{(a^2 - x^2)^{5/2}}{5} - \frac{a^2(a^2 - x^2)^{3/2}}{3}$

17.11.12. $\int \frac{\sqrt{a^2 - x^2}}{x}\, dx = \sqrt{a^2 - x^2} - a\ln\left(\frac{a + \sqrt{a^2 - x^2}}{x}\right)$

17.11.13. $\int \frac{\sqrt{a^2 - x^2}}{x^2}\, dx = -\frac{\sqrt{a^2 - x^2}}{x} - \sin^{-1}\frac{x}{a}$

17.11.14. $\int \frac{\sqrt{a^2 - x^2}}{x^3}\, dx = -\frac{\sqrt{a^2 - x^2}}{2x^2} + \frac{1}{2a}\ln\left(\frac{a + \sqrt{a^2 - x^2}}{x}\right)$

17.11.15. $\int \frac{dx}{(a^2 - x^2)^{3/2}} = \frac{x}{a^2\sqrt{a^2 - x^2}}$

17.11.16. $\int \frac{x\, dx}{(a^2 - x^2)^{3/2}} = \frac{1}{\sqrt{a^2 - x^2}}$

17.11.17. $\int \frac{x^2 dx}{(a^2 - x^2)^{3/2}} = \frac{x}{\sqrt{a^2 - x^2}} - \sin^{-1}\frac{x}{a}$

17.11.18. $\int \frac{x^3 dx}{(a^2 - x^2)^{3/2}} = \sqrt{a^2 - x^2} + \frac{a^2}{\sqrt{a^2 - x^2}}$

17.11.19. $\int \frac{dx}{x(a^2 - x^2)^{3/2}} = \frac{1}{a^2\sqrt{a^2 - x^2}} - \frac{1}{a^3}\ln\left(\frac{a + \sqrt{a^2 - x^2}}{x}\right)$

17.11.20. $\int \frac{dx}{x^2(a^2 - x^2)^{3/2}} = -\frac{\sqrt{a^2 - x^2}}{a^4 x} + \frac{x}{a^4\sqrt{a^2 - x^2}}$

17.11.21. $\int \frac{dx}{x^3(a^2 - x^2)^{3/2}} = \frac{-1}{2a^2x^2\sqrt{a^2 - x^2}} + \frac{3}{2a^4\sqrt{a^2 - x^2}} - \frac{3}{2a^5}\ln\left(\frac{a + \sqrt{a^2 - x^2}}{x}\right)$

17.11.22. $\int (a^2-x^2)^{3/2}dx = \frac{x(a^2-x^2)^{3/2}}{4} + \frac{3a^2x\sqrt{a^2-x^2}}{8} + \frac{3}{8}a^4\sin^{-1}\frac{x}{a}$

17.11.23. $\int x(a^2-x^2)^{3/2}dx = -\frac{(a^2-x^2)^{5/2}}{5}$

17.11.24. $\int x^2(a^2-x^2)^{3/2}dx = -\frac{x(a^2-x^2)^{5/2}}{6} + \frac{a^2x(a^2-x^2)^{3/2}}{24} + \frac{a^4x\sqrt{a^2-x^2}}{16} + \frac{a^6}{16}\sin^{-1}\frac{x}{a}$

17.11.25. $\int x^3(a^2-x^2)^{3/2}dx = \frac{(a^2-x^2)^{7/2}}{7} - \frac{a^2(a^2-x^2)^{5/2}}{5}$

17.11.26. $\int \frac{(a^2-x^2)^{3/2}}{x}dx = \frac{(a^2-x^2)^{3/2}}{3} + a^2\sqrt{a^2-x^2} - a^3\ln\left(\frac{a+\sqrt{a^2-x^2}}{x}\right)$

17.11.27. $\int \frac{(a^2-x^2)^{3/2}}{x^2}dx = -\frac{(a^2-x^2)^{3/2}}{x} - \frac{3x\sqrt{a^2-x^2}}{2} - \frac{3}{2}a^2\sin^{-1}\frac{x}{a}$

17.11.28. $\int \frac{(a^2-x^2)^{3/2}}{x^3}dx = -\frac{(a^2-x^2)^{3/2}}{2x^2} - \frac{3\sqrt{a^2-x^2}}{2} + \frac{3}{2}a\ln\left(\frac{a+\sqrt{a^2-x^2}}{x}\right)$

(12) Integrals Involving $ax^2 + bx + c$

17.12.1. $\int \frac{dx}{ax^2+bx+c} = \begin{cases} \frac{2}{\sqrt{4ac-b^2}}\tan^{-1}\frac{2ax+b}{\sqrt{4ac-b^2}} \\ \frac{1}{\sqrt{b^2-4ac}}\ln\left(\frac{2ax+b-\sqrt{b^2-4ac}}{2ax+b+\sqrt{b^2-4ac}}\right) \end{cases}$

If $b^2 = 4ac$, $ax^2+bx+c = a(x+b/2a)^2$ and the results 17.1.6 to 17.1.10 and 17.1.14 to 17.1.17 can be used. If $b = 0$ use results on page 75. If a or $c = 0$ use results on pages 71–72.

17.12.2. $\int \frac{x\,dx}{ax^2+bx+c} = \frac{1}{2a}\ln(ax^2+bx+c) - \frac{b}{2a}\int \frac{dx}{ax^2+bx+c}$

17.12.3. $\int \frac{x^2dx}{ax^2+bx+c} = \frac{x}{a} - \frac{b}{2a^2}\ln(ax^2+bx+c) + \frac{b^2-2ac}{2a^2}\int \frac{dx}{ax^2+bx+c}$

17.12.4. $\int \frac{x^m dx}{ax^2+bx+c} = \frac{x^{m-1}}{(m-1)a} - \frac{c}{a}\int \frac{x^{m-2}dx}{ax^2+bx+c} - \frac{b}{a}\int \frac{x^{m-1}dx}{ax^2+bx+c}$

17.12.5. $\int \frac{dx}{x(ax^2+bx+c)} = \frac{1}{2c}\ln\left(\frac{x^2}{ax^2+bx+c}\right) - \frac{b}{2c}\int \frac{dx}{ax^2+bx+c}$

17.12.6. $\int \frac{dx}{x^2(ax^2+bx+c)} = \frac{b}{2c^2}\ln\left(\frac{ax^2+bx+c}{x^2}\right) - \frac{1}{cx} + \frac{b^2-2ac}{2c^2}\int \frac{dx}{ax^2+bx+c}$

17.12.7. $\int \frac{dx}{x^n(ax^2+bx+c)} = -\frac{1}{(n-1)cx^{n-1}} - \frac{b}{c}\int \frac{dx}{x^{n-1}(ax^2+bx+c)} - \frac{a}{c}\int \frac{dx}{x^{n-2}(ax^2+bx+c)}$

17.12.8. $\int \frac{dx}{(ax^2+bx+c)^2} = \frac{2ax+b}{(4ac-b^2)(ax^2+bx+c)} + \frac{2a}{4ac-b^2}\int \frac{dx}{ax^2+bx+c}$

17.12.9. $\int \frac{x\,dx}{(ax^2+bx+c)^2} = -\frac{bx+2c}{(4ac-b^2)(ax^2+bx+c)} - \frac{b}{4ac-b^2}\int \frac{dx}{ax^2+bx+c}$

17.12.10. $$\int \frac{x^2 dx}{(ax^2+bx+c)^2} = \frac{(b^2-2ac)x+bc}{a(4ac-b^2)(ax^2+bx+c)} + \frac{2c}{4ac-b^2}\int \frac{dx}{ax^2+bx+c}$$

17.12.11. $$\int \frac{x^m dx}{(ax^2+bx+c)^n} = -\frac{x^{m-1}}{(2n-m-1)a(ax^2+bx+c)^{n-1}} + \frac{(m-1)c}{(2n-m-1)a}\int \frac{x^{m-2}dx}{(ax^2+bx+c)^n} - \frac{(n-m)b}{(2n-m-1)a}\int \frac{x^{m-1}dx}{(ax^2+bx+c)^n}$$

17.12.12. $$\int \frac{x^{2n-1}dx}{(ax^2+bx+c)^n} = \frac{1}{a}\int \frac{x^{2n-3}dx}{(ax^2+bx+c)^{n-1}} - \frac{c}{a}\int \frac{x^{2n-3}dx}{(ax^2+bx+c)^n} - \frac{b}{a}\int \frac{x^{2n-2}dx}{(ax^2+bx+c)^n}$$

17.12.13. $$\int \frac{dx}{x(ax^2+bx+c)^2} = \frac{1}{2c(ax^2+bx+c)} - \frac{b}{2c}\int \frac{dx}{(ax^2+bx+c)^2} + \frac{1}{c}\int \frac{dx}{x(ax^2+bx+c)}$$

17.12.14. $$\int \frac{dx}{x^2(ax^2+bx+c)^2} = -\frac{1}{cx(ax^2+bx+c)} - \frac{3a}{c}\int \frac{dx}{(ax^2+bx+c)^2} - \frac{2b}{c}\int \frac{dx}{x(ax^2+bx+c)^2}$$

17.12.15. $$\int \frac{dx}{x^m(ax^2+bx+c)^n} = -\frac{1}{(m-1)cx^{m-1}(ax^2+bx+c)^{n-1}} - \frac{(m+2n-3)a}{(m-1)c}\int \frac{dx}{x^{m-2}(ax^2+bx+c)^n} - \frac{(m+n-2)b}{(m-1)c}\int \frac{dx}{x^{m-1}(ax^2+bx+c)^n}$$

(13) Integrals Involving $\sqrt{ax^2+bx+c}$

In the following results if $b^2 = 4ac$, $\sqrt{ax^2+bx+c} = \sqrt{a}(x+b/2a)$ and the results 17.1 can be used. If $b = 0$ use the results 17.9. If $a = 0$ or $c = 0$ use the results 17.2 and 17.5.

17.13.1. $$\int \frac{dx}{\sqrt{ax^2+bx+c}} = \begin{cases} \dfrac{1}{\sqrt{a}}\ln(2\sqrt{a}\sqrt{ax^2+bx+c}+2ax+b) \\ -\dfrac{1}{\sqrt{-a}}\sin^{-1}\left(\dfrac{2ax+b}{\sqrt{b^2-4ac}}\right) \quad \text{or} \quad \dfrac{1}{\sqrt{a}}\sinh^{-1}\left(\dfrac{2ax+b}{\sqrt{4ac-b^2}}\right) \end{cases}$$

17.13.2. $$\int \frac{x\,dx}{\sqrt{ax^2+bx+c}} = \frac{\sqrt{ax^2+bx+c}}{a} - \frac{b}{2a}\int \frac{dx}{\sqrt{ax^2+bx+c}}$$

17.13.3. $$\int \frac{x^2 dx}{\sqrt{ax^2+bx+c}} = \frac{2ax-3b}{4a^2}\sqrt{ax^2+bx+c} + \frac{3b^2-4ac}{8a^2}\int \frac{dx}{\sqrt{ax^2+bx+c}}$$

17.13.4. $$\int \frac{dx}{x\sqrt{ax^2+bx+c}} = \begin{cases} -\dfrac{1}{\sqrt{c}}\ln\left(\dfrac{2\sqrt{c}\sqrt{ax^2+bx+c}+bx+2c}{x}\right) \\ \dfrac{1}{\sqrt{-c}}\sin^{-1}\left(\dfrac{bx+2c}{|x|\sqrt{b^2-4ac}}\right) \quad \text{or} \quad -\dfrac{1}{\sqrt{c}}\sinh^{-1}\left(\dfrac{bx+2c}{|x|\sqrt{4ac-b^2}}\right) \end{cases}$$

17.13.5. $$\int \frac{dx}{x^2\sqrt{ax^2+bx+c}} = -\frac{\sqrt{ax^2+bx+c}}{cx} - \frac{b}{2c}\int \frac{dx}{x\sqrt{ax^2+bx+c}}$$

17.13.6. $$\int \sqrt{ax^2+bx+c}\,dx = \frac{(2ax+b)\sqrt{ax^2+bx+c}}{4a} + \frac{4ac-b^2}{8a}\int \frac{dx}{\sqrt{ax^2+bx+c}}$$

17.13.7. $$\int x\sqrt{ax^2+bx+c}\,dx = \frac{(ax^2+bx+c)^{3/2}}{3a} - \frac{b(2ax+b)}{8a^2}\sqrt{ax^2+bx+c} - \frac{b(4ac-b^2)}{16a^2}\int\frac{dx}{\sqrt{ax^2+bx+c}}$$

17.13.8. $$\int x^2\sqrt{ax^2+bx+c}\,dx = \frac{6ax-5b}{24a^2}(ax^2+bx+c)^{3/2} + \frac{5b^2-4ac}{16a^2}\int\sqrt{ax^2+bx+c}\,dx$$

17.13.9. $$\int\frac{\sqrt{ax^2+bx+c}}{x}dx = \sqrt{ax^2+bx+c} + \frac{b}{2}\int\frac{dx}{\sqrt{ax^2+bx+c}} + c\int\frac{dx}{x\sqrt{ax^2+bx+c}}$$

17.13.10. $$\int\frac{\sqrt{ax^2+bx+c}}{x^2}dx = -\frac{\sqrt{ax^2+bx+c}}{x} + a\int\frac{dx}{\sqrt{ax^2+bx+c}} + \frac{b}{2}\int\frac{dx}{x\sqrt{ax^2+bx+c}}$$

17.13.11. $$\int\frac{dx}{(ax^2+bx+c)^{3/2}} = \frac{2(2ax+b)}{(4ac-b^2)\sqrt{ax^2+bx+c}}$$

17.13.12. $$\int\frac{x\,dx}{(ax^2+bx+c)^{3/2}} = \frac{2(bx+2c)}{(b^2-4ac)\sqrt{ax^2+bx+c}}$$

17.13.13. $$\int\frac{x^2dx}{(ax^2+bx+c)^{3/2}} = \frac{(2b^2-4ac)x+2bc}{a(4ac-b^2)\sqrt{ax^2+bx+c}} + \frac{1}{a}\int\frac{dx}{\sqrt{ax^2+bx+c}}$$

17.13.14. $$\int\frac{dx}{x(ax^2+bx+c)^{3/2}} = \frac{1}{c\sqrt{ax^2+bx+c}} + \frac{1}{c}\int\frac{dx}{x\sqrt{ax^2+bx+c}} - \frac{b}{2c}\int\frac{dx}{(ax^2+bx+c)^{3/2}}$$

17.13.15. $$\int\frac{dx}{x^2(ax^2+bx+c)^{3/2}} = -\frac{ax^2+2bx+c}{c^2x\sqrt{ax^2+bx+c}} + \frac{b^2-2ac}{2c^2}\int\frac{dx}{(ax^2+bx+c)^{3/2}} - \frac{3b}{2c^2}\int\frac{dx}{x\sqrt{ax^2+bx+c}}$$

17.13.16. $$\int(ax^2+bx+c)^{n+1/2}\,dx = \frac{(2ax+b)(ax^2+bx+c)^{n+1/2}}{4a(n+1)} + \frac{(2n+1)(4ac-b^2)}{8a(n+1)}\int(ax^2+bx+c)^{n-1/2}dx$$

17.13.17. $$\int x(ax^2+bx+c)^{n+1/2}dx = \frac{(ax^2+bx+c)^{n+3/2}}{a(2n+3)} - \frac{b}{2a}\int(ax^2+bx+c)^{n+1/2}dx$$

17.13.18. $$\int\frac{dx}{(ax^2+bx+c)^{n+1/2}} = \frac{2(2ax+b)}{(2n-1)(4ac-b^2)(ax^2+bx+c)^{n-1/2}} + \frac{8a(n-1)}{(2n-1)(4ac-b^2)}\int\frac{dx}{(ax^2+bx+c)^{n-1/2}}$$

17.13.19. $$\int\frac{dx}{x(ax^2+bx+c)^{n+1/2}} = \frac{1}{(2n-1)c(ax^2+bx+c)^{n-1/2}} + \frac{1}{c}\int\frac{dx}{x(ax^2+bx+c)^{n-1/2}} - \frac{b}{2c}\int\frac{dx}{(ax^2+bx+c)^{n+1/2}}$$

(14) Integrals Involving $x^3 + a^3$

Note that for formulas involving $x^3 - a^3$ replace a with $-a$.

17.14.1. $\displaystyle\int \frac{dx}{x^3+a^3} = \frac{1}{6a^2}\ln\left(\frac{(x+a)^2}{x^2-ax+a^2}\right) + \frac{1}{a^2\sqrt{3}}\tan^{-1}\frac{2x-a}{a\sqrt{3}}$

17.14.2. $\displaystyle\int \frac{x\,dx}{x^3+a^3} = \frac{1}{6a}\ln\left(\frac{x^2-ax+a^2}{(x+a)^2}\right) + \frac{1}{a\sqrt{3}}\tan^{-1}\frac{2x-a}{a\sqrt{3}}$

17.14.3. $\displaystyle\int \frac{x^2dx}{x^3+a^3} = \frac{1}{3}\ln(x^3+a^3)$

17.14.4. $\displaystyle\int \frac{dx}{x(x^3+a^3)} = \frac{1}{3a^3}\ln\left(\frac{x^3}{x^3+a^3}\right)$

17.14.5. $\displaystyle\int \frac{dx}{x^2(x^3+a^3)} = -\frac{1}{a^3x} - \frac{1}{6a^4}\ln\left(\frac{x^2-ax+a^2}{(x+a)^2}\right) - \frac{1}{a^4\sqrt{3}}\tan^{-1}\frac{2x-a}{a\sqrt{3}}$

17.14.6. $\displaystyle\int \frac{dx}{(x^3+a^3)^2} = \frac{x}{3a^3(x^3+a^3)} + \frac{1}{9a^5}\ln\left(\frac{(x+a)^2}{x^2-ax+a^2}\right) + \frac{2}{3a^5\sqrt{3}}\tan^{-1}\frac{2x-a}{a\sqrt{3}}$

17.14.7. $\displaystyle\int \frac{x\,dx}{(x^3+a^3)^2} = \frac{x^2}{3a^3(x^3+a^3)} + \frac{1}{18a^4}\ln\left(\frac{x^2-ax+a^2}{(x+a)^2}\right) + \frac{1}{3a^4\sqrt{3}}\tan^{-1}\frac{2x-a}{a\sqrt{3}}$

17.14.8. $\displaystyle\int \frac{x^2dx}{(x^3+a^3)^2} = -\frac{1}{3(x^3+a^3)}$

17.14.9. $\displaystyle\int \frac{dx}{x(x^3+a^3)^2} = \frac{1}{3a^3(x^3+a^3)} + \frac{1}{3a^6}\ln\left(\frac{x^3}{x^3+a^3}\right)$

17.14.10. $\displaystyle\int \frac{dx}{x^2(x^3+a^3)^2} = -\frac{1}{a^6x} - \frac{x^2}{3a^6(x^3+a^3)} - \frac{4}{3a^6}\int \frac{x\,dx}{x^3+a^3}$ (See 17.14.2.)

17.14.11. $\displaystyle\int \frac{x^m dx}{x^3+a^3} = \frac{x^{m-2}}{m-2} - a^3\int \frac{x^{m-3}dx}{x^3+a^3}$

17.14.12. $\displaystyle\int \frac{dx}{x^n(x^3+a^3)} = \frac{-1}{a^3(n-1)x^{n-1}} - \frac{1}{a^3}\int \frac{dx}{x^{n-3}(x^3+a^3)}$

(15) Integrals Involving $x^4 \pm a^4$

17.15.1. $\displaystyle\int \frac{dx}{x^4+a^4} = \frac{1}{4a^3\sqrt{2}}\ln\left(\frac{x^2+ax\sqrt{2}+a^2}{x^2-ax\sqrt{2}+a^2}\right) - \frac{1}{2a^3\sqrt{2}}\left[\tan^{-1}\left(1-\frac{x\sqrt{2}}{a}\right) - \tan^{-1}\left(1+\frac{x\sqrt{2}}{a}\right)\right]$

17.15.2. $\displaystyle\int \frac{x\,dx}{x^4+a^4} = \frac{1}{2a^2}\tan^{-1}\frac{x^2}{a^2}$

17.15.3. $\displaystyle\int \frac{x^2dx}{x^4+a^4} = \frac{1}{4a\sqrt{2}}\ln\left(\frac{x^2-ax\sqrt{2}+a^2}{x^2+ax\sqrt{2}+a^2}\right) - \frac{1}{2a\sqrt{2}}\left[\tan^{-1}\left(1-\frac{x\sqrt{2}}{a}\right) - \tan^{-1}\left(1+\frac{x\sqrt{2}}{a}\right)\right]$

17.15.4. $\int \frac{x^3 dx}{x^4 + a^4} = \frac{1}{4} \ln (x^4 + a^4)$

17.15.5. $\int \frac{dx}{x(x^4 + a^4)} = \frac{1}{4a^4} \ln\left(\frac{x^4}{x^4 + a^4}\right)$

17.15.6. $\int \frac{dx}{x^2(x^4 + a^4)} = -\frac{1}{a^4 x} - \frac{1}{4a^5\sqrt{2}} \ln\left(\frac{x^2 - ax\sqrt{2} + a^2}{x^2 + ax\sqrt{2} + a^2}\right)$

$$+\frac{1}{2a^5\sqrt{2}}\left[\tan^{-1}\left(1 - \frac{x\sqrt{2}}{a}\right) - \tan^{-1}\left(1 + \frac{x\sqrt{2}}{a}\right)\right]$$

17.15.7. $\int \frac{dx}{x^3(x^4 + a^4)} = -\frac{1}{2a^4 x^2} - \frac{1}{2a^6} \tan^{-1} \frac{x^2}{a^2}$

17.15.8. $\int \frac{dx}{x^4 - a^4} = \frac{1}{4a^3} \ln\left(\frac{x - a}{x + a}\right) - \frac{1}{2a^3} \tan^{-1} \frac{x}{a}$

17.15.9. $\int \frac{x\,dx}{x^4 - a^4} = \frac{1}{4a^2} \ln\left(\frac{x^2 - a^2}{x^2 + a^2}\right)$

17.15.10. $\int \frac{x^2 dx}{x^4 - a^4} = \frac{1}{4a} \ln\left(\frac{x - a}{x + a}\right) + \frac{1}{2a} \tan^{-1} \frac{x}{a}$

17.15.11. $\int \frac{x^3 dx}{x^4 - a^4} = \frac{1}{4} \ln (x^4 - a^4)$

17.15.12. $\int \frac{dx}{x(x^4 - a^4)} = \frac{1}{4a^4} \ln\left(\frac{x^4 - a^4}{x^4}\right)$

17.15.13. $\int \frac{dx}{x^3(x^4 - a^4)} = \frac{1}{a^4 x} + \frac{1}{4a^5} \ln\left(\frac{x - a}{x + a}\right) + \frac{1}{2a^5} \tan^{-1} \frac{x}{a}$

17.15.14. $\int \frac{dx}{x^3(x^4 - a^4)} = \frac{1}{2a^4 x^2} + \frac{1}{4a^6} \ln\left(\frac{x^2 - a^2}{x^2 + a^2}\right)$

(16) Integrals Involving $x^n \pm a^n$

17.16.1. $\int \frac{dx}{x(x^n + a^n)} = \frac{1}{na^n} \ln\left(\frac{x^n}{x^n + a^n}\right)$

17.16.2. $\int \frac{x^{n-1} dx}{x^n + a^n} = \frac{1}{n} \ln (x^n + a^n)$

17.16.3. $\int \frac{x^m dx}{(x^n + a^n)^r} = \int \frac{x^{m-n} dx}{(x^n + a^n)^{r-1}} - a^n \int \frac{x^{m-n} dx}{(x^n + a^n)^r}$

17.16.4. $\int \frac{dx}{x^m (x^n + a^n)^r} = \frac{1}{a^n} \int \frac{dx}{x^m (x^n + a^n)^{r-1}} - \frac{1}{a^n} \int \frac{dx}{x^{m-n} (x^n + a^n)^r}$

17.16.5. $\int \frac{dx}{x\sqrt{x^n + a^n}} = \frac{1}{n\sqrt{a^n}} \ln\left(\frac{\sqrt{x^n + a^n} - \sqrt{a^n}}{\sqrt{x^n + a^n} + \sqrt{a^n}}\right)$

17.16.6. $\int \frac{dx}{x(x^n - a^n)} = \frac{1}{na^n} \ln\left(\frac{x^n - a^n}{x^n}\right)$

17.16.7. $\int \frac{x^{n-1}dx}{x^n - a^n} = \frac{1}{n}\ln(x^n - a^n)$

17.16.8. $\int \frac{x^m dx}{(x^n - a^n)^r} = a^n \int \frac{x^{m-n}dx}{(x^n - a^n)^r} + \int \frac{x^{m-n}dx}{(x^n - a^n)^{r-1}}$

17.16.9. $\int \frac{dx}{x^m(x^n - a^n)^r} = \frac{1}{a^n}\int \frac{dx}{x^{m-n}(x^n - a^n)^r} - \frac{1}{a^n}\int \frac{dx}{x^m(x^n - a^n)^{r-}}$

17.16.10. $\int \frac{dx}{x\sqrt{x^n - a^n}} = \frac{2}{n\sqrt{a^n}}\cos^{-1}\sqrt{\frac{a^n}{x^n}}$

17.16.11.
$$\int \frac{x^{p-1}dx}{x^{2m} + a^{2m}} = \frac{1}{ma^{2m-p}}\sum_{k=1}^{m}\sin\frac{(2k-1)p\pi}{2m}\tan^{-1}\left(\frac{x + a\cos[(2k-1)\pi/2m]}{a\sin[(2k-1)\pi/2m]}\right)$$
$$-\frac{1}{2ma^{2m-p}}\sum_{k=1}^{m}\cos\frac{(2k-1)p\pi}{2m}\ln\left(x^2 + 2ax\cos\frac{(2k-1)\pi}{2m} + a^2\right)$$

where $0 < p \leqq 2m$.

17.16.12.
$$\int \frac{x^{p-1}dx}{x^{2m} - a^{2m}} = \frac{1}{2ma^{2m-p}}\sum_{k=1}^{m-1}\cos\frac{kp\pi}{m}\ln\left(x^2 - 2ax\cos\frac{k\pi}{m} + a^2\right)$$
$$-\frac{1}{ma^{2m-p}}\sum_{k=1}^{m-1}\sin\frac{kp\pi}{m}\tan^{-1}\left(\frac{x - a\cos(k\pi/m)}{a\sin(k\pi/m)}\right)$$
$$+\frac{1}{2ma^{2m-p}}\{\ln(x-a) + (-1)^p\ln(x+a)\}$$

where $0 < p \leqq 2m$.

17.16.13.
$$\int \frac{x^{p-1}dx}{x^{2m+1} + a^{2m+1}} = \frac{2(-1)^{p-1}}{(2m+1)a^{2m-p+1}}\sum_{k=1}^{m}\sin\frac{2kp\pi}{2m+1}\tan^{-1}\left(\frac{x + a\cos[2k\pi/(2m+1)]}{a\sin[2k\pi/(2m+1)]}\right)$$
$$-\frac{(-1)^{p-1}}{(2m+1)a^{2m-p+1}}\sum_{k=1}^{m}\cos\frac{2kp\pi}{2m+1}\ln\left(x^2 + 2ax\cos\frac{2k\pi}{2m+1} + a^2\right)$$
$$+\frac{(-1)^{p-1}\ln(x+a)}{(2m+1)a^{2m-p+1}}$$

where $0 < p \leqq 2m + 1$.

17.16.14.
$$\int \frac{x^{p-1}dx}{x^{2m+1} - a^{2m+1}} = \frac{-2}{(2m+1)a^{2m-p+1}}\sum_{k=1}^{m}\sin\frac{2kp\pi}{2m+1}\tan^{-1}\left(\frac{x - a\cos[2k\pi/(2m+1)]}{a\sin[2k\pi/(2m+1)]}\right)$$
$$+\frac{1}{(2m+1)a^{2m-p+1}}\sum_{k=1}^{m}\cos\frac{2kp\pi}{2m+1}\ln\left(x^2 - 2ax\cos\frac{2k\pi}{2m+1} + a^2\right)$$
$$+\frac{\ln(x-a)}{(2m+1)a^{2m-p+1}}$$

where $0 < p \leqq 2m + 1$.

(17) Integrals Involving sin ax

17.17.1. $\int \sin ax\, dx = -\frac{\cos ax}{a}$

17.17.2. $\int x \sin ax\, dx = \frac{\sin ax}{a^2} - \frac{x \cos ax}{a}$

17.17.3. $\int x^2 \sin ax\, dx = \frac{2x}{a^2}\sin ax + \left(\frac{2}{a^3} - \frac{x^2}{a}\right)\cos ax$

17.17.4. $\int x^3 \sin ax\, dx = \left(\frac{3x^2}{a^2} - \frac{6}{a^4}\right)\sin ax + \left(\frac{6x}{a^3} - \frac{x^3}{a}\right)\cos ax$

17.17.5. $\int \frac{\sin ax}{x}\, dx = ax - \frac{(ax)^3}{3\cdot 3!} + \frac{(ax)^5}{5\cdot 5!} - \cdots$

17.17.6. $\int \frac{\sin ax}{x^2}\, dx = -\frac{\sin ax}{x} + a\int \frac{\cos ax}{x}\, dx$ (See 17.18.5.)

17.17.7. $\int \frac{dx}{\sin ax} = \frac{1}{\alpha}\ln(\csc ax - \cot ax) = \frac{1}{\alpha}\ln\tan\frac{ax}{2}$

17.17.8. $\int \frac{x\, dx}{\sin ax} = \frac{1}{a^2}\left\{ax + \frac{(ax)^3}{18} + \frac{7(ax)^5}{1800} + \cdots + \frac{2(2^{2n-1}-1)B_n(ax)^{2n+1}}{(2n+1)!} + \cdots\right\}$

17.17.9. $\int \sin^2 ax\, dx = \frac{x}{2} - \frac{\sin 2ax}{4a}$

17.17.10. $\int x \sin^2 ax\, dx = \frac{x^2}{4} - \frac{x \sin 2ax}{4a} - \frac{\cos 2ax}{8a^2}$

17.17.11. $\int \sin^3 ax\, dx = -\frac{\cos ax}{a} + \frac{\cos^3 ax}{3a}$

17.17.12. $\int \sin^4 ax\, dx = \frac{3x}{8} - \frac{\sin 2ax}{4a} + \frac{\sin 4ax}{32a}$

17.17.13. $\int \frac{dx}{\sin^2 ax} = -\frac{1}{a}\cot ax$

17.17.14. $\int \frac{dx}{\sin^3 ax} = -\frac{\cos ax}{2a \sin^2 ax} + \frac{1}{2a}\ln\tan\frac{ax}{2}$

17.17.15. $\int \sin px \sin qx\, dx = \frac{\sin(p-q)x}{2(p-q)} - \frac{\sin(p+q)x}{2(p+q)}$ (If $p = \pm q$, see 17.17.9.)

17.17.16. $\int \frac{dx}{1-\sin ax} = \frac{1}{a}\tan\left(\frac{\pi}{4} + \frac{ax}{2}\right)$

17.17.17. $\int \frac{x\, dx}{1-\sin ax} = \frac{x}{a}\tan\left(\frac{\pi}{4} + \frac{ax}{2}\right) + \frac{2}{a^2}\ln\sin\left(\frac{\pi}{4} - \frac{ax}{2}\right)$

17.17.18. $\int \frac{dx}{1+\sin ax} = -\frac{1}{\alpha}\tan\left(\frac{\pi}{4} - \frac{ax}{2}\right)$

17.17.19. $\int \frac{x\, dx}{1+\sin ax} = -\frac{x}{a}\tan\left(\frac{\pi}{4} - \frac{ax}{2}\right) + \frac{2}{a^2}\ln\sin\left(\frac{\pi}{4} + \frac{ax}{2}\right)$

17.17.20. $\int \frac{dx}{(1-\sin ax)^2} = \frac{1}{2a}\tan\left(\frac{\pi}{4}+\frac{ax}{2}\right)+\frac{1}{6a}\tan^3\left(\frac{\pi}{4}+\frac{ax}{2}\right)$

17.17.21. $\int \frac{dx}{(1+\sin ax)^2} = -\frac{1}{2a}\tan\left(\frac{\pi}{4}-\frac{ax}{2}\right)-\frac{1}{6a}\tan^3\left(\frac{\pi}{4}-\frac{ax}{2}\right)$

17.17.22. $\int \frac{dx}{p+q\sin ax} = \begin{cases} \dfrac{2}{a\sqrt{p^2-q^2}}\tan^{-1}\dfrac{p\tan\frac{1}{2}ax+q}{\sqrt{p^2-q^2}} \\ \dfrac{1}{a\sqrt{q^2-p^2}}\ln\left(\dfrac{p\tan\frac{1}{2}ax+q-\sqrt{q^2-p^2}}{p\tan\frac{1}{2}ax+q+\sqrt{q^2-p^2}}\right) \end{cases}$

(If $p=\pm q$, see 17.17.16 and 17.17.18.)

17.17.23. $\int \frac{dx}{(p+q\sin ax)^2} = \frac{q\cos ax}{a(p^2-q^2)(p+q\sin ax)}+\frac{p}{p^2-q^2}\int\frac{dx}{p+q\sin ax}$

(If $p=\pm q$, see 17.17.20 and 17.17.21.)

17.17.24. $\int \frac{dx}{p^2+q^2\sin^2 ax} = \frac{1}{ap\sqrt{p^2+q^2}}\tan^{-1}\frac{\sqrt{p^2+q^2}\tan ax}{p}$

17.17.25. $\int \frac{dx}{p^2-q^2\sin^2 ax} = \begin{cases} \dfrac{1}{ap\sqrt{p^2-q^2}}\tan^{-1}\dfrac{\sqrt{p^2-q^2}\tan ax}{p} \\ \dfrac{1}{2ap\sqrt{q^2-p^2}}\ln\left(\dfrac{\sqrt{q^2-p^2}\tan ax+p}{\sqrt{q^2-p^2}\tan ax-p}\right) \end{cases}$

17.17.26. $\int x^m\sin ax\,dx = -\frac{x^m\cos ax}{a}+\frac{mx^{m-1}\sin ax}{a^2}-\frac{m(m-1)}{a^2}\int x^{m-2}\sin ax\,dx$

17.17.27. $\int \frac{\sin ax}{x^n}dx = -\frac{\sin ax}{(n-1)x^{n-1}}+\frac{a}{n-1}\int\frac{\cos ax}{x^{n-1}}dx$ (See 17.18.30.)

17.17.28. $\int \sin^n ax\,dx = -\frac{\sin^{n-1}ax\cos ax}{an}+\frac{n-1}{n}\int\sin^{n-2}ax\,dx$

17.17.29. $\int \frac{dx}{\sin^n ax} = \frac{-\cos ax}{a(n-1)\sin^{n-1}ax}+\frac{n-2}{n-1}\int\frac{dx}{\sin^{n-2}ax}$

17.17.30. $\int \frac{x\,dx}{\sin^n ax} = \frac{-x\cos ax}{a(n-1)\sin^{n-1}ax}-\frac{1}{a^2(n-1)(n-2)\sin^{n-2}ax}+\frac{n-2}{n-1}\int\frac{x\,dx}{\sin^{n-2}ax}$

(18) Integrals Involving $\cos ax$

17.18.1. $\int \cos ax\,dx = \frac{\sin ax}{a}$

17.18.2. $\int x\cos ax\,dx = \frac{\cos ax}{a^2}+\frac{x\sin ax}{a}$

17.18.3. $\int x^2\cos ax\,dx = \frac{2x}{a^2}\cos ax+\left(\frac{x^2}{a}-\frac{2}{a^3}\right)\sin ax$

17.18.4. $\int x^3 \cos ax\, dx = \left(\frac{3x^2}{a^2} - \frac{6}{a^4}\right)\cos ax + \left(\frac{x^3}{a} - \frac{6x}{a^3}\right)\sin ax$

17.18.5. $\int \frac{\cos ax}{x}\, dx = \ln x - \frac{(ax)^2}{2 \cdot 2!} + \frac{(ax)^4}{4 \cdot 4!} - \frac{(ax)^6}{6 \cdot 6!} + \cdots$

17.18.6. $\int \frac{\cos ax}{x^2}\, dx = -\frac{\cos ax}{x} - a \int \frac{\sin ax}{x}\, dx$ (See 17.17.5.)

17.18.7. $\int \frac{dx}{\cos ax} = \frac{1}{a}\ln(\sec ax + \tan ax) = \frac{1}{a}\ln \tan\left(\frac{\pi}{4} + \frac{ax}{2}\right)$

17.18.8. $\int \frac{x\, dx}{\cos ax} = \frac{1}{a^2}\left\{\frac{(ax)^2}{2} + \frac{(ax)^4}{8} + \frac{5(ax)^6}{144} + \cdots + \frac{E_n (ax)^{2n+2}}{(2n+2)(2n)!} + \cdots\right\}$

17.18.9. $\int \cos^2 ax\, dx = \frac{x}{2} + \frac{\sin 2ax}{4a}$

17.18.10. $\int x \cos^2 ax\, dx = \frac{x^2}{4} + \frac{x \sin 2ax}{4a} + \frac{\cos 2ax}{8a^2}$

17.18.11. $\int \cos^3 ax\, dx = \frac{\sin ax}{a} - \frac{\sin^3 ax}{3a}$

17.18.12. $\int \cos^4 ax\, dx = \frac{3x}{8} + \frac{\sin 2ax}{4a} + \frac{\sin 4ax}{32a}$

17.18.13. $\int \frac{dx}{\cos^2 ax} = \frac{\tan ax}{a}$

17.18.14. $\int \frac{dx}{\cos^3 ax} = \frac{\sin ax}{2a \cos^2 ax} + \frac{1}{2a}\ln \tan\left(\frac{\pi}{4} + \frac{ax}{2}\right)$

17.18.15. $\int \cos ax \cos px\, dx = \frac{\sin(a-p)x}{2(a-p)} + \frac{\sin(a+p)x}{2(a+p)}$ (If $a = \pm p$, see 17.18.9.)

17.18.16. $\int \frac{dx}{1 - \cos ax} = -\frac{1}{a}\cot \frac{ax}{2}$

17.18.17. $\int \frac{x\, dx}{1 - \cos ax} = -\frac{x}{a}\cot \frac{ax}{2} + \frac{2}{a^2}\ln \sin \frac{ax}{2}$

17.18.18. $\int \frac{dx}{1 + \cos ax} = \frac{1}{a}\tan \frac{ax}{2}$

17.18.19. $\int \frac{x\, dx}{1 + \cos ax} = \frac{x}{a}\tan \frac{ax}{2} + \frac{2}{a^2}\ln \cos \frac{ax}{2}$

17.18.20. $\int \frac{dx}{(1 - \cos ax)^2} = -\frac{1}{2a}\cot \frac{ax}{2} - \frac{1}{6a}\cot^3 \frac{ax}{2}$

17.18.21. $\int \frac{dx}{(1 + \cos ax)^2} = \frac{1}{2a}\tan \frac{ax}{2} + \frac{1}{6a}\tan^3 \frac{ax}{2}$

17.18.22. $\int \frac{dx}{p + q\cos ax} = \begin{cases} \frac{2}{a\sqrt{p^2 - q^2}} \tan^{-1} \sqrt{(p-q)/(p+q)}\, \tan \frac{1}{2}ax \\ \frac{1}{a\sqrt{q^2 - p^2}} \ln\left(\frac{\tan \frac{1}{2}ax + \sqrt{(q+p)/(q-p)}}{\tan \frac{1}{2}ax - \sqrt{(q+p)/(q-p)}}\right) \end{cases}$ (If $p = \pm q$, see 17.18.16 and 17.18.18.)

17.18.23. $\displaystyle\int \frac{dx}{(p+q\cos ax)^2} = \frac{q\sin ax}{a(q^2-p^2)(p+q\cos ax)} - \frac{p}{q^2-p^2}\int \frac{dx}{p+q\cos ax}$ (If $p = \pm q$ see 17.18.19 and 17.18.20.)

17.18.24. $\displaystyle\int \frac{dx}{p^2+q^2\cos^2 ax} = \frac{1}{ap\sqrt{p^2+q^2}}\tan^{-1}\frac{p\tan ax}{\sqrt{p^2+q^2}}$

17.18.25. $\displaystyle\int \frac{dx}{p^2-q^2\cos^2 ax} = \begin{cases} \dfrac{1}{ap\sqrt{p^2-q^2}}\tan^{-1}\dfrac{p\tan ax}{\sqrt{p^2-q^2}} \\ \dfrac{1}{2ap\sqrt{q^2-p^2}}\ln\left(\dfrac{p\tan ax-\sqrt{q^2-p^2}}{p\tan ax+\sqrt{q^2-p^2}}\right) \end{cases}$

17.18.26. $\displaystyle\int x^m\cos ax\,dx = \frac{x^m\sin ax}{a} + \frac{mx^{m-1}}{a^2}\cos ax - \frac{m(m-1)}{a^2}\int x^{m-2}\cos ax\,dx$

17.18.27. $\displaystyle\int \frac{\cos ax}{x^n}dx = -\frac{\cos ax}{(n-1)x^{n-1}} - \frac{a}{n-1}\int \frac{\sin ax}{x^{n-1}}dx$ (See 17.17.27.)

17.18.28. $\displaystyle\int \cos^n ax\,dx = \frac{\sin ax\cos^{n-1}ax}{an} + \frac{n-1}{n}\int \cos^{n-2}ax\,dx$

17.18.29. $\displaystyle\int \frac{dx}{\cos^n ax} = \frac{\sin ax}{a(n-1)\cos^{n-1}ax} + \frac{n-2}{b-1}\int \frac{dx}{\cos^{n-2}ax}$

17.18.30. $\displaystyle\int \frac{x\,dx}{\cos^n ax} = \frac{x\sin ax}{a(n-1)\cos^{n-1}ax} - \frac{1}{a^2(n-1)(n-2)\cos^{n-2}ax} + \frac{n-2}{n-1}\int \frac{x\,dx}{\cos^{n-2}ax}$

(19) Integrals Involving sin ax and cos ax

17.19.1. $\displaystyle\int \sin ax\cos ax\,dx = \frac{\sin^2 ax}{2a}$

17.19.2. $\displaystyle \sin px\cos qx\,dx = -\frac{\cos(p-q)x}{2(p-q)} - \frac{\cos(p+q)x}{2(p+q)}$

17.19.3. $\displaystyle\int \sin^n ax\cos ax\,dx = \frac{\sin^{n+1}ax}{(n+1)a}$ (If $n=-1$, see 17.21.1.)

17.19.4. $\displaystyle\int \cos^n ax\sin ax\,dx = -\frac{\cos^{n+1}ax}{(n+1)a}$ (If $n=-1$, see 17.20.1.)

17.19.5. $\displaystyle\int \sin^2 ax\cos^2 ax\,dx = \frac{x}{8} - \frac{\sin 4ax}{32a}$

17.19.6. $\displaystyle\int \frac{dx}{\sin ax\cos ax} = \frac{1}{a}\ln\tan ax$

17.19.7. $\displaystyle\int \frac{dx}{\sin^2 ax\cos ax} = \frac{1}{a}\ln\tan\left(\frac{\pi}{4}+\frac{ax}{2}\right) - \frac{1}{a\sin ax}$

17.19.8. $\displaystyle\int \frac{dx}{\sin ax\cos^2 ax} = \frac{1}{a}\ln\tan\frac{ax}{2} + \frac{1}{a\cos ax}$

17.19.9. $\displaystyle\int \frac{dx}{\sin^2 ax\cos^2 ax} = -\frac{2\cot 2ax}{a}$

17.19.10. $\int \frac{\sin^2 ax}{\cos ax}\,dx = -\frac{\sin ax}{a} + \frac{1}{a}\ln\tan\left(\frac{ax}{2} + \frac{\pi}{4}\right)$

17.19.11. $\int \frac{\cos^2 ax}{\sin ax}\,dx = \frac{\cos ax}{a} + \frac{1}{a}\ln\tan\frac{ax}{2}$

17.19.12. $\int \frac{dx}{\cos ax(1 \pm \sin ax)} = \mp\frac{1}{2a(1 \pm \sin ax)} + \frac{1}{2a}\ln\tan\left(\frac{ax}{2} + \frac{\pi}{4}\right)$

17.19.13. $\int \frac{dx}{\sin ax(1 \pm \cos ax)} = \pm\frac{1}{2a(1 \pm \cos ax)} + \frac{1}{2a}\ln\tan\frac{ax}{2}$

17.19.14. $\int \frac{dx}{\sin ax \pm \cos ax} = \frac{1}{a\sqrt{2}}\ln\tan\left(\frac{ax}{2} \pm \frac{\pi}{8}\right)$

17.19.15. $\int \frac{\sin ax\,dx}{\sin ax \pm \cos ax} = \frac{x}{2} \mp \frac{1}{2a}\ln(\sin ax \pm \cos ax)$

17.19.16. $\int \frac{\cos ax\,dx}{\sin ax \pm \cos ax} = \pm\frac{x}{2} + \frac{1}{2a}\ln(\sin ax \pm \cos ax)$

17.19.17. $\int \frac{\sin ax\,dx}{p + q\cos ax} = -\frac{1}{aq}\ln(p + q\cos ax)$

17.19.18. $\int \frac{\cos ax\,dx}{p + q\sin ax} = \frac{1}{aq}\ln(p + q\sin ax)$

17.19.19. $\int \frac{\sin ax\,dx}{(p + q\cos ax)^n} = \frac{1}{aq(n-1)(p + q\cos ax)^{n-1}}$

17.19.20. $\int \frac{\cos ax\,dx}{(p + q\sin ax)^n} = \frac{-1}{aq(n-1)(p + q\sin ax)^{n-1}}$

17.19.21. $\int \frac{dx}{p\sin ax + q\cos ax} = \frac{1}{a\sqrt{p^2 + q^2}}\ln\tan\left(\frac{ax + \tan^{-1}(q/p)}{2}\right)$

17.19.22. $\int \frac{dx}{p\sin ax + q\cos ax + r} = \begin{cases} \dfrac{2}{a\sqrt{r^2 - p^2 - q^2}}\tan^{-1}\left(\dfrac{p + (r-q)\tan(ax/2)}{\sqrt{r^2 - p^2 - q^2}}\right) \\ \dfrac{1}{a\sqrt{p^2 + q^2 - r^2}}\ln\left(\dfrac{p - \sqrt{p^2 + q^2 - r^2} + (r-q)\tan(ax/2)}{p + \sqrt{p^2 + q^2 - r^2} + (r-q)\tan(ax/2)}\right) \end{cases}$

(If $r = q$ see 17.19.23. If $r^2 = p^2 + q^2$ see 17.19.24.)

17.19.23. $\int \frac{dx}{p\sin ax + q(1 + \cos ax)} = \frac{1}{ap}\ln\left(q + p\tan\frac{ax}{2}\right)$

17.19.24. $\int \frac{dx}{p\sin ax + q\cos ax \pm \sqrt{p^2 + q^2}} = \frac{-1}{a\sqrt{p^2 + q^2}}\tan\left(\frac{\pi}{4} \mp \frac{ax + \tan^{-1}(q/p)}{2}\right)$

17.19.25. $\int \frac{dx}{p^2\sin^2 ax + q^2\cos^2 ax} = \frac{1}{apq}\tan^{-1}\left(\frac{p\tan ax}{q}\right)$

17.19.26. $\int \frac{dx}{p^2\sin^2 ax - q^2\cos^2 ax} = \frac{1}{2apq}\ln\left(\frac{p\tan ax - q}{p\tan ax + q}\right)$

17.19.27. $\int \sin^m ax \cos^n ax\, dx = \begin{cases} -\dfrac{\sin^{m-1} ax \cos^{n+1} ax}{a(m+n)} + \dfrac{m-1}{m+n}\displaystyle\int \sin^{m-2} ax \cos^n ax\, dx \\ \dfrac{\sin^{m+1} ax \cos^{n-1} ax}{a(m+n)} + \dfrac{n-1}{m+n}\displaystyle\int \sin^m ax \cos^{n-2} ax\, dx \end{cases}$

17.19.28. $\int \dfrac{\sin^m ax}{\cos^n ax}\, dx = \begin{cases} \dfrac{\sin^{m-1} ax}{a(n-1)\cos^{n-1} ax} - \dfrac{m-1}{n-1}\displaystyle\int \dfrac{\sin^{m-2} ax}{\cos^{n-2} ax}\, dx \\ \dfrac{\sin^{m+1} ax}{a(n-1)\cos^{n-1} ax} - \dfrac{m-n+2}{n-1}\displaystyle\int \dfrac{\sin^m ax}{\cos^{n-2} ax}\, dx \\ \dfrac{-\sin^{m-1} ax}{a(m-n)\cos^{n-1} ax} + \dfrac{m-1}{m-n}\displaystyle\int \dfrac{\sin^{m-2} ax}{\cos^n ax}\, dx \end{cases}$

17.19.29. $\int \dfrac{\cos^m ax}{\sin^n ax}\, dx = \begin{cases} \dfrac{-\cos^{m-1} ax}{a(n-1)\sin^{n-1} ax} - \dfrac{m-1}{n-1}\displaystyle\int \dfrac{\cos^{m-2} ax}{\sin^{n-2} ax}\, dx \\ \dfrac{-\cos^{m+1} ax}{a(n-1)\sin^{n-1} ax} - \dfrac{m-n+2}{n-1}\displaystyle\int \dfrac{\cos^m ax}{\sin^{n-2} ax}\, dx \\ \dfrac{\cos^{m-1} ax}{a(m-n)\sin^{n-1} ax} + \dfrac{m-1}{m-n}\displaystyle\int \dfrac{\cos^{m-2} ax}{\sin^n ax}\, dx \end{cases}$

17.19.30. $\int \dfrac{dx}{\sin^m ax \cos^n ax} = \begin{cases} \dfrac{1}{a(n-1)\sin^{m-1} ax \cos^{n-1} ax} + \dfrac{m+n-2}{n-1}\displaystyle\int \dfrac{dx}{\sin^m ax \cos^{n-2} ax} \\ \dfrac{-1}{a(m-1)\sin^{m-1} ax \cos^{n-1} ax} + \dfrac{m+n-2}{m-1}\displaystyle\int \dfrac{dx}{\sin^{m-2} ax \cos^n ax} \end{cases}$

(20) Integrals Involving tan *ax*

17.20.1. $\int \tan ax\, dx = -\dfrac{1}{a}\ln\cos ax = \dfrac{1}{a}\ln\sec ax$

17.20.2. $\int \tan^2 ax\, dx = \dfrac{\tan ax}{a} - x$

17.20.3. $\int \tan^3 ax\, dx = \dfrac{\tan^2 ax}{2a} + \dfrac{1}{a}\ln\cos ax$

17.20.4. $\int \tan^n ax \sec^2 ax\, dx = \dfrac{\tan^{n+1} ax}{(n+1)a}$

17.20.5. $\int \dfrac{\sec^2 ax}{\tan ax}\, dx = \dfrac{1}{a}\ln\tan ax$

17.20.6. $\int \dfrac{dx}{\tan ax} = \dfrac{1}{a}\ln\sin ax$

17.20.7. $\int x\tan ax\, dx = \dfrac{1}{a^2}\left\{\dfrac{(ax)^3}{3} + \dfrac{(ax)^5}{15} + \dfrac{2(ax)^7}{105} + \cdots + \dfrac{2^{2n}(2^{2n}-1)B_n(ax)^{2n+1}}{(2n+1)!} + \cdots\right\}$

17.20.8. $\int \dfrac{\tan ax}{x}\, dx = ax + \dfrac{(ax)^3}{9} + \dfrac{2(ax)^5}{75} + \cdots + \dfrac{2^{2n}(2^{2n}-1)B_n(ax)^{2n-1}}{(2n-1)(2n)!} + \cdots$

17.20.9. $\int x\tan^2 ax\, dx = \dfrac{x\tan ax}{a} + \dfrac{1}{a^2}\ln\cos ax - \dfrac{x^2}{2}$

17.20.10. $\int \frac{dx}{p+q\tan ax} = \frac{px}{p^2+q^2} + \frac{q}{a(p^2+q^2)}\ln(q\sin ax + p\cos ax)$

17.20.11. $\int \tan^n ax\,dx = \frac{\tan^{n-1} ax}{(n-1)a} - \int \tan^{n-2} ax\,dx$

(21) Integrals Involving cot ax

17.21.1. $\int \cot ax\,dx = \frac{1}{a}\ln \sin ax$

17.21.2. $\int \cot^2 ax\,dx = -\frac{\cot ax}{a} - x$

17.21.3. $\int \cot^3 ax\,dx = -\frac{\cot^2 ax}{2a} - \frac{1}{a}\ln \sin ax$

17.21.4. $\int \cot^n ax \csc^2 ax\,dx = -\frac{\cot^{n+1} ax}{(n+1)a}$

17.21.5. $\int \frac{\csc^2 ax}{\cot ax}dx = -\frac{1}{a}\ln \cot ax$

17.21.6. $\int \frac{dx}{\cot ax} = -\frac{1}{a}\ln \cos ax$

17.21.7. $\int x\cot ax\,dx = \frac{1}{a^2}\left\{ax - \frac{(ax)^3}{9} - \frac{(ax)^5}{225} - \cdots - \frac{2^{2n}B_n(ax)^{2n+1}}{(2n+1)!} - \cdots\right\}$

17.21.8. $\int \frac{\cot ax}{x}dx = -\frac{1}{ax} - \frac{ax}{3} - \frac{(ax)^3}{135} - \cdots - \frac{2^{2n}B_n(ax)^{2n-1}}{(2n-1)(2n)!} - \cdots$

17.21.9. $\int x\cot^2 ax\,dx = -\frac{x\cot ax}{a} + \frac{1}{a^2}\ln \sin ax - \frac{x^2}{2}$

17.21.10. $\int \frac{dx}{p+q\cot ax} = \frac{px}{p^2+q^2} - \frac{q}{a(p^2+q^2)}\ln(q\sin ax + q\cos ax)$

17.21.11. $\int \cot^n ax\,dx = -\frac{\cot^{n-1} ax}{(n-1)a} - \int \cot^{n-2} ax\,dx$

(22) Integrals Involving sec ax

17.22.1. $\int \sec ax\,dx = \frac{1}{a}\ln(\sec ax + \tan ax) = \frac{1}{a}\ln \tan\left(\frac{ax}{2} + \frac{\pi}{4}\right)$

17.22.2. $\int \sec^2 ax\,dx = \frac{\tan ax}{a}$

17.22.3. $\int \sec^3 ax\,dx = \frac{\sec ax \tan ax}{2a} + \frac{1}{2a}\ln(\sec ax + \tan ax)$

17.22.4. $\int \sec^n ax \tan ax\, dx = \dfrac{\sec^n ax}{na}$

17.22.5. $\int \dfrac{dx}{\sec ax} = \dfrac{\sin ax}{a}$

17.22.6. $\int x \sec ax\, dx = \dfrac{1}{a^2}\left\{\dfrac{(ax)^2}{2} + \dfrac{(ax)^4}{8} + \dfrac{5(ax)^6}{144} + \cdots + \dfrac{E_n(ax)^{2n+2}}{(2n+2)(2n)!} + \cdots\right\}$

17.22.7. $\int \dfrac{\sec ax}{x}\, dx = \ln x + \dfrac{(ax)^2}{4} + \dfrac{5(ax)^4}{96} + \dfrac{61(ax)^6}{4320} + \cdots + \dfrac{E_n(ax)^{2n}}{2n(2n)!} + \cdots$

17.22.8. $\int x \sec^2 ax\, dx = \dfrac{x}{a}\tan ax + \dfrac{1}{a^2}\ln \cos ax$

17.22.9. $\int \dfrac{dx}{q + p\sec ax} = \dfrac{x}{q} - \dfrac{p}{q}\int \dfrac{dx}{p + q\cos ax}$

17.22.10. $\int \sec^n ax\, dx = \dfrac{\sec^{n-2} ax \tan ax}{a(n-1)} + \dfrac{n-2}{n-1}\int \sec^{n-2} ax\, dx$

(23) Integrals Involving csc ax

17.23.1. $\int \csc ax\, dx = \dfrac{1}{a}\ln(\csc ax - \cot ax) = \dfrac{1}{a}\ln \tan \dfrac{ax}{2}$

17.23.2. $\int \csc^2 ax\, dx = -\dfrac{\cot ax}{a}$

17.23.3. $\int \csc^3 ax\, dx = -\dfrac{\csc ax \cot ax}{2a} + \dfrac{1}{2a}\ln \tan \dfrac{ax}{2}$

17.23.4. $\int \csc^n ax \cot ax\, dx = -\dfrac{\csc^n ax}{na}$

17.23.5. $\int \dfrac{dx}{\csc ax} = -\dfrac{\cos ax}{a}$

17.23.6. $\int x \csc ax\, dx = \dfrac{1}{a^2}\left\{ax + \dfrac{(ax)^3}{18} + \dfrac{7(ax)^5}{1800} + \cdots + \dfrac{2(2^{2n-1}-1)B_n(ax)^{2n+1}}{(2n+1)!} + \cdots\right\}$

17.23.7. $\int \dfrac{\csc ax}{x}\, dx = -\dfrac{1}{ax} + \dfrac{ax}{6} + \dfrac{7(ax)^3}{1080} + \cdots + \dfrac{2(2^{2n-1}-1)B_n(ax)^{2n-1}}{(2n-1)(2n)!} + \cdots$

17.23.8. $\int x \csc^2 ax\, dx = -\dfrac{x \cot ax}{a} + \dfrac{1}{a^2}\ln \sin ax$

17.23.9. $\int \dfrac{dx}{q + p\csc ax} = \dfrac{x}{q} - \dfrac{p}{q}\int \dfrac{dx}{p + q\sin ax}$ (See 17.17.22.)

17.23.10. $\int \csc^n ax\, dx = -\dfrac{\csc^{n-2} ax \cot ax}{a(n-1)} + \dfrac{n-2}{n-1}\int \csc^{n-2} ax\, dx$

(24) Integrals Involving Inverse Trigonometric Functions

17.24.1. $\int \sin^{-1}\frac{x}{a}\,dx = x\sin^{-1}\frac{x}{a} + \sqrt{a^2 - x^2}$

17.24.2. $\int x\sin^{-1}\frac{x}{a}\,dx = \left(\frac{x^2}{2} - \frac{a^2}{4}\right)\sin^{-1}\frac{x}{a} + \frac{x\sqrt{a^2 - x^2}}{4}$

17.24.3. $\int x^2\sin^{-1}\frac{x}{a}\,dx = \frac{x^3}{3}\sin^{-1}\frac{x}{a} + \frac{(x^2 + 2a^2)\sqrt{a^2 - x^2}}{9}$

17.24.4. $\int \frac{\sin^{-1}(x/a)}{x}\,dx = \frac{x}{a} + \frac{(x/a)^3}{2\cdot 3\cdot 3} + \frac{1\cdot 3(x/a)^5}{2\cdot 4\cdot 5\cdot 5} + \frac{1\cdot 3\cdot 5(x/a)^7}{2\cdot 4\cdot 6\cdot 7\cdot 7} + \cdots$

17.24.5. $\int \frac{\sin^{-1}(x/a)}{x^2}\,dx = -\frac{\sin^{-1}(x/a)}{x} - \frac{1}{a}\ln\left(\frac{a + \sqrt{a^2 - x^2}}{x}\right)$

17.24.6. $\int \left(\sin^{-1}\frac{x}{a}\right)^2 dx = x\left(\sin^{-1}\frac{x}{a}\right)^2 - 2x + 2\sqrt{a^2 - x^2}\,\sin^{-1}\frac{x}{a}$

17.24.7. $\int \cos^{-1}\frac{x}{a}\,dx = x\cos^{-1}\frac{x}{a} - \sqrt{a^2 - x^2}$

17.24.8. $\int x\cos^{-1}\frac{x}{a}\,dx = \left(\frac{x^2}{2} - \frac{a^2}{4}\right)\cos^{-1}\frac{x}{a} - \frac{x\sqrt{a^2 - x^2}}{4}$

17.24.9. $\int x^2\cos^{-1}\frac{x}{a}\,dx = \frac{x^3}{3}\cos^{-1}\frac{x}{a} - \frac{(x^2 + 2a^2)\sqrt{a^2 - x^2}}{9}$

17.24.10. $\int \frac{\cos^{-1}(x/a)}{x}\,dx = \frac{\pi}{2}\ln x - \int \frac{\sin^{-1}(x/a)}{x}\,dx$ (See 17.24.4.)

17.24.11. $\int \frac{\cos^{-1}(x/a)}{x^2}\,dx = -\frac{\cos^{-1}(x/a)}{x} + \frac{1}{a}\ln\left(\frac{a + \sqrt{a^2 - x^2}}{x}\right)$

17.24.12. $\int \left(\cos^{-1}\frac{x}{a}\right)^2 dx = x\left(\cos^{-1}\frac{x}{a}\right)^2 - 2x - 2\sqrt{a^2 - x^2}\,\cos^{-1}\frac{x}{a}$

17.24.13. $\int \tan^{-1}\frac{x}{a}\,dx = x\tan^{-1}\frac{x}{a} - \frac{a}{2}\ln(x^2 + a^2)$

17.24.14. $\int x\tan^{-1}\frac{x}{a}\,dx = \frac{1}{2}(x^2 + a^2)\tan^{-1}\frac{x}{a} - \frac{ax}{2}$

17.24.15. $\int x^2\tan^{-1}\frac{x}{a}\,dx = \frac{x^3}{3}\tan^{-1}\frac{x}{a} - \frac{ax^2}{6} + \frac{a^3}{6}\ln(x^2 + a^2)$

17.24.16. $\int \frac{\tan^{-1}(x/a)}{x}\,dx = \frac{x}{a} - \frac{(x/a)^3}{3^2} + \frac{(x/a)^5}{5^2} - \frac{(x/a)^7}{7^2} + \cdots$

17.24.17. $\int \frac{\tan^{-1}(x/a)}{x^2}\,dx = -\frac{1}{x}\tan^{-1}\frac{x}{a} - \frac{1}{2a}\ln\left(\frac{x^2 + a^2}{x^2}\right)$

17.24.18. $\int \cot^{-1}\frac{x}{a}\,dx = x\cot^{-1}\frac{x}{a} + \frac{a}{2}\ln(x^2+a^2)$

17.24.19. $\int x\cot^{-1}\frac{x}{a}\,dx = \frac{1}{2}(x^2+a^2)\cot^{-1}\frac{x}{a} + \frac{ax}{2}$

17.24.20. $\int x^2\cot^{-1}\frac{x}{a}\,dx = \frac{x^3}{3}\cot^{-1}\frac{x}{a} + \frac{ax^2}{6} - \frac{a^3}{6}\ln(x^2+a^2)$

17.24.21. $\int \frac{\cot^{-1}(x/a)}{x}\,dx = \frac{\pi}{2}\ln x - \int \frac{\tan^{-1}(x/a)}{x}\,dx$ (See 17.24.16.)

17.24.22. $\int \frac{\cot^{-1}(x/a)}{x^2}\,dx = \frac{\cot^{-1}(x/a)}{x} + \frac{1}{2a}\ln\left(\frac{x^2+a^2}{x^2}\right)$

17.24.23. $\int \sec^{-1}\frac{x}{a}\,dx = \begin{cases} x\sec^{-1}\frac{x}{a} - a\ln(x+\sqrt{x^2-a^2}) & 0<\sec^{-1}\frac{x}{a}<\frac{\pi}{2} \\ x\sec^{-1}\frac{x}{a} + a\ln(x+\sqrt{x^2-a^2}) & \frac{\pi}{2}<\sec^{-1}\frac{x}{a}<\pi \end{cases}$

17.24.24. $\int x\sec^{-1}\frac{x}{a}\,dx = \begin{cases} \frac{x^2}{2}\sec^{-1}\frac{x}{a} - \frac{a\sqrt{x^2-a^2}}{2} & 0<\sec^{-1}\frac{x}{a}<\frac{\pi}{2} \\ \frac{x^2}{2}\sec^{-1}\frac{x}{a} + \frac{a\sqrt{x^2-a^2}}{2} & \frac{\pi}{2}<\sec^{-1}\frac{x}{a}<\pi \end{cases}$

17.24.25. $\int x^2\sec^{-1}\frac{x}{a}\,dx = \begin{cases} \frac{x^3}{3}\sec^{-1}\frac{x}{a} - \frac{ax\sqrt{x^2-a^2}}{6} - \frac{a^3}{6}\ln(x+\sqrt{x^2-a^2}) & 0<\sec^{-1}\frac{x}{a}<\frac{\pi}{2} \\ \frac{x^3}{3}\sec^{-1}\frac{x}{a} + \frac{ax\sqrt{x^2-a^2}}{6} + \frac{a^3}{6}\ln(x+\sqrt{x^2-a^2}) & \frac{\pi}{2}<\sec^{-1}\frac{x}{a}<\pi \end{cases}$

17.24.26. $\int \frac{\sec^{-1}(x/a)}{x}\,dx = \frac{\pi}{2}\ln x + \frac{a}{x} + \frac{(a/x)^3}{2\cdot 3\cdot 3} + \frac{1\cdot 3(a/x)^5}{2\cdot 4\cdot 5\cdot 5} + \frac{1\cdot 3\cdot 5(a/x)^7}{2\cdot 4\cdot 6\cdot 7\cdot 7} + \cdots$

17.24.27. $\int \frac{\sec^{-1}(x/a)}{x^2}\,dx = \begin{cases} -\frac{\sec^{-1}(x/a)}{x} + \frac{\sqrt{x^2-a^2}}{ax} & 0<\sec^{-1}\frac{x}{a}<\frac{\pi}{2} \\ -\frac{\sec^{-1}(x/a)}{x} - \frac{\sqrt{x^2-a^2}}{ax} & \frac{\pi}{2}<\sec^{-1}\frac{x}{a}<\pi \end{cases}$

17.24.28. $\int \csc^{-1}\frac{x}{a}\,dx = \begin{cases} x\csc^{-1}\frac{x}{a} + a\ln(x+\sqrt{x^2-a^2}) & 0<\csc^{-1}\frac{x}{a}<\frac{\pi}{2} \\ x\csc^{-1}\frac{x}{a} - a\ln(x+\sqrt{x^2-a^2}) & -\frac{\pi}{2}<\csc^{-1}\frac{x}{a}<0 \end{cases}$

17.24.29. $\int x\csc^{-1}\frac{x}{a}\,dx = \begin{cases} \frac{x^2}{2}\csc^{-1}\frac{x}{a} + \frac{a\sqrt{x^2-a^2}}{2} & 0<\csc^{-1}\frac{x}{a}<\frac{\pi}{2} \\ \frac{x^2}{2}\csc^{-1}\frac{x}{a} - \frac{a\sqrt{x^2-a^2}}{2} & -\frac{\pi}{2}<\csc^{-1}\frac{x}{a}<0 \end{cases}$

17.24.30. $\int x^2\csc^{-1}\frac{x}{a}\,dx = \begin{cases} \frac{x^3}{3}\csc^{-1}\frac{x}{a} + \frac{ax\sqrt{x^2-a^2}}{6} + \frac{a^3}{6}\ln(x+\sqrt{x^2-a^2}) & 0<\csc^{-1}\frac{x}{a}<\frac{\pi}{2} \\ \frac{x^3}{3}\csc^{-1}\frac{x}{a} - \frac{ax\sqrt{x^2-a^2}}{6} - \frac{a^3}{6}\ln(x+\sqrt{x^2-a^2}) & -\frac{\pi}{2}<\csc^{-1}\frac{x}{a}<0 \end{cases}$

17.24.31. $\displaystyle\int \frac{\csc^{-1}(x/a)}{x}\,dx = -\left(\frac{a}{x} + \frac{(a/x)^3}{2\cdot3\cdot3} + \frac{1\cdot3(a/x)^5}{2\cdot4\cdot5\cdot5} + \frac{1\cdot3\cdot5(a/x)^7}{2\cdot4\cdot6\cdot7\cdot7} + \cdots\right)$

17.24.32. $\displaystyle\int \frac{\csc^{-1}(x/a)}{x^2}\,dx = \begin{cases} -\dfrac{\csc^{-1}(x/a)}{x} - \dfrac{\sqrt{x^2-a^2}}{ax} & 0 < \csc^{-1}\dfrac{x}{a} < \dfrac{\pi}{2} \\ -\dfrac{\csc^{-1}(x/a)}{x} + \dfrac{\sqrt{x^2-a^2}}{ax} & -\dfrac{\pi}{2} < \csc^{-1}\dfrac{x}{a} < 0 \end{cases}$

17.24.33. $\displaystyle\int x^m \sin^{-1}\frac{x}{a}\,dx = \frac{x^{m+1}}{m+1}\sin^{-1}\frac{x}{a} - \frac{1}{m+1}\int \frac{x^{m+1}}{\sqrt{a^2-x^2}}\,dx$

17.24.34. $\displaystyle\int x^m \cos^{-1}\frac{x}{a}\,dx = \frac{x^{m+1}}{m+1}\cos^{-1}\frac{x}{a} + \frac{1}{m+1}\int \frac{x^{m+1}}{\sqrt{a^2-x^2}}\,dx$

17.24.35. $\displaystyle\int x^m \tan^{-1}\frac{x}{a}\,dx = \frac{x^{m+1}}{m+1}\tan^{-1}\frac{x}{a} - \frac{a}{m+1}\int \frac{x^{m+1}}{x^2+a^2}\,dx$

17.24.36. $\displaystyle\int x^m \cot^{-1}\frac{x}{a}\,dx = \frac{x^{m+1}}{m+1}\cot^{-1}\frac{x}{a} + \frac{a}{m+1}\int \frac{x^{m+1}}{x^2+a^2}\,dx$

17.24.37. $\displaystyle\int x^m \sec^{-1}\frac{x}{a}\,dx = \begin{cases} \dfrac{x^{m+1}\sec^{-1}(x/a)}{m+1} - \dfrac{a}{m+1}\displaystyle\int \frac{x^m\,dx}{\sqrt{x^2-a^2}} & 0 < \sec^{-1}\dfrac{x}{a} < \dfrac{\pi}{2} \\ \dfrac{x^{m+1}\sec^{-1}(x/a)}{m+1} + \dfrac{a}{m+1}\displaystyle\int \frac{x^m\,dx}{\sqrt{x^2-a^2}} & \dfrac{\pi}{2} < \sec^{-1}\dfrac{x}{a} < \pi \end{cases}$

17.24.38. $\displaystyle\int x^m \csc^{-1}\frac{x}{a}\,dx = \begin{cases} \dfrac{x^{m+1}\sec^{-1}(x/a)}{m+1} + \dfrac{a}{m+1}\displaystyle\int \frac{x^m\,dx}{\sqrt{x^2-a^2}} & 0 < \csc^{-1}\dfrac{x}{a} < \dfrac{\pi}{2} \\ \dfrac{x^{m+1}\csc^{-1}(x/a)}{m+1} - \dfrac{a}{m+1}\displaystyle\int \frac{x^m\,dx}{\sqrt{x^2-a^2}} & -\dfrac{\pi}{2} < \csc^{-1}\dfrac{x}{a} < 0 \end{cases}$

(25) Integrals Involving e^{ax}

17.25.1. $\displaystyle\int e^{ax}\,dx = \frac{e^{ax}}{a}$

17.25.2. $\displaystyle\int xe^{ax}\,dx = \frac{e^{ax}}{a}\left(x - \frac{1}{a}\right)$

17.25.3. $\displaystyle\int x^2e^{ax}\,dx = \frac{e^{ax}}{a}\left(x^2 - \frac{2x}{a} + \frac{2}{a^2}\right)$

17.25.4. $\displaystyle\int x^ne^{ax}\,dx = \frac{x^ne^{ax}}{a} - \frac{n}{a}\int x^{n-1}e^{ax}\,dx$

$\displaystyle = \frac{e^{ax}}{a}\left(x^n - \frac{nx^{n-1}}{a} + \frac{n(n-1)x^{n-2}}{a^2} - \cdots \frac{(-1)^n n!}{a^n}\right)$ if n = positive integer

17.25.5. $\displaystyle\int \frac{e^{ax}}{x}\,dx = \ln x + \frac{ax}{1\cdot1!} + \frac{(ax)^2}{2\cdot2!} + \frac{(ax)^3}{3\cdot3!} + \cdots$

17.25.6. $\displaystyle\int \frac{e^{ax}}{x^n}\,dx = \frac{-e^{ax}}{(n-1)x^{n-1}} + \frac{a}{n-1}\int \frac{e^{ax}}{x^{n-1}}\,dx$

17.25.7. $\displaystyle\int \frac{dx}{p+qe^{ax}} = \frac{x}{p} - \frac{1}{ap}\ln(p+qe^{ax})$

17.25.8. $\displaystyle\int \frac{dx}{(p+qe^{ax})^2} = \frac{x}{p^2} + \frac{1}{ap(p+qe^{ax})} - \frac{1}{ap^2}\ln(p+qe^{ax})$

17.25.9. $$\int \frac{dx}{pe^{ax}+qe^{-ax}} = \begin{cases} \dfrac{1}{a\sqrt{pq}}\tan^{-1}\left(\sqrt{\dfrac{p}{q}}\,e^{ax}\right) \\ \dfrac{1}{2a\sqrt{-pq}}\ln\left(\dfrac{e^{ax}-\sqrt{-q/p}}{e^{ax}+\sqrt{-q/p}}\right) \end{cases}$$

17.25.10. $\displaystyle\int e^{ax}\sin bx\,dx = \frac{e^{ax}(a\sin bx - b\cos bx)}{a^2+b^2}$

17.25.11. $\displaystyle\int e^{ax}\cos bx\,dx = \frac{e^{ax}(a\cos bx + b\sin bx)}{a^2+b^2}$

17.25.12. $\displaystyle\int xe^{ax}\sin bx\,dx = \frac{xe^{ax}(a\sin bx - b\cos bx)}{a^2+b^2} - \frac{e^{ax}\{(a^2-b^2)\sin bx - 2ab\cos bx\}}{(a^2+b^2)^2}$

17.25.13. $\displaystyle\int xe^{ax}\cos bx\,dx = \frac{xe^{ax}(a\cos bx + b\sin bx)}{a^2+b^2} - \frac{e^{ax}\{(a^2-b^2)\cos bx + 2ab\sin bx\}}{(a^2+b^2)^2}$

17.25.14. $\displaystyle\int e^{ax}\ln x\,dx = \frac{e^{ax}\ln x}{a} - \frac{1}{a}\int \frac{e^{ax}}{x}dx$

17.25.15. $\displaystyle\int e^{ax}\sin^n bx\,dx = \frac{e^{ax}\sin^{n-1}bx}{a^2+n^2b^2}(a\sin bx - nb\cos bx) + \frac{n(n-1)b^2}{a^2+n^2b^2}\int e^{ax}\sin^{n-2}bx\,dx$

17.25.16. $\displaystyle\int e^{ax}\cos^n bx\,dx = \frac{e^{ax}\cos^{n-1}bx}{a^2+n^2b^2}(a\cos bx + nb\sin bx) + \frac{n(n-1)b^2}{a^2+n^2b^2}\int e^{ax}\cos^{n-2}bx\,dx$

(26) Integrals Involving $\ln x$

17.26.1. $\displaystyle\int \ln x\,dx = x\ln x - x$

17.26.2. $\displaystyle\int x\ln x\,dx = \frac{x^2}{2}\left(\ln x - \frac{1}{2}\right)$

17.26.3. $\displaystyle\int x^m\ln x\,dx = \frac{x^{m+1}}{m+1}\left(\ln x - \frac{1}{m+1}\right)$ (If $m=-1$, see 17.26.4.)

17.26.4. $\displaystyle\int \frac{\ln x}{x}dx = \frac{1}{2}\ln^2 x$

17.26.5. $\displaystyle\int \frac{\ln x}{x^2}dx = -\frac{\ln x}{x} - \frac{1}{x}$

17.26.6. $\displaystyle\int \ln^2 x\,dx = x\ln^2 x - 2x\ln x + 2x$

17.26.7. $\displaystyle\int \frac{\ln^n x\,dx}{x} = \frac{\ln^{n+1}x}{n+1}$ (If $n=-1$, see 17.26.8.)

17.26.8. $\displaystyle\int \frac{dx}{x\ln x} = \ln(\ln x)$

17.26.9. $\int \frac{dx}{\ln x} = \ln(\ln x) + \ln x + \frac{\ln^2 x}{2 \cdot 2!} + \frac{\ln^3 x}{3 \cdot 3!} + \cdots$

17.26.10. $\int \frac{x^m dx}{\ln x} = \ln(\ln x) + (m+1)\ln x + \frac{(m+1)^2 \ln^2 x}{2 \cdot 2!} + \frac{(m+1)^3 \ln^3 x}{3 \cdot 3!} + \cdots$

17.26.11. $\int \ln^n x\, dx = x \ln^n x - n \int \ln^{n-1} x\, dx$

17.26.12. $\int x^m \ln^n x\, dx = \frac{x^{m+1} \ln^n x}{m+1} - \frac{n}{m+1} \int x^m \ln^{n-1} x dx$

If $m = -1$, see 17.26.7.

17.26.13. $\int \ln(x^2 + a^2)\, dx = x \ln(x^2 + a^2) - 2x + 2a \tan^{-1} \frac{x}{a}$

17.26.14. $\int \ln(x^2 - a^2)\, dx = x \ln(x^2 - a^2) - 2x + a \ln\left(\frac{x+a}{x-a}\right)$

17.26.15. $\int x^m \ln(x^2 \pm a^2)\, dx = \frac{x^{m+1} \ln(x^2 \pm a^2)}{m+1} - \frac{2}{m+1} \int \frac{x^{m+2}}{x^2 \pm a^2} dx$

(27) Integrals Involving sinh *ax*

17.27.1. $\int \sinh ax\, dx = \frac{\cosh ax}{a}$

17.27.2. $\int x \sinh ax\, dx = \frac{x \cosh ax}{a} - \frac{\sinh ax}{a^2}$

17.27.3. $\int x^2 \sinh ax\, dx = \left(\frac{x^2}{a} + \frac{2}{a^3}\right) \cosh ax - \frac{2x}{a^2} \sinh ax$

17.27.4. $\int \frac{\sinh ax}{x} dx = ax + \frac{(ax)^3}{3 \cdot 3!} + \frac{(ax)^5}{5 \cdot 5!} + \cdots$

17.27.5. $\int \frac{\sinh ax}{x^2} dx = -\frac{\sinh ax}{x} + a \int \frac{\cosh ax}{x} dx$ (See 17.28.4.)

17.27.6. $\int \frac{dx}{\sinh ax} = \frac{1}{a} \ln \tanh \frac{ax}{2}$

17.27.7. $\int \frac{x\, dx}{\sinh ax} = \frac{1}{a^2} \left\{ ax - \frac{(ax)^3}{18} + \frac{7(ax)^5}{1800} - \cdots + \frac{2(-1)^n (2^{2n} - 1) B_n (ax)^{2n+1}}{(2n+1)!} + \cdots \right\}$

17.27.8. $\int \sinh^2 ax\, dx = \frac{\sinh ax \cosh ax}{2a} - \frac{x}{2}$

17.27.9. $\int x \sinh^2 ax\, dx = \frac{x \sinh 2ax}{4a} - \frac{\cosh 2ax}{8a^2} - \frac{x^2}{4}$

17.27.10. $\int \frac{dx}{\sinh^2 ax} = -\frac{\coth ax}{a}$

17.27.11. $\int \sinh ax \sinh px\, dx = \frac{\sinh (a+p)x}{2(a+p)} - \frac{\sinh (a-p)x}{2(a-p)}$

For $a = \pm p$ see 17.27.8.

17.27.12. $\int x^m \sinh ax\, dx = \frac{x^m \cosh ax}{a} - \frac{m}{a}\int x^{m-1} \cosh ax\, dx$ (See 17.28.12.)

17.27.13. $\int \sinh^n ax\, dx = \frac{\sinh^{n-1} ax \cosh ax}{an} - \frac{n-1}{n}\int \sinh^{n-2} ax\, dx$

17.27.14. $\int \frac{\sinh ax}{x^n} dx = \frac{-\sinh ax}{(n-1)x^{n-1}} + \frac{a}{n-1}\int \frac{\cosh ax}{x^{n-1}} dx$ (See 17.28.14.)

17.27.15. $\int \frac{dx}{\sinh^n ax} = \frac{-\cosh ax}{a(n-1)\sinh^{n-1} ax} - \frac{n-2}{n-1}\int \frac{dx}{\sinh^{n-2} ax}$

17.27.16. $\int \frac{x\,dx}{\sinh^n ax} = \frac{-x\cosh ax}{a(n-1)\sinh^{n-1} ax} - \frac{1}{a^2(n-1)(n-2)\sinh^{n-2} ax} - \frac{n-2}{n-1}\int \frac{x\,dx}{\sinh^{n-2} ax}$

(28) Integrals Involving cosh *ax*

17.28.1. $\int \cosh ax\, dx = \frac{\sinh ax}{a}$

17.28.2. $\int x \cosh ax\, dx = \frac{x \sinh ax}{a} - \frac{\cosh ax}{a^2}$

17.28.3. $\int x^2 \cosh ax\, dx = -\frac{2x\cosh ax}{a^2} + \left(\frac{x^2}{a} + \frac{2}{a^3}\right)\sinh ax$

17.28.4. $\int \frac{\cosh ax}{x} dx = \ln x + \frac{(ax)^2}{2 \cdot 2!} + \frac{(ax)^4}{4 \cdot 4!} + \frac{(ax)^6}{6 \cdot 6!} + \cdots$

17.28.5. $\int \frac{\cosh ax}{x^2} dx = -\frac{\cosh ax}{x} + a\int \frac{\sinh ax}{x} dx$ (See 17.27.4.)

17.28.6. $\int \frac{dx}{\cosh ax} = \frac{2}{a}\tan^{-1} e^{ax}$

17.28.7. $\int \frac{x\,dx}{\cosh ax} = \frac{1}{a^2}\left\{\frac{(ax)^2}{2} - \frac{(ax)^4}{8} + \frac{5(ax)^6}{144} + \cdots + \frac{(-1)^n E_n (ax)^{2n+2}}{(2n+2)(2n)!} + \cdots\right\}$

17.28.8. $\int \cosh^2 ax\, dx = \frac{x}{2} + \frac{\sinh ax \cosh ax}{2a}$

17.28.9. $\int x\cosh^2 ax\, dx = \frac{x^2}{4} + \frac{x \sinh 2ax}{4a} - \frac{\cosh 2ax}{8a^2}$

17.28.10. $\displaystyle\int \frac{dx}{\cosh^2 ax} = \frac{\tanh ax}{a}$

17.28.11. $\displaystyle\int \cosh ax \cosh px\, dx = \frac{\sinh(a-p)x}{2(a-p)} + \frac{\sinh(a+p)x}{2(a+p)}$

17.28.12. $\displaystyle\int x^m \cosh ax\, dx = \frac{x^m \sinh ax}{a} - \frac{m}{a}\int x^{m-1} \sinh ax\, dx$ (See 17.27.12.)

17.28.13. $\displaystyle\int \cosh^n ax\, dx = \frac{\cosh^{n-1} ax \sinh ax}{an} + \frac{n-1}{n}\int \cosh^{n-2} ax\, dx$

17.28.14. $\displaystyle\int \frac{\cosh ax}{x^n} dx = \frac{-\cosh ax}{(n-1)x^{n-1}} + \frac{a}{n-1}\int \frac{\sinh ax}{x^{n-1}} dx$ (See 17.27.14.)

17.28.15. $\displaystyle\int \frac{dx}{\cosh^n ax} = \frac{\sinh ax}{a(n-1)\cosh^{n-1} ax} + \frac{n-2}{n-1}\int \frac{dx}{\cosh^{n-2} ax}$

17.28.16. $\displaystyle\int \frac{x\, dx}{\cosh^n ax} = \frac{x \sinh ax}{a(n-1)\cosh^{n-1} ax} + \frac{1}{(n-1)(n-2)a^2 \cosh^{n-2} ax} + \frac{n-2}{n-1}\int \frac{x\, dx}{\cosh^{n-2} ax}$

(29) Integrals Involving sinh ax and cosh ax

17.29.1. $\displaystyle\int \sinh ax \cosh ax\, dx = \frac{\sinh^2 ax}{2a}$

17.29.2. $\displaystyle\int \sinh px \cosh qx\, dx = \frac{\cosh(p+q)x}{2(p+q)} + \frac{\cosh(p-q)x}{2(p-q)}$

17.29.3. $\displaystyle\int \sinh^2 ax \cosh^2 ax\, dx = \frac{\sinh 4ax}{32a} - \frac{x}{8}$

17.29.4. $\displaystyle\int \frac{dx}{\sinh ax \cosh ax} = \frac{1}{a} \ln \tanh ax$

17.29.5. $\displaystyle\int \frac{dx}{\sinh^2 ax \cosh^2 ax} = -\frac{2\coth 2ax}{a}$

17.29.6. $\displaystyle\int \frac{\sinh^2 ax}{\cosh ax} dx = \frac{\sinh ax}{a} \frac{1}{a} \tan^{-1} \sinh ax$

17.29.7. $\displaystyle\int \frac{\cosh^2 ax}{\sinh ax} dx = \frac{\cosh ax}{a} + \frac{1}{a} \ln \tanh \frac{ax}{2}$

(30) Integrals Involving tanh ax

17.30.1. $\int \tanh ax\,dx = \frac{1}{a}\ln\cosh ax$

17.30.2. $\int \tanh^2 ax\,dx = x - \frac{\tanh ax}{a}$

17.30.3. $\int \tanh^3 ax\,dx = \frac{1}{a}\ln\cosh ax - \frac{\tanh^2 ax}{2a}$

17.30.4. $\int x\tanh ax\,dx = \frac{1}{a^2}\left\{\frac{(ax)^3}{3} - \frac{(ax)^5}{15} + \frac{2(ax)^7}{105} - \cdots \frac{(-1)^{n-1}2^{2n}(2^{2n}-1)B_n(ax)^{2n+1}}{(2n+1)!} + \cdots\right\}$

17.30.5. $\int x\tanh^2 ax\,dx = \frac{x^2}{2} - \frac{x\tanh ax}{a} + \frac{1}{a^2}\ln\cosh ax$

17.30.6. $\int \frac{\tanh ax}{x}dx = ax - \frac{(ax)^3}{9} + \frac{2(ax)^5}{75} - \cdots \frac{(-1)^{n-1}2^{2n}(2^{2n}-1)B_n(ax)^{2n-1}}{(2n-1)(2n)!} + \cdots$

17.30.7. $\int \frac{dx}{p+q\tanh ax} = \frac{px}{p^2-q^2} - \frac{q}{a(p^2-q^2)}\ln(q\sinh ax + p\cosh ax)$

17.30.8. $\int \tanh^n ax\,dx = \frac{-\tanh^{n-1} ax}{a(a-1)} + \int \tanh^{n-2} ax\,dx$

(31) Integrals Involving coth ax

17.31.1. $\int \coth ax\,dx = \frac{1}{a}\ln\sinh ax$

17.31.2. $\int \coth^2 ax\,dx = x - \frac{\coth ax}{a}$

17.31.3. $\int \coth^3 ax\,dx = \frac{1}{a}\ln\sinh ax - \frac{\coth^2 ax}{2a}$

17.31.4. $\int x\coth ax\,dx = \frac{1}{a^2}\left\{ax + \frac{(ax)^3}{9} - \frac{(ax)^5}{225} + \cdots \frac{(-1)^{n-1}2^{2n}B_n(ax)^{2n+1}}{(2n+1)!} + \cdots\right\}$

17.31.5. $\int x\coth^2 ax\,dx = \frac{x^2}{2} - \frac{x\coth ax}{a} + \frac{1}{a^2}\ln\sinh ax$

17.31.6. $\int \frac{\coth ax}{x}dx = -\frac{1}{ax} + \frac{ax}{3} - \frac{(ax)^3}{135} + \cdots \frac{(-1)^n 2^{2n}B_n(ax)^{2n-1}}{(2n-1)(2n)!} + \cdots$

17.31.7. $\int \frac{dx}{p+q\coth ax} = \frac{px}{p^2-q^2} - \frac{q}{a(p^2-q^2)}\ln(p\sinh ax + q\cosh ax)$

17.31.8. $\int \coth^n ax\,dx = -\frac{\coth^{n-1} ax}{a(n-1)} + \int \coth^{n-2} ax\,dx$

(32) Integrals Involving sech *ax*

17.32.1. $\int \operatorname{sech} ax\, dx = \frac{2}{a} \tan^{-1} e^{ax}$

17.32.2. $\int \operatorname{sech}^2 ax\, dx = \frac{\tanh ax}{a}$

17.32.3. $\int \operatorname{sech}^3 ax\, dx = \frac{\operatorname{sech} ax \tanh ax}{2a} + \frac{1}{2a} \tan^{-1} \sinh ax$

17.32.4. $\int x \operatorname{sech} ax\, dx = \frac{1}{a^2}\left\{\frac{(ax)^2}{2} - \frac{(ax)^4}{8} + \frac{5(ax)^6}{144} + \cdots \frac{(-1)^n E_n (ax)^{2n+2}}{(2n+2)(2n)!} + \cdots\right\}$

17.32.5. $\int x \operatorname{sech}^2 ax\, dx = \frac{x \tanh ax}{a} - \frac{1}{a^2} \ln \cosh ax$

17.32.6. $\int \frac{\operatorname{sech} ax}{x} dx = \ln x - \frac{(ax)^2}{4} + \frac{5(ax)^4}{96} - \frac{61(ax)^6}{4320} + \cdots \frac{(-1)^n E_n (ax)^{2n}}{2n(2n)!} + \cdots$

17.32.7. $\int \operatorname{sech}^n ax\, dx = \frac{\operatorname{sech}^{n-2} ax \tanh ax}{a(n-1)} + \frac{n-2}{n-1} \int \operatorname{sech}^{n-2} ax\, dx$

(33) Integrals Involving csch *ax*

17.33.1. $\int \operatorname{csch} ax\, dx = \frac{1}{a} \ln \tanh \frac{ax}{2}$

17.33.2. $\int \operatorname{csch}^2 ax\, dx = -\frac{\coth ax}{a}$

17.33.3. $\int \operatorname{csch}^3 ax\, dx = -\frac{\operatorname{csch} ax \coth ax}{2a} - \frac{1}{2a} \ln \tanh \frac{ax}{2}$

17.33.4. $\int x \operatorname{csch} ax\, dx = \frac{1}{a^2}\left\{ax - \frac{(ax)^3}{18} + \frac{7(ax)^5}{1800} + \cdots + \frac{2(-1)^n (2^{2n-1} - 1) B_n (ax)^{2n+1}}{(2n+1)!} + \cdots\right\}$

17.33.5. $\int x \operatorname{csch}^2 ax\, dx = -\frac{x \coth ax}{a} + \frac{1}{a^2} \ln \sinh ax$

17.33.6. $\int \frac{\operatorname{csch} ax}{x} dx = -\frac{1}{ax} - \frac{ax}{6} + \frac{7(ax)^3}{1080} + \cdots \frac{(-1)^n 2(2^{2n-1} - 1) B_n (ax)^{2n-1}}{(2n-1)(2n)!} + \cdots$

17.33.7. $\int \operatorname{csch}^n ax\, dx = \frac{-\operatorname{csch}^{n-2} ax \coth ax}{a(n-1)} - \frac{n-2}{n-1} \int \operatorname{csch}^{n-2} ax\, dx$

(34) Integrals Involving Inverse Hyperbolic Functions

17.34.1. $\int \sinh^{-1}\frac{x}{a}\,dx = x\sinh^{-1}\frac{x}{a} - \sqrt{x^2+a^2}$

17.34.2. $\int x\sinh^{-1}\frac{x}{a}\,dx = \left(\frac{x^2}{2}+\frac{a^2}{4}\right)\sinh^{-1}\frac{x}{a} - x\frac{\sqrt{x^2+a^2}}{4}$

17.34.3.
$$\int \frac{\sinh^{-1}(x/a)}{x}\,dx = \begin{cases} \frac{x}{a} - \frac{(x/a)^3}{2\cdot 3\cdot 3} + \frac{1\cdot 3(x/a)^5}{2\cdot 4\cdot 5\cdot 5} - \frac{1\cdot 3\cdot 5(x/a)^7}{2\cdot 4\cdot 6\cdot 7\cdot 7} + \cdots & |x| < a \\ \frac{\ln^2(2x/a)}{2} - \frac{(a/x)^2}{2\cdot 2\cdot 2} + \frac{1\cdot 3(a/x)^4}{2\cdot 4\cdot 4\cdot 4} - \frac{1\cdot 3\cdot 5(a/x)^6}{2\cdot 4\cdot 6\cdot 6\cdot 6} + \cdots & x > a \\ -\frac{\ln^2(-2x/a)}{2} + \frac{(a/x)^2}{2\cdot 2\cdot 2} - \frac{1\cdot 3(a/x)^4}{2\cdot 4\cdot 4\cdot 4} + \frac{1\cdot 3\cdot 5(a/x)^6}{2\cdot 4\cdot 6\cdot 6\cdot 6} - \cdots & x < -a \end{cases}$$

17.34.4.
$$\int \cosh^{-1}\frac{x}{a}\,dx = \begin{cases} x\cosh^{-1}(x/a) - \sqrt{x^2-a^2}, & \cosh^{-1}(x/a) > 0 \\ x\cosh^{-1}(x/a) + \sqrt{x^2-a^2}, & \cosh^{-1}(x/a) < 0 \end{cases}$$

17.34.5.
$$\int x\cosh^{-1}\frac{x}{a}\,dx = \begin{cases} \frac{1}{4}(2x^2-a^2)\cosh^{-1}(x/a) - \frac{1}{4}x\sqrt{x^2-a^2}, & \cosh^{-1}(x/a) > 0 \\ \frac{1}{4}(2x^2-a^2)\cosh^{-1}(x/a) + \frac{1}{4}x\sqrt{x^2-a^2}, & \cosh^{-1}(x/a) < 0 \end{cases}$$

17.34.6.
$$\int \frac{\cosh^{-1}(x/a)}{x}\,dx = \pm\left[\frac{1}{2}\ln^2(2x/a) + \frac{(a/x)^2}{2\cdot 2\cdot 2} + \frac{1\cdot 3(a/x)^4}{2\cdot 4\cdot 4\cdot 4} + \frac{1\cdot 3\cdot 5(a/x)^6}{2\cdot 4\cdot 6\cdot 6\cdot 6} + \cdots\right]$$

$+$ if $\cosh^{-1}(x/a) > 0$, $-$ if $\cosh^{-1}(x/a) < 0$

17.34.7. $\int \tanh^{-1}\frac{x}{a}\,dx = x\tanh^{-1}\frac{x}{a} + \frac{a}{2}\ln(a^2-x^2)$

17.34.8. $\int x\tanh^{-1}\frac{x}{a}\,dx = \frac{ax}{2} + \frac{1}{2}(x^2-a^2)\tanh^{-1}\frac{x}{a}$

17.34.9. $\int \frac{\tanh^{-1}(x/a)}{x}\,dx = \frac{x}{a} + \frac{(x/a)^3}{3^2} + \frac{(x/a)^5}{5^2} + \cdots$

17.34.10. $\int \coth^{-1}\frac{x}{a}\,dx = x\coth^{-1}x + \frac{a}{2}\ln(x^2-a^2)$

17.34.11. $\int x\coth^{-1}\frac{x}{a}\,dx = \frac{ax}{2} + \frac{1}{2}(x^2-a^2)\coth^{-1}\frac{x}{a}$

17.34.12. $\int \frac{\coth^{-1}(x/a)}{x}\,dx = -\left(\frac{a}{x} + \frac{(a/x)^3}{3^2} + \frac{(a/x)^5}{5^2} + \cdots\right)$

17.34.13.
$$\int \operatorname{sech}^{-1}\frac{x}{a}\,dx = \begin{cases} x\operatorname{sech}^{-1}(x/a) + a\sin^{-1}(x/a), & \operatorname{sech}^{-1}(x/a) > 0 \\ x\operatorname{sech}^{-1}(x/a) - a\sin^{-1}(x/a), & \operatorname{sech}^{-1}(x/a) < 0 \end{cases}$$

17.34.14. $\int \operatorname{csch}^{-1}\frac{x}{a}\,dx = x\operatorname{csch}^{-1}\frac{x}{a} \pm a\sinh^{-1}\frac{x}{a}$ $\quad$ ($+$ if $x > 0$, $-$ if $x < 0$)

17.34.15. $\int x^m \sinh^{-1}\frac{x}{a}\,dx = \frac{x^{m+1}}{m+1}\sinh^{-1}\frac{x}{a} - \frac{1}{m+1}\int \frac{x^{m+1}}{\sqrt{x^2+a^2}}\,dx$

17.34.16. $\int x^m \cosh^{-1}\frac{x}{a}\,dx = \begin{cases} \frac{x^{m+1}}{m+1}\cosh^{-1}\frac{x}{a} - \frac{1}{m+1}\int \frac{x^{m+1}}{\sqrt{x^2-a^2}}\,dx & \cosh^{-1}(x/a) > 0 \\ \frac{x^{m+1}}{m+1}\cosh^{-1}\frac{x}{a} + \frac{1}{m+1}\int \frac{x^{m+1}}{\sqrt{x^2-a^2}}\,dx & \cosh^{-1}(x/a) < 0 \end{cases}$

17.34.17. $\int x^m \tanh^{-1}\frac{x}{a}\,dx = \frac{x^{m+1}}{m+1}\tanh^{-1}\frac{x}{a} - \frac{a}{m+1}\int \frac{x^{m+1}}{a^2-x^2}\,dx$

17.34.18. $\int x^m \coth^{-1}\frac{x}{a}\,dx = \frac{x^{m+1}}{m+1}\coth^{-1}\frac{x}{a} - \frac{1}{m+1}\int \frac{x^{m+1}}{a^2-x^2}\,dx$

17.34.19. $\int x^m \operatorname{sech}^{-1}\frac{x}{a}\,dx = \begin{cases} \frac{x^{m+1}}{m+1}\operatorname{sech}^{-1}\frac{x}{a} + \frac{a}{m+1}\int \frac{x^m\,dx}{\sqrt{a^2-x^2}} & \operatorname{sech}^{-1}(x/a) > 0 \\ \frac{x^{m+1}}{m+1}\operatorname{sech}^{-1}\frac{x}{a} - \frac{1}{m+1}\int \frac{x^m\,dx}{\sqrt{a^2-x^2}} & \operatorname{sech}^{-1}(x/a) < 0 \end{cases}$

17.34.20. $\int x^m \operatorname{csch}^{-1}\frac{x}{a}\,dx = \frac{x^{m+1}}{m+1}\operatorname{csch}^{-1}\frac{x}{a} \pm \frac{a}{m+1}\int \frac{x^m\,dx}{\sqrt{x^2+a^2}}$ $(+\text{ if } x > 0,\ -\text{ if } x < 0)$

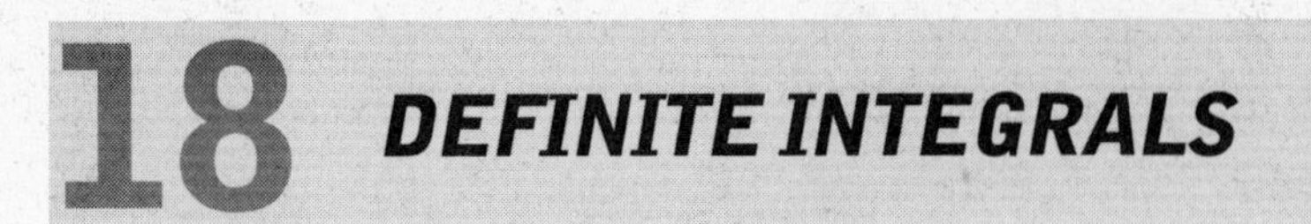

18 DEFINITE INTEGRALS

Definition of a Definite Integral

Let $f(x)$ be defined in an interval $a \leqq x \leqq b$. Divide the interval into n equal parts of length $\Delta x = (b - a)/n$. Then the definite integral of $f(x)$ between $x = a$ and $x = b$ is defined as

18.1. $$\int_a^b f(x)\,dx = \lim_{n\to\infty}\{f(a)\,\Delta x + f(a+\Delta x)\,\Delta x + f(a+2\Delta x)\,\Delta x + \cdots + f(a+(n-1)\,\Delta x)\,\Delta x\}$$

The limit will certainly exist if $f(x)$ is piecewise continuous.

If $f(x) = \dfrac{d}{dx}g(x)$, then by the fundamental theorem of the integral calculus the above definite integral can be evaluated by using the result

18.2. $$\int_a^b f(x)\,dx = \int_a^b \frac{d}{dx}g(x)\,dx = g(x)\Big|_a^b = g(b) - g(a)$$

If the interval is infinite or if $f(x)$ has a singularity at some point in the interval, the definite integral is called an *improper integral* and can be defined by using appropriate limiting procedures. For example,

18.3. $$\int_a^\infty f(x)\,dx = \lim_{b\to\infty}\int_a^b f(x)\,dx$$

18.4. $$\int_{-\infty}^\infty f(x)\,dx = \lim_{\substack{a\to-\infty\\ b\to\infty}}\int_a^b f(x)\,dx$$

18.5. $$\int_a^b f(x)\,dx = \lim_{\epsilon\to 0}\int_\alpha^{b-\epsilon} f(x)\,dx \quad \text{if } b \text{ is a singular point.}$$

18.6. $$\int_a^b f(x)\,dx = \lim_{\epsilon\to 0}\int_{a+\epsilon}^{b} f(x)\,dx \quad \text{if } a \text{ is a singular point.}$$

General Formulas Involving Definite Integrals

18.7. $$\int_a^b \{f(x) \pm g(x) \pm h(x) \pm \cdots\}\,dx = \int_a^b f(x)\,dx \pm \int_a^b g(x)\,dx \pm \int_a^b h(x)\,dx \pm \cdots$$

18.8. $$\int_a^b cf(x)\,dx = c\int_a^b f(x)\,dx \quad \text{where } c \text{ is any constant.}$$

18.9. $$\int_a^a f(x)\,dx = 0$$

18.10. $$\int_a^b f(x)\,dx = -\int_b^a f(x)\,dx$$

18.11. $$\int_a^b f(x)\,dx = \int_a^c f(x)\,dx + \int_c^b f(x)\,dx$$

18.12. $\int_a^b f(x)\,dx = (b-a)\,f(c)$ where c is between a and b.

This is called the *mean value theorem* for definite integrals and is valid if $f(x)$ is continuous in $a \leqq x \leqq b$.

18.13. $\int_a^b f(x)\,g(x)\,dx = f(c)\int_a^b g(x)\,dx$ where c is between a and b

This is a generalization of 18.12 and is valid if $f(x)$ and $g(x)$ are continuous in $a \leqq x \leqq b$ and $g(x) \geqq 0$.

Leibnitz's Rules for Differentiation of Integrals

18.14. $$\frac{d}{d\alpha}\int_{\phi_1(\alpha)}^{\phi_2(\alpha)} F(x,\alpha)\,dx = \int_{\phi_1(\alpha)}^{\phi_2(\alpha)} \frac{\partial F}{\partial \alpha}\,dx + F(\phi_2,\alpha)\frac{d\phi_2}{d\alpha} - F(\phi_1,\alpha)\frac{d\phi_1}{d\alpha}$$

Approximate Formulas for Definite Integrals

In the following the interval from $x = a$ to $x = b$ is subdivided into n equal parts by the points $a = x_0, x_1, x_2, \ldots, x_{n-1}, x_n = b$ and we let $y_0 = f(x_0)$, $y_1 = f(x_1)$, $y_2 = f(x_2)$, $\ldots$, $y_n = f(x_n)$, $h = (b-a)/n$.

Rectangular formula:

18.15. $$\int_a^b f(x)\,dx \approx h(y_0 + y_1 + y_2 + \cdots + y_{n-1})$$

Trapezoidal formula:

18.16. $$\int_a^b f(x)\,dx \approx \frac{h}{2}(y_0 + 2y_1 + 2y_2 + \cdots + 2y_{n-1} + y_n)$$

Simpson's formula (or parabolic formula) for n even:

18.17. $$\int_a^b f(x)\,dx \approx \frac{h}{3}(y_0 + 4y_1 + 2y_2 + 4y_3 + \cdots + 2y_{n-2} + 4y_{n-1} + y_n)$$

Definite Integrals Involving Rational or Irrational Expressions

18.18. $$\int_0^\infty \frac{dx}{x^2 + a^2} = \frac{\pi}{2a}$$

18.19. $$\int_0^\infty \frac{x^{p-1}\,dx}{1+x} = \frac{\pi}{\sin p\pi}, \quad 0 < p < 1$$

18.20. $$\int_0^\infty \frac{x^m\,dx}{x^n + a^n} = \frac{\pi a^{m+1-n}}{n \sin[(m+1)\pi/n]}, \quad 0 < m+1 < n$$

18.21. $$\int_0^\infty \frac{x^m\,dx}{1 + 2x\cos\beta + x^2} = \frac{\pi}{\sin m\pi}\,\frac{\sin m\beta}{\sin\beta}$$

18.22. $$\int_0^a \frac{dx}{\sqrt{a^2 - x^2}} = \frac{\pi}{2}$$

18.23. $$\int_0^a \sqrt{a^2 - x^2}\,dx = \frac{\pi a^2}{4}$$

18.24. $\int_0^a x^m (a^n - x^n)^p\, dx = \dfrac{a^{m+1+np}\Gamma[(m+1)/n]\Gamma(p+1)}{n\Gamma[(m+1)/n + p+1]}$

18.25. $\int_0^\infty \dfrac{x^m dx}{(x^n + a^n)^r} = \dfrac{(-1)^{r-1}\pi a^{m+1-nr}\Gamma[(m+1)/n]}{n\sin[(m+1)\pi/n](r-1)!\Gamma[(m+1)/n - r+1]}, \quad 0 < m+1 < nr$

Definite Integrals Involving Trigonometric Functions

All letters are considered positive unless otherwise indicated.

18.26. $\int_0^\pi \sin mx \sin nx\, dx = \begin{cases} 0 & m, n \text{ integers and } m \neq n \\ \pi/2 & m, n \text{ integers and } m = n \end{cases}$

18.27. $\int_0^\pi \cos mx \cos nx\, dx = \begin{cases} 0 & m, n \text{ integers and } m \neq n \\ \pi/2 & m, n \text{ integers and } m = n \end{cases}$

18.28. $\int_0^\pi \sin mx \cos nx\, dx = \begin{cases} 0 & m, n \text{ integers and } m+n \text{ even} \\ 2m/(m^2 - n^2) & m, n \text{ integers and } m+n \text{ odd} \end{cases}$

18.29. $\int_0^{\pi/2} \sin^2 x\, dx = \int_0^{\pi/2} \cos^2 x\, dx = \dfrac{\pi}{4}$

18.30. $\int_0^{\pi/2} \sin^{2m} x\, dx = \int_0^{\pi/2} \cos^{2m} x\, dx = \dfrac{1\cdot 3\cdot 5\cdots 2m-1}{2\cdot 4\cdot 6\cdots 2m}\dfrac{\pi}{2}, \quad m = 1, 2, \ldots$

18.31. $\int_0^{\pi/2} \sin^{2m+1} x\, dx = \int_0^{\pi/2} \cos^{2m+1} x\, dx = \dfrac{2\cdot 4\cdot 6\cdots 2m}{1\cdot 3\cdot 5\cdots 2m+1}, \quad m = 1, 2, \ldots$

18.32. $\int_0^{\pi/2} \sin^{2p-1} x \cos^{2q-1} x\, dx = \dfrac{\Gamma(p)\Gamma(q)}{2\Gamma(p+q)}$

18.33. $\int_0^\infty \dfrac{\sin px}{x}\, dx = \begin{cases} \pi/2 & p > 0 \\ 0 & p = 0 \\ -\pi/2 & p < 0 \end{cases}$

18.34. $\int_0^\infty \dfrac{\sin px \cos qx}{x}\, dx = \begin{cases} 0 & p > q > 0 \\ \pi/2 & 0 < p < q \\ \pi/4 & p = q > 0 \end{cases}$

18.35. $\int_0^\infty \dfrac{\sin px \sin qx}{x^2}\, dx = \begin{cases} \pi p/2 & 0 < p \leqq q \\ \pi q/2 & p \geqq q > 0 \end{cases}$

18.36. $\int_0^\infty \dfrac{\sin^2 px}{x^2}\, dx = \dfrac{\pi p}{2}$

18.37. $\int_0^\infty \dfrac{1 - \cos px}{x^2}\, dx = \dfrac{\pi p}{2}$

18.38. $\int_0^\infty \frac{\cos px - \cos qx}{x}\,dx = \ln\frac{q}{p}$

18.39. $\int_0^\infty \frac{\cos px - \cos qx}{x^2}\,dx = \frac{\pi(q-p)}{2}$

18.40. $\int_0^\infty \frac{\cos mx}{x^2+a^2}\,dx = \frac{\pi}{2a}e^{-ma}$

18.41. $\int_0^\infty \frac{x\sin mx}{x^2+a^2}\,dx = \frac{\pi}{2}e^{-ma}$

18.42. $\int_0^\infty \frac{\sin mx}{x(x^2+a^2)}\,dx = \frac{\pi}{2a^2}(1-e^{-ma})$

18.43. $\int_0^{2\pi} \frac{dx}{a+b\sin x} = \frac{2\pi}{\sqrt{a^2-b^2}}$

18.44. $\int_0^{2\pi} \frac{dx}{a+b\cos x} = \frac{2\pi}{\sqrt{a^2-b^2}}$

18.45. $\int_0^{\pi/2} \frac{dx}{a+b\cos x} = \frac{\cos^{-1}(b/a)}{\sqrt{a^2-b^2}}$

18.46. $\int_0^{2\pi} \frac{dx}{(a+b\sin x)^2} = \int_0^{2\pi} \frac{dx}{(a+b\,\cos x)^2} = \frac{2\pi a}{(a^2-b^2)^{3/2}}$

18.47. $\int_0^{2\pi} \frac{dx}{1-2a\cos x+a^2} = \frac{2\pi}{1-a^2}, \quad 0<a<1$

18.48. $\int_0^{\pi} \frac{x\sin x\,dx}{1-2a\cos x+a^2} = \begin{cases} (\pi/a)\ln(1+a), & |a|<1 \\ \pi\ln(1+1/a), & |a|>1 \end{cases}$

18.49. $\int_0^{\pi} \frac{\cos mx\,dx}{1-2a\cos x+a^2} = \frac{\pi a^m}{1-a^2}, \quad a^2<1, \quad m = 0, 1, 2, \ldots$

18.50. $\int_0^\infty \sin ax^2\,dx = \int_0^\infty \cos ax^2 dx = \frac{1}{2}\sqrt{\frac{\pi}{2a}}$

18.51. $\int_0^\infty \sin ax^n\,dx = \frac{1}{na^{1/n}}\Gamma(1/n)\sin\frac{\pi}{2n}, \quad n>1$

18.52. $\int_0^\infty \cos ax^n\,dx = \frac{1}{na^{1/n}}\Gamma(1/n)\cos\frac{\pi}{2n}, \quad n>1$

18.53. $\int_0^\infty \frac{\sin x}{\sqrt{x}}\,dx = \int_0^\infty \frac{\cos x}{\sqrt{x}}\,dx = \sqrt{\frac{\pi}{2}}$

18.54. $\int_0^\infty \frac{\sin x}{x^p}\,dx = \frac{\pi}{2\Gamma(p)\sin(p\pi/2)}, \quad 0<p<1$

18.55. $\int_0^\infty \frac{\cos x}{x^p}\,dx = \frac{\pi}{2\Gamma(p)\cos(p\pi/2)}, \quad 0<p<1$

18.56. $\int_0^\infty \sin ax^2\cos 2bx\,dx = \frac{1}{2}\sqrt{\frac{\pi}{2a}}\left(\cos\frac{b^2}{a} - \sin\frac{b^2}{a}\right)$

18.57. $\int_0^\infty \cos ax^2 \cos 2bx\, dx = \frac{1}{2}\sqrt{\frac{\pi}{2a}}\left(\cos\frac{b^2}{a} + \sin\frac{b^2}{a}\right)$

18.58. $\int_0^\infty \frac{\sin^3 x}{x^3}\, dx = \frac{3\pi}{8}$

18.59. $\int_0^\infty \frac{\sin^4 x}{x^4}\, dx = \frac{\pi}{3}$

18.60. $\int_0^\infty \frac{\tan x}{x}\, dx = \frac{\pi}{2}$

18.61. $\int_0^{\pi/2} \frac{dx}{1+\tan^m x} = \frac{\pi}{4}$

18.62. $\int_0^{\pi/2} \frac{x}{\sin x}\, dx = 2\left\{\frac{1}{1^2} - \frac{1}{3^2} + \frac{1}{5^2} - \frac{1}{7^2} + \cdots\right\}$

18.63. $\int_0^1 \frac{\tan^{-1} x}{x}\, dx = \frac{1}{1^2} - \frac{1}{3^2} + \frac{1}{5^2} - \frac{1}{7^2} + \cdots$

18.64. $\int_0^1 \frac{\sin^{-1} x}{x}\, dx = \frac{\pi}{2}\ln 2$

18.65. $\int_0^1 \frac{1-\cos x}{x}\, dx - \int_1^\infty \frac{\cos x}{x}\, dx = \gamma$

18.66. $\int_0^\infty \left(\frac{1}{1+x^2} - \cos x\right)\frac{dx}{x} = \gamma$

18.67. $\int_0^\infty \frac{\tan^{-1} px - \tan^{-1} qx}{x}\, dx = \frac{\pi}{2}\ln\frac{p}{q}$

Definite Integrals Involving Exponential Functions

Some integrals contain Euler's constant $\gamma = 0.5772156\ldots$ (see 1.3, page 3).

18.68. $\int_0^\infty e^{-ax}\cos bx\, dx = \frac{a}{a^2+b^2}$

18.69. $\int_0^\infty e^{-ax}\sin bx\, dx = \frac{b}{a^2+b^2}$

18.70. $\int_0^\infty \frac{e^{-ax}\sin bx}{x}\, dx = \tan^{-1}\frac{b}{a}$

18.71. $\int_0^\infty \frac{e^{-ax} - e^{-bx}}{x}\, dx = \ln\frac{b}{a}$

18.72. $\int_0^\infty e^{-ax^2}\, dx = \frac{1}{2}\sqrt{\frac{\pi}{a}}$

18.73. $\int_0^\infty e^{-ax^2}\cos bx\, dx = \frac{1}{2}\sqrt{\frac{\pi}{a}}\, e^{-b^2/4a}$

18.74. $$\int_0^\infty e^{-(ax^2+bx+c)}dx = \frac{1}{2}\sqrt{\frac{\pi}{a}}e^{(b^2-4ac)/4a}\operatorname{erfc}\frac{b}{2\sqrt{a}}$$

where $\operatorname{erfc}(p) = \frac{2}{\sqrt{\pi}}\int_p^\infty e^{-x^2}dx$

18.75. $$\int_{-\infty}^\infty e^{-(ax^2+bx+c)}dx = \sqrt{\frac{\pi}{a}}e^{(b^2-4ac)/4a}$$

18.76. $$\int_0^\infty x^n e^{-ax}dx = \frac{\Gamma(n+1)}{a^{n+1}}$$

18.77. $$\int_0^\infty x^m e^{-ax^2}dx = \frac{\Gamma[(m+1)/2]}{2a^{(m+1)/2}}$$

18.78. $$\int_0^\infty e^{-(ax^2+b/x^2)}dx = \frac{1}{2}\sqrt{\frac{\pi}{a}}e^{-2\sqrt{ab}}$$

18.79. $$\int_0^\infty \frac{x\,dx}{e^x-1} = \frac{1}{1^2}+\frac{1}{2^2}+\frac{1}{3^2}+\frac{1}{4^2}+\cdots = \frac{\pi^2}{6}$$

18.80. $$\int_0^\infty \frac{x^{n-1}}{e^x-1}dx = \Gamma(n)\left(\frac{1}{1^n}+\frac{1}{2^n}+\frac{1}{3^n}+\cdots\right)$$

For even n this can be summed in terms of Bernoulli numbers (see pages 142–143).

18.81. $$\int_0^\infty \frac{x\,dx}{e^x+1} = \frac{1}{1^2}-\frac{1}{2^2}+\frac{1}{3^2}-\frac{1}{4^2}+\cdots = \frac{\pi^2}{12}$$

18.82. $$\int_0^\infty \frac{x^{n-1}}{e^x+1}dx = \Gamma(n)\left(\frac{1}{1^n}-\frac{1}{2^n}+\frac{1}{3^n}-\cdots\right)$$

For some positive integer values of n the series can be summed (see 23.10).

18.83. $$\int_0^\infty \frac{\sin mx}{e^{2\pi x}-1}dx = \frac{1}{4}\coth\frac{m}{2}-\frac{1}{2m}$$

18.84. $$\int_0^\infty\left(\frac{1}{1+x}-e^{-x}\right)\frac{dx}{x} = \gamma$$

18.85. $$\int_0^\infty \frac{e^{-x^2}-e^{-x}}{x}dx = \tfrac{1}{2}\gamma$$

18.86. $$\int_0^\infty\left(\frac{1}{e^x-1}-\frac{e^{-x}}{x}\right)dx = \gamma$$

18.87. $$\int_0^\infty \frac{e^{-ax}-e^{-bx}}{x\sec px}dx = \frac{1}{2}\ln\left(\frac{b^2+p^2}{a^2+p^2}\right)$$

18.88. $$\int_0^\infty \frac{e^{-ax}-e^{-bx}}{x\csc px}dx = \tan^{-1}\frac{b}{p}-\tan^{-1}\frac{a}{p}$$

18.89. $$\int_0^\infty \frac{e^{-ax}(1-\cos x)}{x^2}dx = \cot^{-1}a-\frac{a}{2}\ln(a^2+1)$$

Definite Integrals Involving Logarithmic Functions

18.90. $\int_0^1 x^m (\ln x)^n \, dx = \frac{(-1)^n n!}{(m+1)^{n+1}} \qquad m > -1, \quad n = 0, 1, 2, \ldots$

If $n \neq 0, 1, 2, \ldots$ replace $n!$ by $\Gamma(n+1)$.

18.91. $\int_0^1 \frac{\ln x}{1+x} \, dx = -\frac{\pi^2}{12}$

18.92. $\int_0^1 \frac{\ln x}{1-x} \, dx = -\frac{\pi^2}{6}$

18.93. $\int_0^1 \frac{\ln(1+x)}{x} \, dx = \frac{\pi^2}{12}$

18.94. $\int_0^1 \frac{\ln(1-x)}{x} \, dx = -\frac{\pi^2}{6}$

18.95. $\int_0^1 \ln x \ln(1+x) \, dx = 2 - 2\ln 2 - \frac{\pi^2}{12}$

18.96. $\int_0^1 \ln x \ln(1-x) \, dx = 2 - \frac{\pi^2}{6}$

18.97. $\int_0^\infty \frac{x^{p-1} \ln x}{1+x} \, dx = -\pi^2 \csc p\pi \cot p\pi \qquad 0 < p < 1$

18.98. $\int_0^1 \frac{x^m - x^n}{\ln x} \, dx = \ln \frac{m+1}{n+1}$

18.99. $\int_0^\infty e^{-x} \ln x \, dx = -\gamma$

18.100. $\int_0^\infty e^{-x^2} \ln x \, dx = -\frac{\sqrt{\pi}}{4}(\gamma + 2\ln 2)$

18.101. $\int_0^\infty \ln\left(\frac{e^x + 1}{e^x - 1}\right) dx = \frac{\pi^2}{4}$

18.102. $\int_0^{\pi/2} \ln \sin x \, dx = \int_0^{\pi/2} \ln \cos x \, dx = -\frac{\pi}{2} \ln 2$

18.103. $\int_0^{\pi/2} (\ln \sin x)^2 \, dx = \int_0^{\pi/2} (\ln \cos x)^2 \, dx = \frac{\pi}{2}(\ln 2)^2 + \frac{\pi^3}{24}$

18.104. $\int_0^\pi x \ln \sin x \, dx = -\frac{\pi^2}{2} \ln 2$

18.105. $\int_0^{\pi/2} \sin x \ln \sin x \, dx = \ln 2 - 1$

18.106. $\int_0^{2\pi} \ln(a + b \sin x) \, dx = \int_0^{2\pi} \ln(a + b\cos x) \, dx = 2\pi \ln(a + \sqrt{a^2 - b^2})$

18.107. $\int_0^\pi \ln(a + b\cos x) \, dx = \pi \ln\left(\frac{a + \sqrt{a^2 - b^2}}{2}\right)$

18.108. $\int_0^{\pi} \ln(a^2 - 2ab\cos x + b^2)\,dx = \begin{cases} 2\pi \ln a, & a \geqq b > 0 \\ 2\pi \ln b, & b \geqq a > 0 \end{cases}$

18.109. $\int_0^{\pi/4} \ln(1 + \tan x)\,dx = \frac{\pi}{8} \ln 2$

18.110. $\int_0^{\pi/2} \sec x \ln\left(\frac{1 + b\cos x}{1 + a\cos x}\right) dx = \frac{1}{2}\{(\cos^{-1} a)^2 - (\cos^{-1} b)^2\}$

18.111. $\int_0^{a} \ln\left(2\sin\frac{x}{2}\right) dx = -\left(\frac{\sin a}{1^2} + \frac{\sin 2a}{2^2} + \frac{\sin 3a}{3^2} + \cdots\right)$

See also 18.102.

Definite Integrals Involving Hyperbolic Functions

18.112. $\int_0^{\infty} \frac{\sin ax}{\sinh bx}\,dx = \frac{\pi}{2b} \tanh \frac{a\pi}{2b}$

18.113. $\int_0^{\infty} \frac{\cos ax}{\cosh bx}\,dx = \frac{\pi}{2b} \operatorname{sech} \frac{a\pi}{2b}$

18.114. $\int_0^{\infty} \frac{x\,dx}{\sinh ax} = \frac{\pi^2}{4a^2}$

18.115. $\int_0^{\infty} \frac{x^n\,dx}{\sinh ax} = \frac{2^{n+1} - 1}{2^n a^{n+1}} \Gamma(n+1)\left\{\frac{1}{1^{n+1}} + \frac{1}{2^{n+1}} + \frac{1}{3^{n+1}} + \cdots\right\}$

If n is an odd positive integer, the series can be summed.

18.116. $\int_0^{\infty} \frac{\sinh ax}{e^{bx} + 1}\,dx = \frac{\pi}{2b} \csc \frac{a\pi}{b} - \frac{1}{2a}$

18.117. $\int_0^{\infty} \frac{\sinh ax}{e^{bx} - 1}\,dx = \frac{1}{2a} - \frac{\pi}{2b} \cot \frac{a\pi}{b}$

Miscellaneous Definite Integrals

18.118. $\int_0^{\infty} \frac{f(ax) - f(bx)}{x}\,dx = \{f(0) - f(\infty)\} \ln \frac{b}{a}$

This is called *Frullani's integral*. It holds if $f'(x)$ is continuous and $\int_0^{\infty} \frac{f(x) - f(\infty)}{x}\,dx$ converges.

18.119. $\int_0^{1} \frac{dx}{x^x} = \frac{1}{1^1} + \frac{1}{2^2} + \frac{1}{3^3} + \cdots$

18.120. $\int_{-a}^{a} (a + x)^{m-1}(a - x)^{n-1}\,dx = (2a)^{m+n-1} \frac{\Gamma(m)\,\Gamma(n)}{\Gamma(m+n)}$

BASIC DIFFERENTIAL EQUATIONS and SOLUTIONS

DIFFERENTIAL EQUATION	SOLUTION
19.1. Separation of variables	
$f_1(x)\, g_1(y)\, dx + f_2(x)\, g_2(y)\, dy = 0$	$\int \frac{f_1(x)}{f_2(x)} dx + \int \frac{g_2(y)}{g_1(y)} dy = c$
19.2. Linear first order equation	
$\frac{dy}{dx} + p(x)y = Q(x)$	$ye^{\int P dx} = \int Qe^{\int P dx} dx + c$
19.3. Bernoulli's equation	
$\frac{dy}{dx} + P(x)y = Q(x)y^n$	$v e^{(1-n)\int P dx} = (1-n)\int Qe^{(1-n)\int P dx} dx + c$ where $v = y^{1-n}$. If $n = 1$, the solution is $\ln y = \int (Q - P)\, dx + c$
19.4. Exact equation	
$M(x, y)dx + N(x, y)dy = 0$ where $\partial M/\partial y = \partial N/\partial x$.	$\int M\, \partial x + \int \left(N - \frac{\partial}{\partial y} \int M\, \partial x \right) dy = c$ where ∂x indicates that the integration is to be performed with respect to x keeping y constant.
19.5 Homogeneous equation	
$\frac{dy}{dx} = F\left(\frac{y}{x}\right)$	$\ln x = \int \frac{dv}{F(v) - v} + c$ where $v = y/x$. If $F(v) = v$, the solution is $y = cx$.

<table>
<tr><td>19.6.

$y\,F(xy)\,dx + x\,G(xy)\,dy = 0$</td><td>$$\ln x = \int \frac{G(v)\,dv}{v\{G(v) - F(v)\}} + c$$
where $v = xy$. If $G(v) = F(v)$, the solution is $xy = c$.</td></tr>
<tr><td>19.7. Linear, homogeneous second order equation

$$\frac{d^2y}{dx^2} + a\frac{dy}{dx} + by = 0$$
a, b are real constants.</td><td>Let m_1, m_2 be the roots of $m^2 + am + b = 0$. Then there are 3 cases.
Case 1. m_1, m_2 real and distinct:
$$y = c_1e^{m_1x} + c_2e^{m_2x}$$
Case 2. m_1, m_2 real and equal:
$$y = c_1\,e^{m_1x} + c_2x\,e^{m_1x}$$
Case 3. $m_1 = p + qi,\quad m_2 = p - qi$:
$$y = e^{px}(c_1\cos qx + c_2\sin qx)$$
where $p = -a/2,\ q = \sqrt{b - a^2/4}$.</td></tr>
<tr><td>19.8. Linear, nonhomogeneous second order equation

$$\frac{d^2y}{dx^2} + a\frac{dy}{dx} + by = R(x)$$
a, b are real constants.</td><td>There are 3 cases corresponding to those of entry 19.7 above.
Case 1.
$$y = c_1e^{m_1x} + c_2\,e^{m_2x} + \frac{e^{m_1x}}{m_1 - m_2}\int e^{-m_1x}R(x)\,dx + \frac{e^{m_2x}}{m_2 - m_1}\int e^{-m_2x}R(x)\,dx$$
Case 2.
$$y = c_1\,e^{m_1x} + c_2x\,e^{m_1x} + xe^{m_1x}\int e^{-m_1x}R(x)\,dx - e^{m_1x}\int xe^{-m_1x}R(x)\,dx$$
Case 3.
$$y = e^{px}(c_1\cos qx + c_2\sin qx) + \frac{e^{px}\sin qx}{q}\int e^{-px}R(x)\cos qx\,dx - \frac{e^{px}\cos qx}{q}\int e^{-px}R(x)\sin qx\,dx$$</td></tr>
</table>

19.9. Euler or Cauchy equation	Putting $x = e^t$, the equation becomes
$$x^2 \frac{d^2y}{dx^2} + ax\frac{dy}{dx} + by = S(x)$$	$$\frac{d^2y}{dt^2} + (a-1)\frac{dy}{dt} + by = S(e^t)$$ and can then be solved as in entries 19.7 and 19.8 above.
19.10. Bessel's equation	
$$x^2 \frac{d^2y}{dx^2} + x\frac{dy}{dx} + (\lambda^2x^2 - n^2)y = 0$$	$$y = c_1J_n(\lambda x) + c_2Y_n(\lambda x)$$ See 27.1 to 27.15.
19.11. Transformed Bessel's equation	
$$x^2 \frac{d^2y}{dx^2} + (2p+1)x\frac{dy}{dx} + (a^2x^{2r} + \beta^2)y = 0$$	$$y = x^{-p}\left\{c_1J_{q/r}\left(\frac{\alpha}{r}x^r\right) + c_2Y_{q/r}\left(\frac{\alpha}{r}x^r\right)\right\}$$ where $q = \sqrt{p^2 - \beta^2}$.
19.12. Legendre's equation	
$$(1-x^2)\frac{d^2y}{dx^2} - 2x\frac{dy}{dx} + n(n+1)y = 0$$	$$y = c_1P_n(x) + c_2Q_n(x)$$ See 28.1 to 28.48.

20 FORMULAS from VECTOR ANALYSIS

Vectors and Scalars

Various quantities in physics such as temperature, volume, and speed can be specified by a real number. Such quantities are called *scalars*.

Other quantities such as force, velocity, and momentum require for their specification a direction as well as magnitude. Such quantities are called *vectors*. A vector is represented by an arrow or directed line segment indicating direction. The magnitude of the vector is determined by the length of the arrow, using an appropriate unit.

Notation for Vectors

A vector is denoted by a bold faced letter such as **A** (Fig. 20.1). The magnitude is denoted by $|\mathbf{A}|$ or A. The tail end of the arrow is called the *initial point,* while the head is called the *terminal point.*

Fundamental Definitions

1. **Equality of vectors.** Two vectors are equal if they have the same magnitude and direction. Thus, $\mathbf{A} = \mathbf{B}$ in (Fig. 20-1).
2. **Multiplication of a vector by a scalar.** If m is any real number (scalar), then $m\mathbf{A}$ is a vector whose magnitude is $|m|$ times the magnitude of **A** and whose direction is the same as or opposite to **A** according as $m > 0$ or $m < 0$. If $m = 0$, then $m\mathbf{A} = \mathbf{0}$ is called the *zero* or *null vector.*

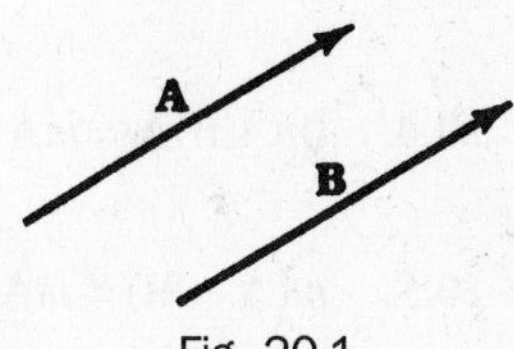

Fig. 20-1

3. **Sums of vectors.** The sum or resultant of **A** and **B** is a vector $\mathbf{C} = \mathbf{A} + \mathbf{B}$ formed by placing the initial point **B** on the terminal point **A** and joining the initial point of **A** to the terminal point of **B** as in Fig. 20-2b. This definition is equivalent to the parallelogram law for vector addition as indicated in Fig. 20-2c. The vector $\mathbf{A} - \mathbf{B}$ is defined as $\mathbf{A} + (-\mathbf{B})$.

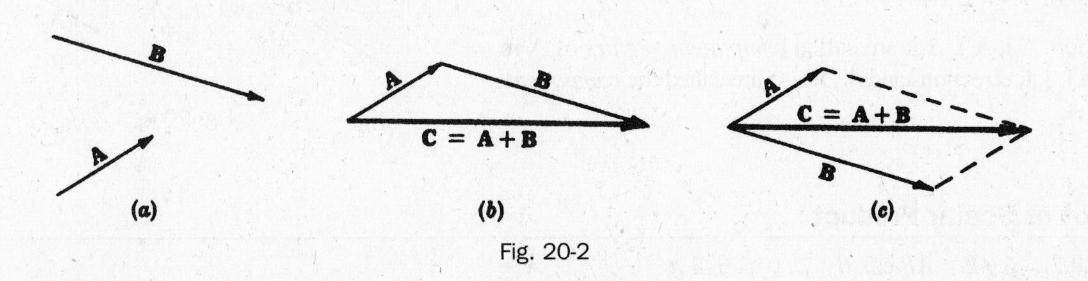

Fig. 20-2

Extension to sums of more than two vectors are immediate. Thus, Fig. 20-3 shows how to obtain the sum **E** of the vectors **A**, **B**, **C**, and **D**.

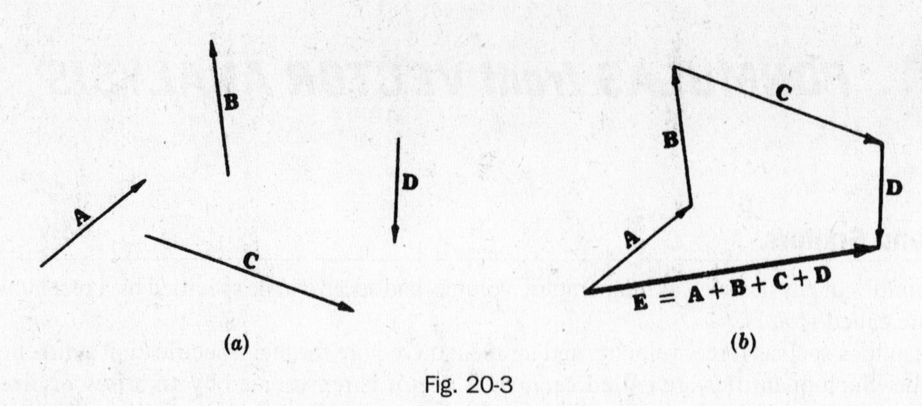

Fig. 20-3

4. **Unit vectors.** A *unit vector* is a vector with unit magnitude. If **A** is a vector, then a unit vector in the direction of **A** is $\mathbf{a} = \mathbf{A}/A$ where $A > 0$.

Laws of Vector Algebra

If **A**, **B**, **C** are vectors and m, n are scalars, then:

20.1.	$\mathbf{A} + \mathbf{B} = \mathbf{B} + \mathbf{A}$	Commutative law for addition
20.2.	$\mathbf{A} + (\mathbf{B} + \mathbf{C}) = (\mathbf{A} + \mathbf{B}) + \mathbf{C}$	Associative law for addition
20.3.	$m(n\mathbf{A}) = (mn)\mathbf{A} = n(m\mathbf{A})$	Associative law for scalar multiplication
20.4.	$(m + n)\mathbf{A} = m\mathbf{A} + n\mathbf{A}$	Distributive law
20.5.	$m(\mathbf{A} + \mathbf{B}) = m\mathbf{A} + m\mathbf{B}$	Distributive law

Components of a Vector

A vector **A** can be represented with initial point at the origin of a rectangular coordinate system. If **i**, **j**, **k** are unit vectors in the directions of the positive x, y, z axes, then

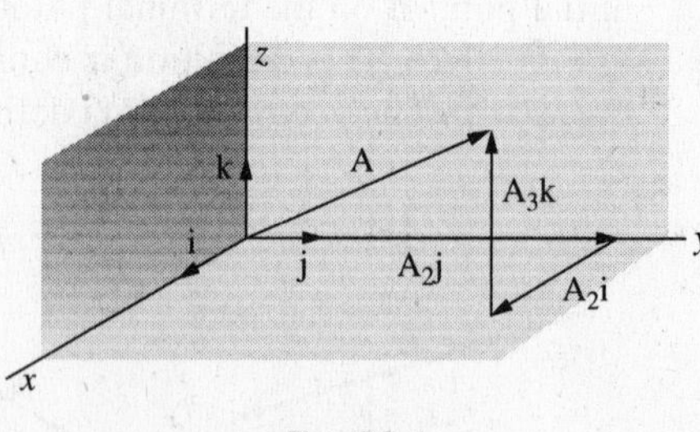

Fig. 20-4

20.6. $\mathbf{A} = A_1\mathbf{i} + A_2\mathbf{j} + A_3\mathbf{k}$

where $A_1\mathbf{i}$, $A_2\mathbf{j}$, $A_3\mathbf{k}$ are called *component vectors* of **A** in the **i**, **j**, **k** directions and A_1, A_2, A_3 are called the *components* of **A**.

Dot or Scalar Product

20.7. $\mathbf{A} \cdot \mathbf{B} = AB\cos\theta \qquad 0 \leqq \theta \leqq \pi$

where θ is the angle between **A** and **B**.

Fundamental results follow:

20.8. $\mathbf{A} \cdot \mathbf{B} = \mathbf{B} \cdot \mathbf{A}$ Commutative law

20.9. $\mathbf{A} \cdot (\mathbf{B} + \mathbf{C}) = \mathbf{A} \cdot \mathbf{B} + \mathbf{A} \cdot \mathbf{C}$ Distributive law

20.10. $\mathbf{A} \cdot \mathbf{B} = A_1B_1 + A_2B_2 + A_3B_3$

where $\mathbf{A} = A_1\mathbf{i} + A_2\mathbf{j} + A_3\mathbf{k}$, $\mathbf{B} = B_1\mathbf{i} + B_2\mathbf{j} + B_3\mathbf{k}$.

Cross or Vector Product

20.11. $\mathbf{A} \times \mathbf{B} = AB \sin\theta\, \mathbf{u} \qquad 0 \leqq \theta \leqq \pi$

where θ is the angle between **A** and **B** and **u** is a unit vector perpendicular to the plane of **A** and **B** such that **A**, **B**, **u** form a *right-handed system* (i.e., a right-threaded screw rotated through an angle less than 180° from **A** to **B** will advance in the direction of **u** as in Fig. 20-5).

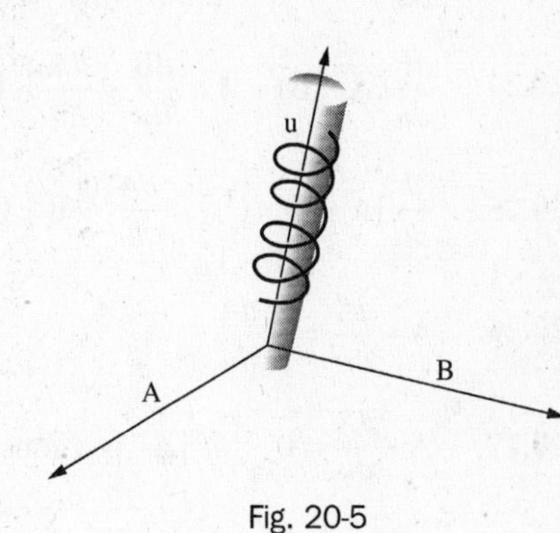

Fig. 20-5

Fundamental results follow:

20.12.
$$\mathbf{A} \times \mathbf{B} = \begin{vmatrix} \mathbf{i} & \mathbf{j} & \mathbf{k} \\ A_1 & A_2 & A_3 \\ B_1 & B_2 & B_3 \end{vmatrix}$$
$$= (A_2B_3 - A_3B_2)\mathbf{i} + (A_3B_1 - A_1B_3)\mathbf{j} + (A_1B_2 - A_2B_1)\mathbf{k}$$

20.13. $\mathbf{A} \times \mathbf{B} = -(\mathbf{B} \times \mathbf{A})$

20.14. $\mathbf{A} \times (\mathbf{B} + \mathbf{C}) = \mathbf{A} \times \mathbf{B} + \mathbf{A} \times \mathbf{C}$

20.15. $|\mathbf{A} \times \mathbf{B}|$ = area of parallelogram having sides **A** and **B**

Miscellaneous Formulas Involving Dot and Cross Products

20.16.
$$\mathbf{A} \cdot (\mathbf{B} \times \mathbf{C}) = \begin{vmatrix} A_1 & A_2 & A_3 \\ B_1 & B_2 & B_3 \\ C_1 & C_2 & C_3 \end{vmatrix} = A_1B_2C_3 + A_2B_3C_1 + A_3B_1C_2 - A_3B_2C_1 - A_2B_1C_3 - A_1B_3C_2$$

20.17. $|\mathbf{A} \cdot (\mathbf{B} \times \mathbf{C})|$ = volume of parallelepiped with sides **A, B, C**

20.18. $\mathbf{A} \times (\mathbf{B} \times \mathbf{C}) = \mathbf{B}(\mathbf{A} \cdot \mathbf{C}) - \mathbf{C}(\mathbf{A} \cdot \mathbf{B})$

20.19. $(\mathbf{A} \times \mathbf{B}) \times \mathbf{C} = \mathbf{B}(\mathbf{A} \cdot \mathbf{C}) - \mathbf{A}(\mathbf{B} \cdot \mathbf{C})$

20.20. $(\mathbf{A} \times \mathbf{B}) \cdot (\mathbf{C} \times \mathbf{D}) = (\mathbf{A} \cdot \mathbf{C})(\mathbf{B} \cdot \mathbf{D}) - (\mathbf{A} \cdot \mathbf{D})(\mathbf{B} \cdot \mathbf{C})$

20.21.
$$(\mathbf{A} \times \mathbf{B}) \times (\mathbf{C} \times \mathbf{D}) = \mathbf{C}\{\mathbf{A} \cdot (\mathbf{B} \times \mathbf{D})\} - \mathbf{D}\{\mathbf{A} \cdot (\mathbf{B} \times \mathbf{C})\}$$
$$= \mathbf{B}\{\mathbf{A} \cdot (\mathbf{C} \times \mathbf{D})\} - \mathbf{A}\{\mathbf{B} \cdot (\mathbf{C} \times \mathbf{D})\}$$

Derivatives of Vectors

The derivative of a vector function $\mathbf{A}(u) = A_1(u)\mathbf{i} + A_2(u)\mathbf{j} + A_3(u)\mathbf{k}$ of the scalar variable u is given by

20.22. $$\frac{d\mathbf{A}}{du} = \lim_{\Delta u \to 0} \frac{\mathbf{A}(u+\Delta u) - \mathbf{A}(u)}{\Delta u} = \frac{dA_1}{du}\mathbf{i} + \frac{dA_2}{du}\mathbf{j} + \frac{dA_3}{du}\mathbf{k}$$

Partial derivatives of a vector function $\mathbf{A}(x, y, z)$ are similarly defined. We assume that all derivatives exist unless otherwise specified.

Formulas Involving Derivatives

20.23. $$\frac{d}{du}(\mathbf{A} \bullet \mathbf{B}) = \mathbf{A} \bullet \frac{d\mathbf{B}}{du} + \frac{d\mathbf{A}}{du} \bullet \mathbf{B}$$

20.24. $$\frac{d}{du}(\mathbf{A} \times \mathbf{B}) = \mathbf{A} \times \frac{d\mathbf{B}}{du} + \frac{d\mathbf{A}}{du} \times \mathbf{B}$$

20.25. $$\frac{d}{du}\{\mathbf{A} \bullet (\mathbf{B} \times C)\} = \frac{d\mathbf{A}}{du} \bullet (\mathbf{B} \times \mathbf{C}) + \mathbf{A} \bullet \left(\frac{d\mathbf{B}}{du} \times \mathbf{C}\right) + \mathbf{A} \bullet \left(\mathbf{B} \times \frac{d\mathbf{C}}{du}\right)$$

20.26. $$\mathbf{A} \bullet \frac{d\mathbf{A}}{du} = A\frac{dA}{du}$$

20.27. $$\mathbf{A} \bullet \frac{d\mathbf{A}}{du} = 0 \quad \textit{if } |\mathbf{A}| \text{ is a constant}$$

The Del Operator

The operator *del* is defined by

20.28. $$\nabla = \mathbf{i}\frac{\partial}{\partial x} + \mathbf{j}\frac{\partial}{\partial y} + \mathbf{k}\frac{\partial}{\partial z}$$

In the following results we assume that $U = U(x, y, z)$, $V = V(x, y, z)$, $\mathbf{A} = \mathbf{A}(x, y, z)$ and $\mathbf{B} = \mathbf{B}(x, y, z)$ have partial derivatives.

The Gradient

20.29. Gradient of U = grad U $$= \nabla U = \left(\mathbf{i}\frac{\partial}{\partial x} + \mathbf{j}\frac{\partial}{\partial y} + \mathbf{k}\frac{\partial}{\partial z}\right)U = \frac{\partial U}{\partial x}\mathbf{i} + \frac{\partial U}{\partial y}\mathbf{j} + \frac{\partial U}{\partial z}\mathbf{k}$$

The Divergence

20.30. Divergence of $\mathbf{A}$ = div $\mathbf{A}$ $$= \nabla \bullet \mathbf{A} = \left(\mathbf{i}\frac{\partial}{\partial x} + \mathbf{j}\frac{\partial}{\partial y} + \mathbf{k}\frac{\partial}{\partial z}\right) \bullet (A_1\mathbf{i} + A_2\mathbf{j} + A_3\mathbf{k})$$

$$= \frac{\partial A_1}{\partial x} + \frac{\partial A_2}{\partial y} + \frac{\partial A_3}{\partial z}$$

The Curl

20.31. Curl of $\mathbf{A}$ = curl $\mathbf{A} = \nabla \times \mathbf{A}$

$$= \left(\mathbf{i}\frac{\partial}{\partial x} + \mathbf{j}\frac{\partial}{\partial y} + \mathbf{k}\frac{\partial}{\partial z}\right) \times (A_1\mathbf{i} + A_2\mathbf{j} + A_3\mathbf{k})$$

$$= \begin{vmatrix} \mathbf{i} & \mathbf{j} & \mathbf{k} \\ \frac{\partial}{\partial x} & \frac{\partial}{\partial y} & \frac{\partial}{\partial z} \\ A_1 & A_2 & A_3 \end{vmatrix}$$

$$= \left(\frac{\partial A_3}{\partial y} - \frac{\partial A_2}{\partial z}\right)\mathbf{i} + \left(\frac{\partial A_1}{\partial z} - \frac{\partial A_3}{\partial x}\right)\mathbf{j} + \left(\frac{\partial A_2}{\partial x} - \frac{\partial A_1}{\partial y}\right)\mathbf{k}$$

The Laplacian

20.32. Laplacian of $U = \nabla^2 U = \nabla \cdot (\nabla U) = \dfrac{\partial^2 U}{\partial x^2} + \dfrac{\partial^2 U}{\partial y^2} + \dfrac{\partial^2 U}{\partial z^2}$

20.33. Laplacian of $\mathbf{A} = \nabla^2 \mathbf{A} = \dfrac{\partial^2 \mathbf{A}}{\partial x^2} + \dfrac{\partial^2 \mathbf{A}}{\partial y^2} + \dfrac{\partial^2 \mathbf{A}}{\partial z^2}$

The Biharmonic Operator

20.34. Biharmonic operator on $U = \nabla^4 U = \nabla^2(\nabla^2 U)$

$$= \frac{\partial^4 U}{\partial x^4} + \frac{\partial^4 U}{\partial y^4} + \frac{\partial^4 U}{\partial z^4} + 2\frac{\partial^4 U}{\partial x^2 \partial y^2} + 2\frac{\partial^4 U}{\partial y^2 \partial z^2} + 2\frac{\partial^4 U}{\partial x^2 \partial z^2}$$

Miscellaneous Formulas Involving ∇

20.35. $\nabla(U + V) = \nabla U + \nabla V$

20.36. $\nabla \cdot (\mathbf{A} + \mathbf{B}) = \nabla \cdot \mathbf{A} + \nabla \cdot \mathbf{B}$

20.37. $\nabla \times (\mathbf{A} + \mathbf{B}) = \nabla \times \mathbf{A} + \nabla \times \mathbf{B}$

20.38. $\nabla \cdot (U\mathbf{A}) = (\nabla U) \cdot \mathbf{A} + U(\nabla \cdot \mathbf{A})$

20.39. $\nabla \times (U\mathbf{A}) = (\nabla U) \times \mathbf{A} + U(\nabla \times \mathbf{A})$

20.40. $\nabla \cdot (\mathbf{A} \times \mathbf{B}) = \mathbf{B} \cdot (\nabla \times \mathbf{A}) - \mathbf{A} \cdot (\nabla \times \mathbf{B})$

20.41. $\nabla \times (\mathbf{A} \times \mathbf{B}) = (\mathbf{B} \cdot \nabla)\,\mathbf{A} - \mathbf{B}(\nabla \cdot \mathbf{A}) - (\mathbf{A} \cdot \nabla)\mathbf{B} + \mathbf{A}(\nabla \cdot \mathbf{B})$

20.42. $\nabla(\mathbf{A} \cdot \mathbf{B}) = (\mathbf{B} \cdot \nabla)\mathbf{A} + (\mathbf{A} \cdot \nabla)\mathbf{B} + \mathbf{B} \times (\nabla \times \mathbf{A}) + \mathbf{A} \times (\nabla \times \mathbf{B})$

20.43. $\nabla \times (\nabla U) = 0$, that is, the curl of the gradient of U is zero.

20.44. $\nabla \cdot (\nabla \times \mathbf{A}) = 0$, that is, the divergence of the curl of $\mathbf{A}$ is zero.

20.45. $\nabla \times (\nabla \times \mathbf{A}) = \nabla(\nabla \cdot \mathbf{A}) - \nabla^2 \mathbf{A}$

Integrals Involving Vectors

If $\mathbf{A}(u) = \dfrac{d}{du}\mathbf{B}(u)$. then the *indefinite integral* of A(u) is as follows:

20.46. $\int \mathbf{A}(u)du = \mathbf{B}(u) + \mathbf{c}, \qquad c = \text{ constant vector}$

The *definite integral* of $\mathbf{A}(u)$ from $u = a$ to $u = b$ in this case is given by

20.47. $\int_a^b \mathbf{A}(u)\,du = \mathbf{B}(b) - \mathbf{B}(a)$

The definite integral can be defined as in 18.1.

Line Integrals

Consider a space curve C joining two points $P_1(a_1, a_2, a_3)$ and $P_2(b_1, b_2, b_3)$ as in Fig. 20-6. Divide the curve into n parts by points of subdivision $(x_1, y_1, z_1), \ldots, (x_{n-1}, y_{n-1}, z_{n-1})$. Then the *line integral* of a vector $\mathbf{A}(x, y, z)$ along C is defined as

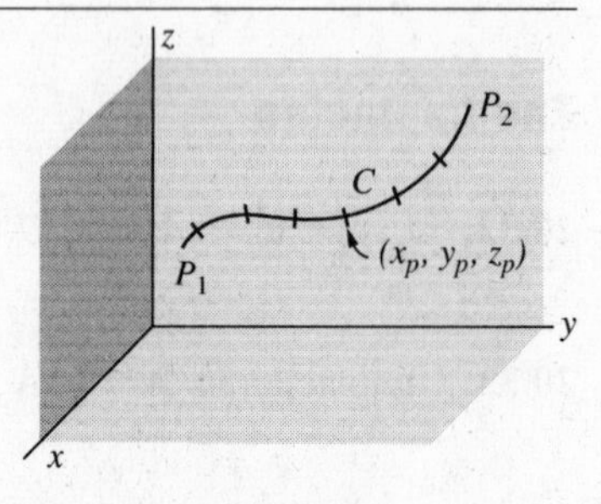

Fig. 20-6

20.48. $\int_C \mathbf{A} \cdot d\mathbf{r} = \int_{P_1}^{P_2} \mathbf{A} \cdot d\mathbf{r} = \lim_{n\to\infty} \sum_{p=1}^{n} \mathbf{A}(x_p, y_p, z_p) \cdot \Delta\mathbf{r}_p$

where $\Delta\mathbf{r}_p = \Delta x_p\mathbf{i} + \Delta y_p\mathbf{j} + \Delta z_p\mathbf{k}$, $\Delta x_p = x_{p+1} - x_p$, $\Delta y_p = y_{p+1} - y_p$, $\Delta z_p = z_{p+1} - z_p$ and where it is assumed that as $n \to \infty$ the largest of the magnitudes $|\Delta\mathbf{r}_p|$ approaches zero. The result 20.48 is a generalization of the ordinary definite integral (see 18.1).

The line integral 20.48 can also be written as

20.49. $\int_C \mathbf{A} \cdot d\mathbf{r} = \int_C (A_1 dx + A_2 dy + A_3\, dz)$

using $\mathbf{A} = A_1\mathbf{i} + A_2\mathbf{j} + A_3\mathbf{k}$ and $d\mathbf{r} = dx\mathbf{i} + dy\mathbf{j} + dz\mathbf{k}$.

Properties of Line Integrals

20.50. $\int_{P_1}^{P_2} \mathbf{A} \cdot d\mathbf{r} = -\int_{P_2}^{P_1} \mathbf{A} \cdot d\mathbf{r}$

20.51. $\int_{P_1}^{P_2} \mathbf{A} \cdot d\mathbf{r} = \int_{P_1}^{P_3} \mathbf{A} \cdot d\mathbf{r} + \int_{P_3}^{P_2} \mathbf{A} \cdot d\mathbf{r}$

Independence of the Path

In general, a line integral has a value that depends on the particular path C joining points P_1 and P_2 in a region $\mathcal{R}$. However, in the case of $\mathbf{A} = \nabla\phi$ or $\nabla \times \mathbf{A} = 0$ where ϕ and its partial derivatives are continuous in $\mathcal{R}$, the line integral $\int_C \mathbf{A} \cdot d\mathbf{r}$ is independent of the path. In such a case,

20.52. $\int_C \mathbf{A} \cdot d\mathbf{r} = \int_{P_1}^{P_2} \mathbf{A} \cdot d\mathbf{r} = \phi(P_2) - \phi(P_1)$

where $\phi(P_1)$ and $\phi(P_2)$ denote the values of ϕ at P_1 and P_2, respectively. In particular if C is a closed curve,

20.53. $\int_C \mathbf{A} \cdot d\mathbf{r} = \oint_C \mathbf{A} \cdot d\mathbf{r} = 0$

where the circle on the integral sign is used to emphasize that C is closed.

Multiple Integrals

Let $F(x, y)$ be a function defined in a region $\mathcal{R}$ of the xy plane as in Fig. 20-7. Subdivide the region into n parts by lines parallel to the x and y axes as indicated. Let $\Delta A_p = \Delta x_p \, \Delta y_p$ denote an area of one of these parts. Then the integral of $F(x, y)$ over $\mathcal{R}$ is defined as

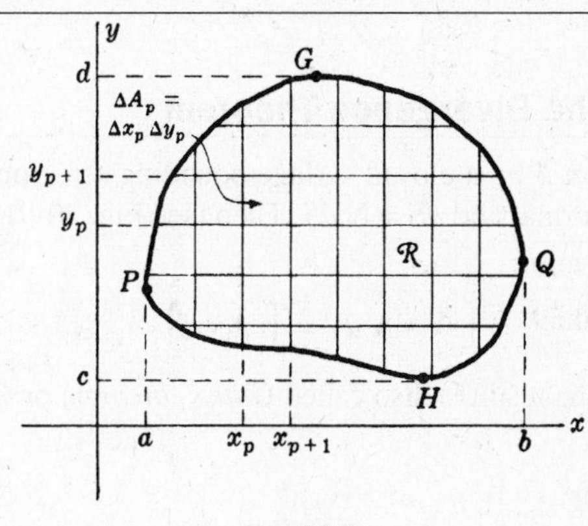

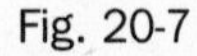
Fig. 20-7

20.54. $\int_{\mathcal{R}} F(x, y)\, dA = \lim_{n \to \infty} \sum_{p=1}^{n} F(x_p, y_p) \Delta A_p$

provided this limit exists.

In such a case, the integral can also be written as

20.55. $\int_{x=a}^{b} \int_{y=f_1(x)}^{f_2(x)} F(x, y)\, dy\, dx$

$$= \int_{x=a}^{b} \left\{ \int_{y=f_1(x)}^{f_2(x)} F(x, y)\, dy \right\} dx$$

where $y = f_1(x)$ and $y = f_2(x)$ are the equations of curves PHQ and PGQ, respectively, and a and b are the x coordinates of points P and Q. The result can also be written as

20.56. $\int_{y=c}^{d} \int_{x=g_1(y)}^{g_2(y)} F(x, y)\, dx\, dy = \int_{y=c}^{d} \left\{ \int_{x=g_1(y)}^{g_2(y)} F(x, y)\, dx \right\} dy$

where $x = g_1(y)$, $x = g_2(y)$ are the equations of curves HPG and HQG, respectively, and c and d are the y coordinates of H and G.

These are called *double integrals* or *area integrals.* The ideas can be similarly extended to *triple* or *volume integrals* or to higher *multiple integrals.*

Surface Integrals

Subdivide the surface S (see Fig. 20-8) into n elements of area ΔS_p, $p = 1, 2, \ldots, n$, Let $\mathbf{A}(x_p, y_p, z_p) = \mathbf{A}_p$ where (x_p, y_p, z_p) is a point P in ΔS_p. Let $\mathbf{N}_p$ be a unit normal to ΔS_p at P. Then the surface integral of the normal component of $\mathbf{A}$ over S is defined as

20.57. $\int_S \mathbf{A} \cdot \mathbf{N}\, dS = \lim_{n \to \infty} \sum_{p=1}^{n} \mathbf{A}_p \cdot \mathbf{N}_p \Delta S_p$

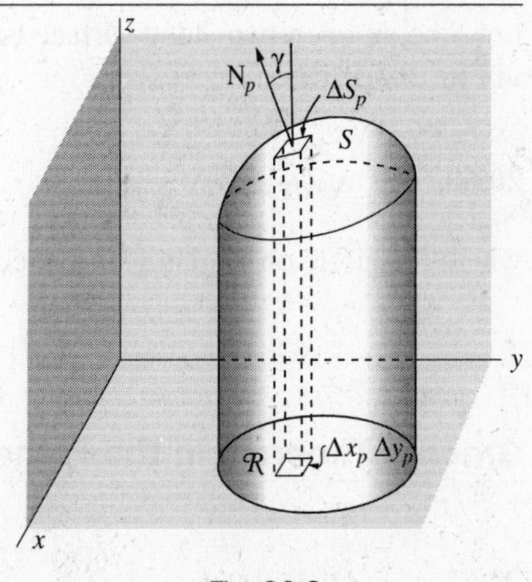

Fig. 20-8

Relation Between Surface and Double Integrals

If $\mathcal{R}$ is the projection of S on the xy plane, then (see Fig. 20-8)

20.58. $$\int_S \mathbf{A} \cdot \mathbf{N}\, dS = \int_{\mathcal{R}} \int \mathbf{A} \cdot \mathbf{N} \frac{dx\, dy}{|\mathbf{N} \cdot \mathbf{k}|}$$

The Divergence Theorem

Let S be a closed surface bounding a region of volume V; and suppose $\mathbf{N}$ is the positive (outward drawn) normal and $d\mathbf{S} = \mathbf{N}\, dS$. Then (see Fig. 20-9)

20.59. $$\int_V \nabla \cdot \mathbf{A}\, dV = \int_S \mathbf{A} \cdot d\mathbf{S}$$

The result is also called *Gauss' theorem* or *Green's theorem.*

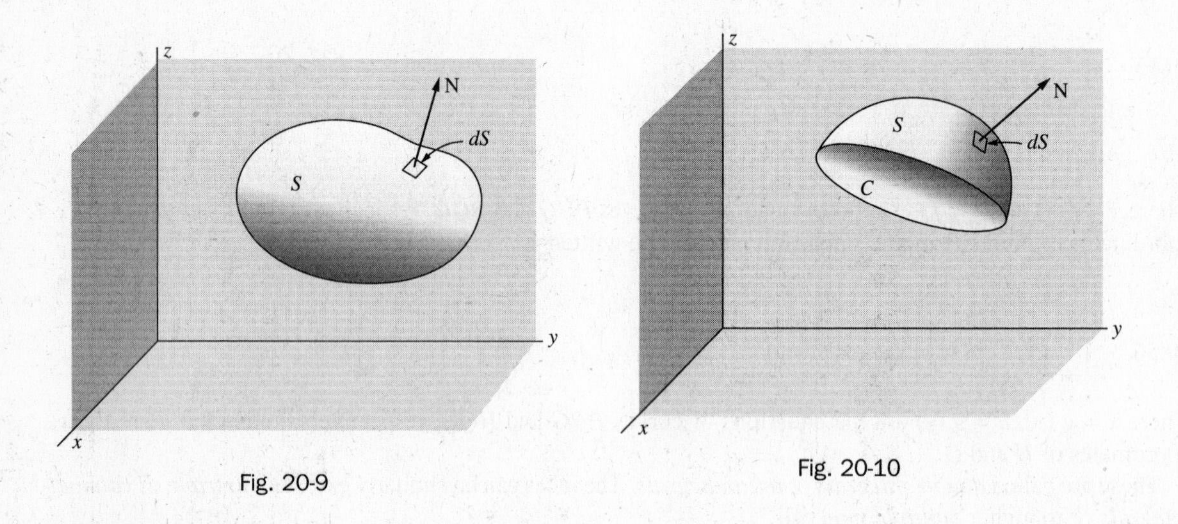

Fig. 20-9

Fig. 20-10

Stokes' Theorem

Let S be an open two-sided surface bounded by a closed non-intersecting curve C(simple closed curve) as in Fig. 20-10. Then

20.60. $$\oint_C \mathbf{A} \cdot d\mathbf{r} = \int_S (\nabla \times \mathbf{A}) \cdot d\mathbf{S}$$

where the circle on the integral is used to emphasize that C is closed.

Green's Theorem in the Plane

20.61. $$\oint_C (P\, dx + Q\, dy) = \int_R \left(\frac{\partial Q}{\partial x} - \frac{\partial P}{\partial y} \right) dx\, dy$$

where R is the area bounded by the closed curve C. This result is a special case of the divergence theorem or Stokes' theorem.

Green's First Identity

20.62. $\int_V \{(\phi \nabla^2 \psi + (\nabla \phi) \cdot (\nabla \psi)\} dV = \int (\phi \nabla \psi) \cdot d\mathbf{S}$

where ϕ and ψ are scalar functions.

Green's Second Identity

20.63. $\int_V (\phi \nabla^2 \psi - \psi \nabla^2 \phi)\, dV = \int_S (\phi \nabla \psi - \psi \nabla \phi) \cdot d\mathbf{S}$

Miscellaneous Integral Theorems

20.64. $\int_V \nabla \times \mathbf{A}\, dV = \int_S dS \times \mathbf{A}$

20.65. $\int_C \phi\, d\mathbf{r} = \int_S d\mathbf{S} \times \nabla \phi$

Curvilinear Coordinates

A point P in space (see Fig. 20-11) can be located by rectangular coordinates $(x, y, z,)$ or curvilinear coordinates (u_1, u_2, u_3) where the transformation equations from one set of coordinates to the other are given by

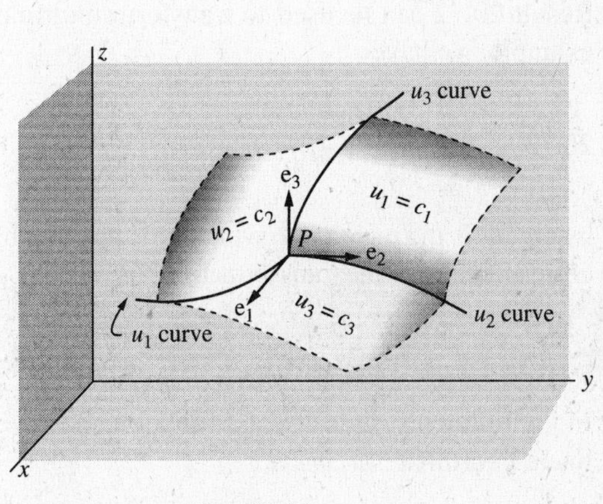

Fig. 20-11

20.66. $x = x(u_1, u_2, u_3)$

$y = y(u_1, u_2, u_3)$

$z = z(u_1, u_2, u_3)$

If u_2 and u_3 are constant, then as u_1 varies, the position vector $\mathbf{r} = x\mathbf{i} + y\mathbf{j} + z\mathbf{k}$ of P describes a curve called the u_1 coordinate curve. Similarly, we define the u_2 and u_3 coordinate curves through P. The vectors $\partial \mathbf{r}/\partial u_1$, $\partial \mathbf{r}/\partial u_2$, $\partial \mathbf{r}/\partial u_3$ represent tangent vectors to the u_1, u_2, u_3 coordinate curves. Letting $\mathbf{e}_1$, $\mathbf{e}_2$, $\mathbf{e}_3$ be unit tangent vectors to these curves, we have

20.67. $\dfrac{\partial \mathbf{r}}{\partial u_1} = h_1 \mathbf{e}_1, \quad \dfrac{\partial \mathbf{r}}{\partial u_2} = h_2 \mathbf{e}_2, \quad \dfrac{\partial \mathbf{r}}{\partial u_3} = h_3 \mathbf{e}_3$

where

20.68. $h_1 = \left| \dfrac{\partial \mathbf{r}}{\partial u_1} \right|, \quad h_2 = \left| \dfrac{\partial \mathbf{r}}{\partial u_2} \right|, \quad h_3 = \left| \dfrac{\partial \mathbf{r}}{\partial u_3} \right|$

are called *scale factors*. If $\mathbf{e}_1$, $\mathbf{e}_2$, $\mathbf{e}_3$ are mutually perpendicular, the curvilinear coordinate system is called *orthogonal*.

Formulas Involving Orthogonal Curvilinear Coordinates

20.69. $d\mathbf{r} = \frac{\partial \mathbf{r}}{\partial u_1} du_1 + \frac{\partial \mathbf{r}}{\partial u_2} du_2 + \frac{\partial \mathbf{r}}{\partial u_3} du_3 = h_1\, du_1 \mathbf{e}_1 + h_2 du_2\, \mathbf{e}_2 + h_3 du_3\, \mathbf{e}_3$

20.70. $ds^2 = d\mathbf{r} \cdot d\mathbf{r} = h_1^2\, du_1^2 + h_2^2\, du_2^2 + h_3^2\, du_3^2$

where ds is the element of are length.

If dV is the element of volume, then

20.71. $dV = |(h_1 \mathbf{e}_1 du_1) \cdot (h_2 \mathbf{e}_2 du_2) \times (h_3 \mathbf{e}_3 du_3)| = h_1 h_2 h_3\, du_1 du_2 du_3$

$$= \left| \frac{\partial \mathbf{r}}{\partial u_1} \cdot \frac{\partial \mathbf{r}}{\partial u_2} \times \frac{\partial \mathbf{r}}{\partial u_3} \right| du_1\, du_2\, du_3 = \left| \frac{\partial(x, y, z)}{\partial(u_1, u_2, u_3)} \right| du_1\, du_2\, du_3$$

where

20.72. $\frac{\partial(x, y, z)}{\partial(u_1, u_2, u_3)} = \begin{vmatrix} \partial x/\partial u_1 & \partial x/\partial u_2 & \partial x/\partial u_3 \\ \partial y/\partial u_1 & \partial y/\partial u_2 & \partial y/\partial u_3 \\ \partial z/\partial u_1 & \partial z/\partial u_2 & \partial z/du_3 \end{vmatrix}$

sometimes written $J(x, y, z; u_1, u_2, u_3)$, is called the *Jacobian* of the transformation.

Transformation of Multiple Integrals

Result 20.72 can be used to transform multiple integrals from rectangular to curvilinear coordinates. For example, we have

20.73. $\iiint_{\mathscr{R}} F(x, y, z)\, dx\, dy\, dz = \iiint_{\mathscr{R}'} G(u_1, u_2, u_3) \left| \frac{\partial(x, y, z)}{\partial(u_1, u_2, u_3)} \right| du_1\, du_2\, du_3$

where $\mathscr{R}'$ is the region into which $\mathscr{R}$ is mapped by the transformation and $G(u_1, u_2, u_3)$ is the value of $F(x, y, z)$ corresponding to the transformation.

Gradient, Divergence, Curl, and Laplacian

In the following, Φ is a scalar function and $\mathbf{A} = A_1\mathbf{e}_1 + A_2\mathbf{e}_2 + A_3\mathbf{e}_3$ is a vector function of orthogonal curvilinear coordinates u_1, u_2, u_3.

20.74. Gradient of $\Phi = \text{grad}\, \Phi = \nabla\Phi = \frac{\mathbf{e}_1}{h_1} \frac{\partial \Phi}{\partial u_1} + \frac{\mathbf{e}_2}{h_2} \frac{\partial \Phi}{du_2} + \frac{\mathbf{e}_3}{h_3} \frac{\partial \Phi}{\partial u_3}$

20.75. Divergence of $\mathbf{A} = \text{div}\, \mathbf{A} = \nabla \cdot \mathbf{A} = \frac{1}{h_1 h_2 h_3} \left[\frac{\partial}{\partial u_1}(h_2 h_3 A_1) + \frac{\partial}{\partial u_2}(h_3 h_1 A_2) + \frac{\partial}{\partial u_3}(h_1 h_2 A_3) \right]$

20.76. Curl of $\mathbf{A} = \text{curl}\, \mathbf{A} = \nabla \times \mathbf{A} = \frac{1}{h_1 h_2 h_3} \begin{vmatrix} h_1 \mathbf{e}_1 & h_2 \mathbf{e}_2 & h_3 \mathbf{e}_3 \\ \frac{\partial}{\partial u_1} & \frac{\partial}{\partial u_2} & \frac{\partial}{\partial u_3} \\ h_1 A_1 & h_2 A_2 & h_3 A_3 \end{vmatrix}$

$$= \frac{1}{h_2 h_3} \left[\frac{\partial}{\partial u_2}(h_3 A_3) - \frac{\partial}{\partial u_3}(h_2 A_2) \right] \mathbf{e}_1 + \frac{1}{h_1 h_3} \left[\frac{\partial}{\partial u_3}(h_1 A_1) - \frac{\partial}{\partial u_1}(h_3 A_3) \right] \mathbf{e}_2$$

$$+ \frac{1}{h_1 h_2} \left[\frac{\partial}{\partial u_1}(h_2 A_2) - \frac{\partial}{\partial u_2}(h_1 A_1) \right] \mathbf{e}_3$$

20.77. Laplacian of $\Phi = \nabla^2\Phi = \dfrac{1}{h_1h_2h_3}\left[\dfrac{\partial}{\partial u_1}\left(\dfrac{h_2h_3}{h_1}\dfrac{\partial\Phi}{\partial u_1}\right) + \dfrac{\partial}{\partial u_2}\left(\dfrac{h_3h_1}{h_2}\dfrac{\partial\Phi}{\partial u_2}\right) + \dfrac{\partial}{\partial u_3}\left(\dfrac{h_1h_2}{h_3}\dfrac{\partial\Phi}{\partial u_3}\right)\right]$

Note that the biharmonic operator $\nabla^4\Phi = \nabla^2(\nabla^2\Phi)$ can be obtained from 20.77.

Special Orthogonal Coordinate Systems

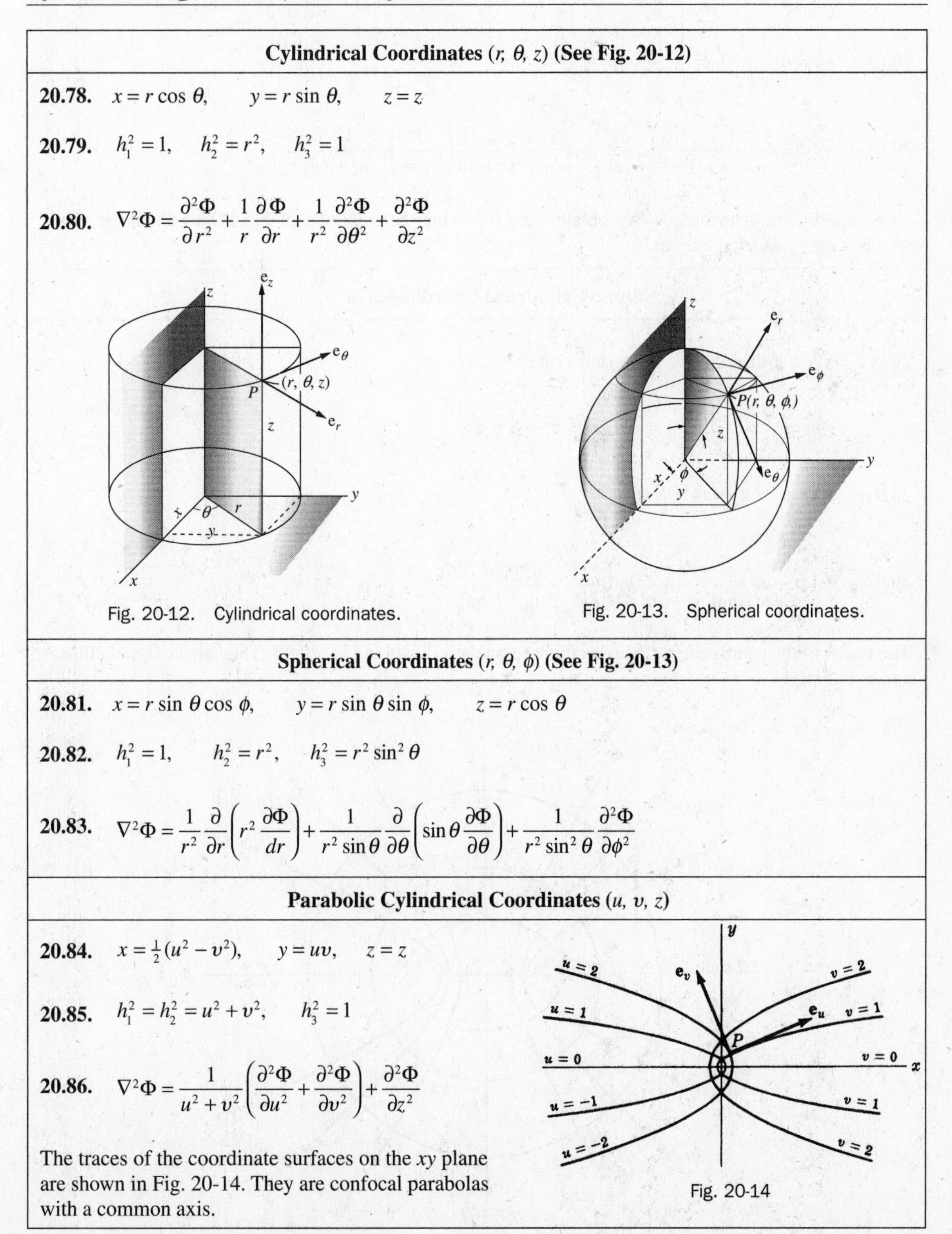

Cylindrical Coordinates (r, θ, z) (See Fig. 20-12)

20.78. $x = r\cos\theta, \quad y = r\sin\theta, \quad z = z$

20.79. $h_1^2 = 1, \quad h_2^2 = r^2, \quad h_3^2 = 1$

20.80. $\nabla^2\Phi = \dfrac{\partial^2\Phi}{\partial r^2} + \dfrac{1}{r}\dfrac{\partial\Phi}{\partial r} + \dfrac{1}{r^2}\dfrac{\partial^2\Phi}{\partial\theta^2} + \dfrac{\partial^2\Phi}{\partial z^2}$

Fig. 20-12. Cylindrical coordinates.

Fig. 20-13. Spherical coordinates.

Spherical Coordinates (r, θ, ϕ) (See Fig. 20-13)

20.81. $x = r\sin\theta\cos\phi, \quad y = r\sin\theta\sin\phi, \quad z = r\cos\theta$

20.82. $h_1^2 = 1, \quad h_2^2 = r^2, \quad h_3^2 = r^2\sin^2\theta$

20.83. $\nabla^2\Phi = \dfrac{1}{r^2}\dfrac{\partial}{\partial r}\left(r^2\dfrac{\partial\Phi}{dr}\right) + \dfrac{1}{r^2\sin\theta}\dfrac{\partial}{\partial\theta}\left(\sin\theta\dfrac{\partial\Phi}{\partial\theta}\right) + \dfrac{1}{r^2\sin^2\theta}\dfrac{\partial^2\Phi}{\partial\phi^2}$

Parabolic Cylindrical Coordinates (u, v, z)

20.84. $x = \frac{1}{2}(u^2 - v^2), \quad y = uv, \quad z = z$

20.85. $h_1^2 = h_2^2 = u^2 + v^2, \quad h_3^2 = 1$

20.86. $\nabla^2\Phi = \dfrac{1}{u^2+v^2}\left(\dfrac{\partial^2\Phi}{\partial u^2} + \dfrac{\partial^2\Phi}{\partial v^2}\right) + \dfrac{\partial^2\Phi}{\partial z^2}$

The traces of the coordinate surfaces on the xy plane are shown in Fig. 20-14. They are confocal parabolas with a common axis.

Fig. 20-14

Paraboloidal Coordinates (u, v, ϕ)

20.87. $x = uv\cos\phi, \quad y = uv\sin\phi, \quad z = \frac{1}{2}(u^2 - v^2)$

where $u \geqq 0, \quad v \geqq 0, \quad 0 \leqq \phi < 2\pi$

20.88. $h_1^2 = h_2^2 = u^2 + v^2, \quad h_3^2 = u^2v^2$

20.89. $$\nabla^2\Phi = \frac{1}{u(u^2+v^2)}\frac{\partial}{\partial u}\left(u\frac{\partial\Phi}{\partial u}\right) + \frac{1}{v(u^2+v^2)}\frac{\partial}{\partial v}\left(v\frac{\partial\Phi}{\partial v}\right) + \frac{1}{u^2v^2}\frac{\partial^2\Phi}{\partial\phi^2}$$

Two sets of coordinate surfaces are obtained by revolving the parabolas of Fig. 20-14 about the x axis which is then relabeled the z axis.

Elliptic Cylindrical Coordinates (u, v, z)

20.90. $x = a\cosh u\cos v, \quad y = a\sinh u\sin v, \quad z = z$

where $u \geqq 0, \quad 0 \leqq v < 2\pi, \quad -\infty < z < \infty$

20.91. $h_1^2 = h_2^2 = a^2(\sinh^2 u + \sin^2 v), \quad h_3^2 = 1$

20.92. $$\nabla^2\Phi = \frac{1}{a^2(\sinh^2 u + \sin^2 v)}\left(\frac{\partial^2\Phi}{\partial u^2} + \frac{\partial^2\Phi}{\partial v^2}\right) + \frac{\partial^2\Phi}{\partial z^2}$$

The traces of the coordinate surfaces on the xy plane are shown in Fig. 20-15. They are confocal ellipses and hyperbolas.

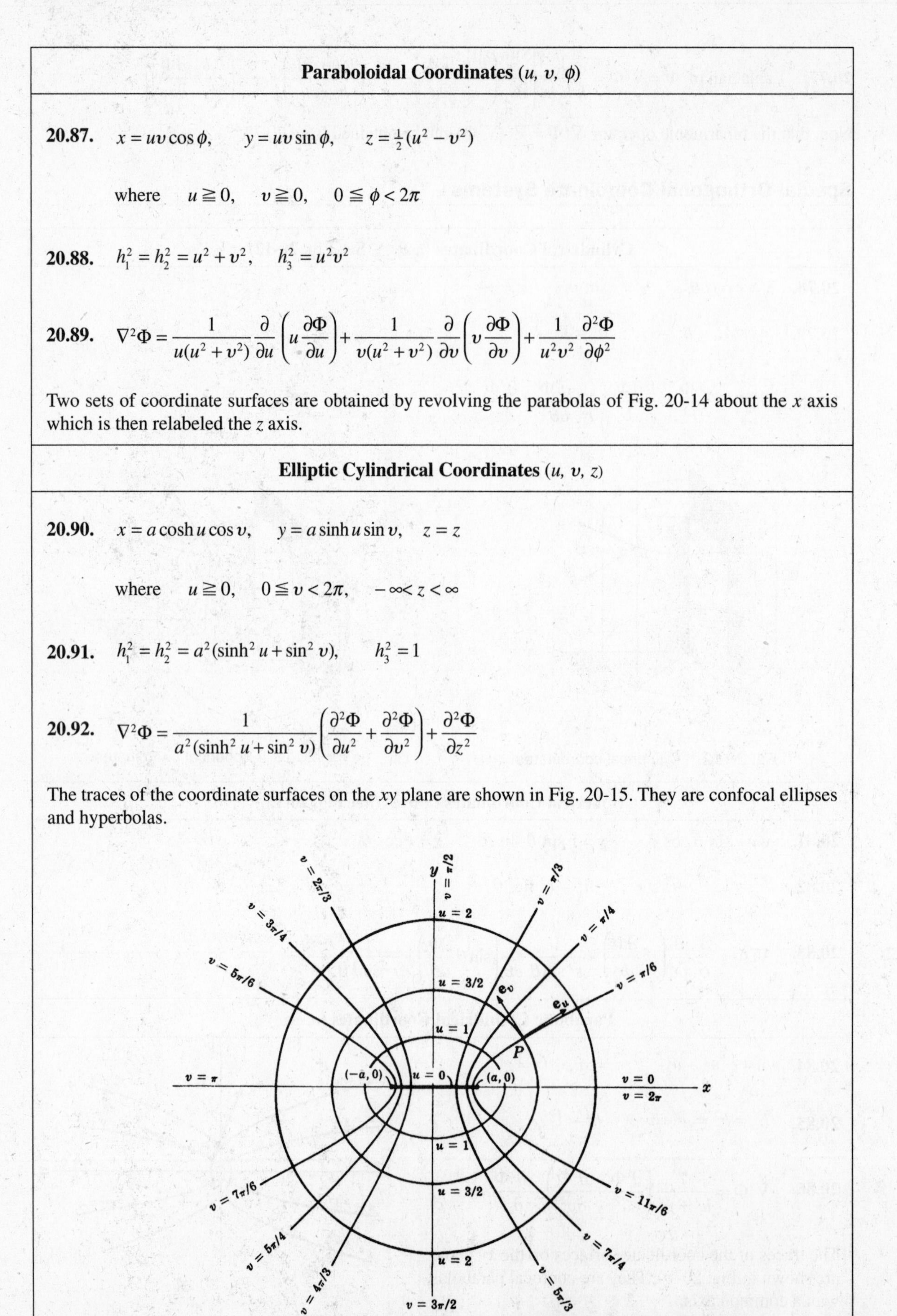

Fig. 20-15. Elliptic cylindrical coordinates.

Prolate Spheroidal Coordinates (ξ, η, ϕ)

20.93. $x = a \sinh \xi \sin \eta \cos \phi, \quad y = a \sinh \xi \sin \eta \sin \phi, \quad z = a \cosh \xi \cos \eta$

where $\quad \xi \geqq 0, \quad 0 \leqq \eta \leqq \pi, \quad 0 \leqq \phi < 2\pi$

20.94. $h_1^2 = h_2^2 = a^2(\sinh^2 \xi \sin^2 \eta), \quad h_3^2 = a^2 \sinh^2 \xi \sin^2 \eta$

20.95.
$$\nabla^2 \Phi = \frac{1}{a^2(\sinh^2 \xi + \sin^2 \eta) \sinh \xi} \frac{\partial}{\partial \xi}\left(\sinh \xi \frac{\partial \Phi}{\partial \xi}\right) + \frac{1}{a^2(\sinh^2 \xi + \sin^2 \eta) \sin \eta} \frac{\partial}{\partial \eta}\left(\sin \eta \frac{\partial \Phi}{\partial \eta}\right) + \frac{1}{a^2 \sinh^2 \xi \sin^2 \eta} \frac{\partial^2 \Phi}{\partial \phi^2}$$

Two sets of coordinate surfaces are obtained by revolving the curves of Fig. 20-15 about the x axis which is relabeled the z axis. The third set of coordinate surfaces consists of planes passing through this axis.

Oblate Spheroidal Coordinates (ξ, η, ϕ)

20.96. $x = a \cosh \xi \cos \eta \cos \phi, \quad y = a \cosh \xi \cos \eta \sin \phi, \quad z = a \sinh \xi \sin \eta$

where $\quad \xi \geqq 0, \quad -\pi/2 \leqq \eta \leqq \pi/2, \quad 0 \leqq \phi < 2\pi$

20.97. $h_1^2 = h_2^2 = a^2(\sinh^2 \xi + \sin^2 \eta), \quad h_3^2 = a^2 \cosh^2 \xi \cos^2 \eta$

20.98.
$$\nabla^2 \Phi = \frac{1}{a^2(\sinh^2 \xi + \sin^2 \eta) \cosh \xi} \frac{\partial}{\partial \xi}\left(\cosh \xi \frac{\partial \Phi}{\partial \xi}\right) + \frac{1}{a^2(\sinh^2 \xi + \sin^2 \eta) \cos \eta} \frac{\partial}{\partial \eta}\left(\cos \eta \frac{\partial \Phi}{\partial \eta}\right) + \frac{1}{a^2 \cosh^2 \xi \cos^2 \eta} \frac{\partial^2 \Phi}{\partial \phi^2}$$

Two sets of coordinate surfaces are obtained by revolving the curves of Fig. 20-15 about the y axis which is relabeled the z axis. The third set of coordinate surfaces are planes passing through this axis.

Bipolar Coordinates (u, v, z)

20.99. $x = \dfrac{a \sinh v}{\cosh v - \cos u}, \quad y = \dfrac{a \sin u}{\cosh v - \cos u}, \quad z = z$

where $\quad 0 \leqq u < 2\pi, \quad -\infty < v < \infty, \quad -\infty < z < \infty$

or

20.100. $x^2 + (y - a \cot u)^2 = a^2 \csc^2 u, \quad (x - a \coth v)^2 + y^2 = a^2 \operatorname{csch}^2 v, \quad z = z$

20.101. $h_1^2 = h_2^2 = \dfrac{a^2}{(\cosh v - \cos u)^2}, \qquad h_3^2 = 1$

20.102. $\nabla^2\Phi = \dfrac{(\cosh v - \cos u)^2}{a^2}\left(\dfrac{\partial^2\Phi}{\partial u^2} + \dfrac{\partial^2\Phi}{\partial v^2}\right) + \dfrac{\partial^2\Phi}{\partial z^2}$

The traces of the coordinate surfaces on the xy plane are shown in Fig. 20-16.

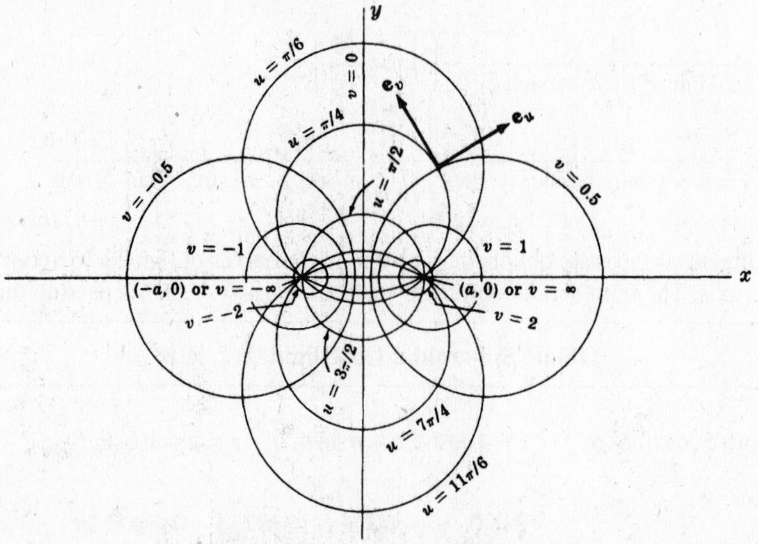

Fig. 20-16. Bipolar coordinates.

Toroidal Coordinates (u, v, ϕ)

20.103. $x = \dfrac{a \sinh v \cos\phi}{\cosh v - \cos u}, \quad y = \dfrac{a \sinh v \sin\phi}{\cosh v - \cos u}, \quad z = \dfrac{a \sin u}{\cosh v - \cos u}$

20.104. $h_1^2 = h_2^2 = \dfrac{a^2}{(\cosh v - \cos u)^2}, \quad h_3^2 = \dfrac{a^2 \sinh^2 v}{(\cosh v - \cos u)^2}$

20.105. $\nabla^2\Phi = \dfrac{(\cosh v - \cos u)^3}{a^2}\dfrac{\partial}{\partial u}\left(\dfrac{1}{\cosh v - \cos u}\dfrac{\partial\Phi}{\partial u}\right)$

$$+ \frac{(\cosh v - \cos u)^3}{a^2 \sinh v}\frac{\partial}{\partial v}\left(\frac{\sinh v}{\cosh v - \cos u}\frac{\partial\Phi}{\partial v}\right) + \frac{(\cosh v - \cos u)^2}{a^2 \sinh^2 v}\frac{\partial^2\Phi}{\partial\phi^2}$$

The coordinate surfaces are obtained by revolving the curves of Fig. 20.16 about the y axis which is relabeled the z axis.

Conical Coordinates (λ, μ, ν)

20.106. $x = \dfrac{\lambda\mu\nu}{ab}, \quad y = \dfrac{\lambda}{a}\sqrt{\dfrac{(\mu^2 - a^2)(\nu^2 - a^2)}{a^2 - b^2}}, \quad z = \dfrac{\lambda}{b}\sqrt{\dfrac{(\mu^2 - b^2)(\nu^2 - b^2)}{b^2 - a^2}}$

20.107. $h_1^2 = 1, \quad h_2^2 = \dfrac{\lambda^2(\mu^2 - \nu^2)}{(\mu^2 - a^2)(b^2 - \mu^2)}, \quad h_3^2 = \dfrac{\lambda^2(\mu^2 - \nu^2)}{(\nu^2 - a^2)(\nu^2 - b^2)}$

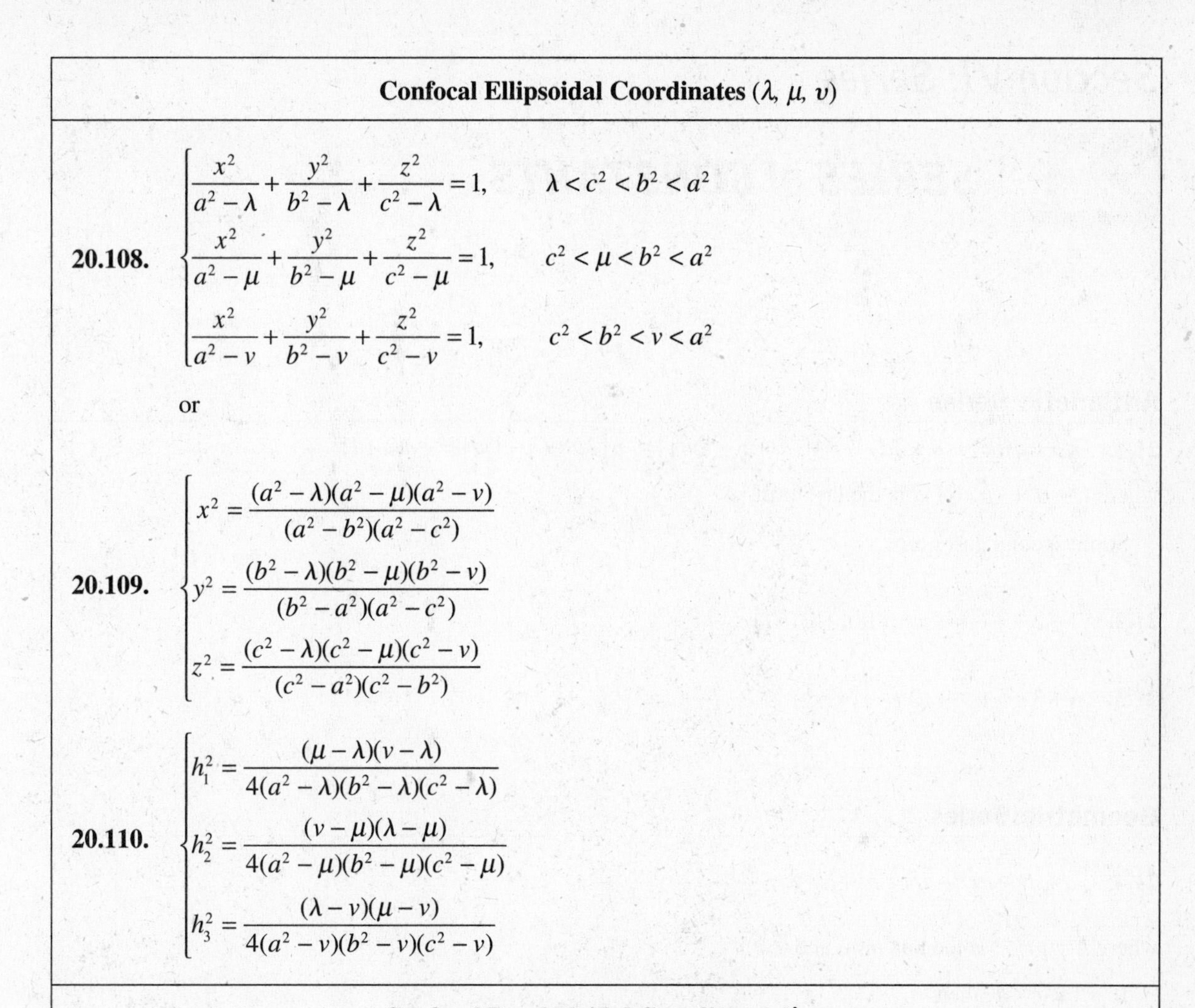

Confocal Ellipsoidal Coordinates (λ, μ, v)

20.108.
$$\begin{cases} \dfrac{x^2}{a^2-\lambda}+\dfrac{y^2}{b^2-\lambda}+\dfrac{z^2}{c^2-\lambda}=1, & \lambda < c^2 < b^2 < a^2 \\ \dfrac{x^2}{a^2-\mu}+\dfrac{y^2}{b^2-\mu}+\dfrac{z^2}{c^2-\mu}=1, & c^2 < \mu < b^2 < a^2 \\ \dfrac{x^2}{a^2-v}+\dfrac{y^2}{b^2-v}+\dfrac{z^2}{c^2-v}=1, & c^2 < b^2 < v < a^2 \end{cases}$$

or

20.109.
$$\begin{cases} x^2 = \dfrac{(a^2-\lambda)(a^2-\mu)(a^2-v)}{(a^2-b^2)(a^2-c^2)} \\ y^2 = \dfrac{(b^2-\lambda)(b^2-\mu)(b^2-v)}{(b^2-a^2)(a^2-c^2)} \\ z^2 = \dfrac{(c^2-\lambda)(c^2-\mu)(c^2-v)}{(c^2-a^2)(c^2-b^2)} \end{cases}$$

20.110.
$$\begin{cases} h_1^2 = \dfrac{(\mu-\lambda)(v-\lambda)}{4(a^2-\lambda)(b^2-\lambda)(c^2-\lambda)} \\ h_2^2 = \dfrac{(v-\mu)(\lambda-\mu)}{4(a^2-\mu)(b^2-\mu)(c^2-\mu)} \\ h_3^2 = \dfrac{(\lambda-v)(\mu-v)}{4(a^2-v)(b^2-v)(c^2-v)} \end{cases}$$

Confocal Paraboloidal Coordinates (λ, μ, v)

20.111.
$$\begin{cases} \dfrac{x^2}{a^2-\lambda}+\dfrac{h^2}{b^2-\lambda}=z-\lambda, & -\infty < \lambda < b^2 \\ \dfrac{x^2}{a^2-\mu}+\dfrac{y^2}{b^2-\mu}=z-\mu, & b^2 < \mu < a^2 \\ \dfrac{x^2}{a^2-v}+\dfrac{y^2}{b^2-v}=z-v, & a^2 < v < \infty \end{cases}$$

or

20.112.
$$\begin{cases} x^2 = \dfrac{(a^2-\lambda)(a^2-\mu)(a^2-v)}{b^2-a^2} \\ y^2 = \dfrac{(b^2-\lambda)(b^2-\mu)(b^2-v)}{a^2-b^2} \\ z = \lambda+\mu+v-a^2-b^2 \end{cases}$$

20.113.
$$\begin{cases} h_1^2 = \dfrac{(\mu-\lambda)(v-\lambda)}{4(a^2-\lambda)(b^2-\lambda)} \\ h_2^2 = \dfrac{(v-\mu)(\lambda-\mu)}{4(a^2-\mu)(b^2-\mu)} \\ h_3^2 = \dfrac{(\lambda-v)(\mu-v)}{16(a^2-v)(b^2-v)} \end{cases}$$

Section VI: Series

21 SERIES of CONSTANTS

Arithmetic Series

21.1. $a+(a+d)+(a+2d)+\cdots+\{a+(n-1)d\}=\frac{1}{2}n\{2a+(n-1)d\}=\frac{1}{2}n(a+l)$

where $l = a + (n-1)d$ is the last term.

Some special cases are

21.2. $1+2+3+\cdots+n=\frac{1}{2}n(n+1)$

21.3. $1+3+5+\cdots+(2n-1)=n^2$

Geometric Series

21.4. $a+ar+ar^2+ar^3+\cdots+ar^{n-1}=\dfrac{a(1-r^n)}{1-r}=\dfrac{a-rl}{1-r}$

where $l = ar^{n-1}$ is the last term and $r \neq 1$.

If $-1<r<1$, then

21.5. $a+ar+ar^2+ar^3+\cdots=\dfrac{a}{1-r}$

Arithmetic-Geometric Series

21.6. $a+(a+d)r+(a+2d)r^2+\cdots+\{a+(n-1)d\}r^{n-1}=\dfrac{a(1-r^n)}{1-r}+\dfrac{rd\{1-nr^{n-1}+(n-1)r^n\}}{(1-r)^2}$

where $r \neq 1$.

If $-1<r<1$, then

21.7. $a+(a+d)r+(d+2d)r^2+\cdots=\dfrac{a}{1-r}+\dfrac{rd}{(1-r)^2}$

Sums of Powers of Positive Integers

21.8. $1^p+2^p+3^p+\cdots+n^p=\dfrac{n^{p+1}}{p+1}+\frac{1}{2}n^p+\dfrac{B_1pn^{p-1}}{2!}-\dfrac{B_2p(p-1)(p-2)n^{p-3}}{4!}+\cdots$

where the series terminates at n^2 or n according as p is odd or even, and B_k are the *Bernoulli numbers* (see page 142).

Some special cases are

21.9. $1+2+3+\cdots+n=\dfrac{n(n+1)}{2}$

21.10. $1^2+2^2+3^2+\cdots+n^2=\dfrac{n(n+1)(2n+1)}{6}$

21.11. $1^3+2^3+3^3+\cdots+n^3=\dfrac{n^2(n+1)^2}{4}=(1+2+3+\cdots+n)^2$

21.12. $1^4+2^4+3^4+\cdots+n^4=\dfrac{n(n+1)(2n+1)(3n^2+3n-1)}{30}$

If $S_k=1^k+2^k+3^k+\cdots+n^k$ where k and n are positive integers, then

21.13. $\binom{k+1}{1}S_1+\binom{k+1}{2}S_2+\cdots+\binom{k+1}{k}S_k=(n+1)^{k+1}-(n+1)$

Series Involving Reciprocals of Powers of Positive Integers

21.14. $1-\dfrac{1}{2}+\dfrac{1}{3}-\dfrac{1}{4}+\dfrac{1}{5}-\cdots=\ln 2$

21.15. $1-\dfrac{1}{3}+\dfrac{1}{5}-\dfrac{1}{7}+\dfrac{1}{9}-\cdots=\dfrac{\pi}{4}$

21.16. $1-\dfrac{1}{4}+\dfrac{1}{7}-\dfrac{1}{10}+\dfrac{1}{13}-\cdots=\dfrac{\pi\sqrt{3}}{9}+\dfrac{1}{3}\ln 2$

21.17. $1-\dfrac{1}{5}+\dfrac{1}{9}-\dfrac{1}{13}+\dfrac{1}{17}-\cdots=\dfrac{\pi\sqrt{2}}{8}+\dfrac{\sqrt{2}\ln(1+\sqrt{2})}{4}$

21.18. $\dfrac{1}{2}-\dfrac{1}{5}+\dfrac{1}{8}-\dfrac{1}{11}+\dfrac{1}{14}-\cdots=\dfrac{\pi\sqrt{3}}{9}+\dfrac{1}{3}\ln 2$

21.19. $\dfrac{1}{1^2}+\dfrac{1}{2^2}+\dfrac{1}{3^2}+\dfrac{1}{4^2}+\cdots=\dfrac{\pi^2}{6}$

21.20. $\dfrac{1}{1^4}+\dfrac{1}{2^4}+\dfrac{1}{3^4}+\dfrac{1}{4^4}+\cdots=\dfrac{\pi^4}{90}$

21.21. $\dfrac{1}{1^6}+\dfrac{1}{2^6}+\dfrac{1}{3^6}+\dfrac{1}{4^6}+\cdots=\dfrac{\pi^6}{945}$

21.22. $\dfrac{1}{1^2}-\dfrac{1}{2^2}+\dfrac{1}{3^2}-\dfrac{1}{4^2}+\cdots=\dfrac{\pi^2}{12}$

21.23. $\dfrac{1}{1^4}-\dfrac{1}{2^4}+\dfrac{1}{3^4}-\dfrac{1}{4^4}+\cdots=\dfrac{7\pi^4}{720}$

21.24. $\dfrac{1}{1^6}-\dfrac{1}{2^6}+\dfrac{1}{3^6}-\dfrac{1}{4^6}+\cdots=\dfrac{31\pi^6}{30{,}240}$

21.25. $\dfrac{1}{1^2}+\dfrac{1}{3^2}+\dfrac{1}{5^2}+\dfrac{1}{7^2}+\cdots=\dfrac{\pi^2}{8}$

21.26. $\dfrac{1}{1^4}+\dfrac{1}{3^4}+\dfrac{1}{5^4}+\dfrac{1}{7^4}+\cdots=\dfrac{\pi^4}{96}$

21.27. $\frac{1}{1^6}+\frac{1}{3^6}+\frac{1}{5^6}+\frac{1}{7^6}+\cdots=\frac{\pi^6}{960}$

21.28. $\frac{1}{1^3}-\frac{1}{3^3}+\frac{1}{5^3}-\frac{1}{7^3}+\cdots=\frac{\pi^3}{32}$

21.29. $\frac{1}{1^3}+\frac{1}{3^3}-\frac{1}{5^3}-\frac{1}{7^3}+\cdots=\frac{3\pi^3\sqrt{2}}{128}$

21.30. $\frac{1}{1\cdot 3}+\frac{1}{3\cdot 5}+\frac{1}{5\cdot 7}+\frac{1}{7\cdot 9}+\cdots=\frac{1}{2}$

21.31. $\frac{1}{1\cdot 3}+\frac{1}{2\cdot 4}+\frac{1}{3\cdot 5}+\frac{1}{4\cdot 6}+\cdots=\frac{3}{4}$

21.32. $\frac{1}{1^2\cdot 3^2}+\frac{1}{3^2\cdot 5^2}+\frac{1}{5^2\cdot 7^2}+\frac{1}{7^2\cdot 9^2}+\cdots=\frac{\pi^2-8}{16}$

21.33. $\frac{1}{1^2\cdot 2^2\cdot 3^2}+\frac{1}{2^2\cdot 3^2\cdot 4^2}+\frac{1}{3^2\cdot 4^2\cdot 5^2}+\cdots=\frac{4\pi^2-39}{16}$

21.34. $\frac{1}{a}-\frac{1}{a+d}+\frac{1}{a+2d}-\frac{1}{a+3d}+\cdots=\int_0^1\frac{u^{a-1}du}{1+u^d}$

21.35. $\frac{1}{1^{2p}}+\frac{1}{2^{2p}}+\frac{1}{3^{2p}}+\frac{1}{4^{2p}}+\cdots=\frac{2^{2p-1}\pi^{2p}B_p}{(2p)!}$

21.36. $\frac{1}{1^{2p}}+\frac{1}{3^{2p}}+\frac{1}{5^{2p}}+\frac{1}{7^{2p}}+\cdots=\frac{(2^{2p}-1)\pi^{2p}B_p}{2(2p)!}$

21.37. $\frac{1}{1^{2p}}-\frac{1}{2^{2p}}+\frac{1}{3^{2p}}-\frac{1}{4^{2p}}+\cdots=\frac{(2^{2p-1}-1)\pi^{2p}B_p}{(2p)!}$

21.38. $\frac{1}{1^{2p+1}}-\frac{1}{3^{2p+1}}+\frac{1}{5^{2p+1}}-\frac{1}{7^{2p+1}}+\cdots=\frac{\pi^{2p+1}E_p}{2^{2p+2}(2p)!}$

Miscellaneous Series

21.39. $\frac{1}{2}+\cos\alpha+\cos 2\alpha+\cdots+\cos n\alpha=\frac{\sin(n+1/2)\alpha}{2\sin(\alpha/2)}$

21.40. $\sin\alpha+\sin 2\alpha+\sin 3\alpha+\cdots+\sin n\alpha=\frac{\sin[1/2(n+1)]\alpha\sin 1/2n\alpha}{\sin(\alpha/2)}$

21.41. $1+r\cos\alpha+r^2\cos 2\alpha+r^3\cos 3\alpha+\cdots=\frac{1-r\cos\alpha}{1-2r\cos\alpha+r^2},\ |r|<1$

21.42. $r\sin\alpha+r^2\sin 2\alpha+r^3\sin 3\alpha+\cdots=\frac{r\sin\alpha}{1-2r\cos\alpha+r^2},\ |r|<1$

21.43. $1+r\cos\alpha+r^2\cos 2\alpha+\cdots+r^n\cos n\alpha=\frac{r^{n+2}\cos n\alpha-r^{n+1}\cos(n+1)\alpha-r\cos\alpha+1}{1-2r\cos\alpha+r^2}$

21.44. $r\sin\alpha+r^2\sin 2\alpha+\cdots+r^n\sin n\alpha=\frac{r\sin\alpha-r^{n+1}\sin(n+1)\alpha+r^{n+2}\sin n\alpha}{1-2r\cos\alpha+r^2}$

The Euler-Maclaurin Summation Formula

21.45. $$\sum_{k=1}^{n-1} F(k) = \int_0^n F(k)dk - \frac{1}{2}\{F(0) + F(n)\}$$

$$+\frac{1}{12}\{F'(n) - F(0)\} - \frac{1}{720}\{F'''(n) - F'''(0)\}$$

$$+\frac{1}{30,240}\{F^{(\text{v})}(n) - F^{(\text{v})}(0)\} - \frac{1}{1,209,600}\{F^{(\text{vii})}(n) - F^{(\text{vii})}(0)\}$$

$$+\cdots(-1)^{p-1}\frac{B_p}{(2p)!}\{F^{(2p-1)}(n) - F^{(2p-1)}(0)\} + \cdots$$

The Poisson Summation Formula

21.46. $$\sum_{k=-\infty}^{\infty} F(k) = \sum_{m=-\infty}^{\infty}\left\{\int_{-\infty}^{\infty} e^{2\pi imx} F(x)dx\right\}$$

22 TAYLOR SERIES

Taylor Series for Functions of One Variable

22.1. $f(x)=f(a)+f'(a)(x-a)+\dfrac{f''(a)(x-a)^2}{2!}+\cdots+\dfrac{f^{(n-1)}(a)(x-a)^{n-1}}{(n-1)!}+R_n$

where R_n, the remainder after n terms, is given by either of the following forms:

22.2. **Lagrange's form:** $R_n=\dfrac{f^{(n)}(\xi)(x-a)^n}{n!}$

22.3. **Cauchy's form:** $R_n=\dfrac{f^{(n)}(\xi)(x-\xi)^{n-1}(x-a)}{(n-1)!}$

The value ξ, which may be different in the two forms, lies between a and x. The result holds if $f(x)$ has continuous derivatives of order n at least.

If $\lim_{n\to\infty} R_n=0$, the infinite series obtained is called the *Taylor series* for $f(x)$ about $x = a$. If $a = 0$, the series is often called a *Maclaurin series*. These series, often called power series, generally converge for all values of x in some interval called the *interval of convergence* and diverge for all x outside this interval.

Some series contain the Bernoulli numbers B_n and the Euler numbers E_n defined in Chapter 23, pages 142–143.

Binomial Series

22.4. $(a+x)^n=a^n+na^{n-1}x+\dfrac{n(n-1)}{2!}a^{n-2}x^2+\dfrac{n(n-1)(n-2)}{3!}a^{n-3}x^3+\cdots$

$$=a^n+\binom{n}{1}a^{n-1}x+\binom{n}{2}a^{n-2}x^2+\binom{n}{3}a^{n-3}x^3+\cdots$$

Special cases are

22.5. $(a+x)^2=a^2+2ax+x^2$

22.6. $(a+x)^3=a^3+3a^2x+3ax^2+x^3$

22.7. $(a+x)^4=a^4+4a^3x+6a^2x^2+4ax^3+x^4$

22.8. $(1+x)^{-1}=1-x+x^2-x^3+x^4-\cdots \qquad -1<x<1$

22.9. $(1+x)^{-2}=1-2x+3x^2-4x^3+5x^4-\cdots \qquad -1<x<1$

22.10. $(1+x)^{-3}=1-3x+6x^2-10x^3+15x^4-\cdots \qquad -1<x<1$

22.11. $(1+x)^{-1/2} = 1 - \frac{1}{2}x + \frac{1 \cdot 3}{2 \cdot 4}x^2 - \frac{1 \cdot 3 \cdot 5}{2 \cdot 4 \cdot 6}x^3 + \cdots \qquad -1 < x \leqq 1$

22.12. $(1+x)^{1/2} = 1 + \frac{1}{2}x - \frac{1}{2 \cdot 4}x^2 + \frac{1 \cdot 3}{2 \cdot 4 \cdot 6}x^3 - \cdots \qquad -1 < x \leqq 1$

22.13. $(1+x)^{-1/3} = 1 - \frac{1}{3}x + \frac{1 \cdot 4}{3 \cdot 6}x^2 - \frac{1 \cdot 4 \cdot 7}{3 \cdot 6 \cdot 9}x^3 + \cdots \qquad -1 < x \leqq 1$

22.14. $(1+x)^{1/3} = 1 + \frac{1}{3}x - \frac{2}{3 \cdot 6}x^2 + \frac{2 \cdot 5}{3 \cdot 6 \cdot 9}x^3 - \cdots \qquad -1 < x \leqq 1$

Series for Exponential and Logarithmic Functions

22.15. $e^x = 1 + x + \frac{x^2}{2!} + \frac{x^3}{3!} + \cdots \qquad -\infty < x < \infty$

22.16. $a^x = e^{x \ln a} = 1 + x \ln a + \frac{(x \ln a)^2}{2!} + \frac{(x \ln a)^3}{3!} + \cdots \qquad -\infty < x < \infty$

22.17. $\ln(1+x) = x - \frac{x^2}{2} + \frac{x^3}{3} - \frac{x^4}{4} + \cdots \qquad -1 < x \leqq 1$

22.18. $\frac{1}{2}\ln\left(\frac{1+x}{1-x}\right) = x + \frac{x^3}{3} + \frac{x^5}{5} + \frac{x^7}{7} + \cdots \qquad -1 < x < 1$

22.19. $\ln x = 2\left\{\left(\frac{x-1}{x+1}\right) + \frac{1}{3}\left(\frac{x-1}{x+1}\right)^3 + \frac{1}{5}\left(\frac{x-1}{x+1}\right)^5 + \cdots\right\} \qquad x > 0$

22.20. $\ln x = \left(\frac{x-1}{x}\right) + \frac{1}{2}\left(\frac{x-1}{x}\right)^2 + \frac{1}{3}\left(\frac{x-1}{x}\right)^3 + \cdots \qquad x \geqq \frac{1}{2}$

Series for Trigonometric Functions

22.21. $\sin x = x - \frac{x^3}{3!} + \frac{x^5}{5!} - \frac{x^7}{7!} + \cdots \qquad -\infty < x < \infty$

22.22. $\cos x = 1 - \frac{x^2}{2!} + \frac{x^4}{4!} - \frac{x^6}{6!} + \cdots \qquad -\infty < x < \infty$

22.23. $\tan x = x + \frac{x^3}{3} + \frac{2x^5}{15} + \frac{17x^7}{315} + \cdots + \frac{2^{2n}(2^{2n}-1)B_n x^{2n-1}}{(2n)!} + \cdots \qquad |x| < \frac{\pi}{2}$

22.24. $\cot x = \frac{1}{x} - \frac{x}{3} - \frac{x^3}{45} - \frac{2x^5}{945} - \cdots - \frac{2^{2n}B_n x^{2n-1}}{(2n)!} - \cdots \qquad 0 < |x| < \pi$

22.25. $\sec x = 1 + \frac{x^2}{2} + \frac{5x^4}{24} + \frac{61x^6}{720} + \cdots + \frac{E_n x^{2n}}{(2n)!} + \cdots \qquad |x| < \frac{\pi}{2}$

22.26. $\csc x = \frac{1}{x} + \frac{x}{6} + \frac{7x^3}{360} + \frac{31x^5}{15{,}120} + \cdots + \frac{2(2^{2n-1}-1)B_n x^{2n-1}}{(2n)!} + \cdots \qquad 0 < |x| < \pi$

22.27. $\sin^{-1} x = x + \frac{1}{2}\frac{x^3}{3} + \frac{1 \cdot 3}{2 \cdot 4}\frac{x^5}{5} + \frac{1 \cdot 3 \cdot 5}{2 \cdot 4 \cdot 6}\frac{x^7}{7} + \cdots \qquad |x| < 1$

22.28. $\cos^{-1} x = \frac{\pi}{2} - \sin^{-1} x = \frac{\pi}{2} - \left(x + \frac{1}{2}\frac{x^3}{3} + \frac{1 \cdot 3}{2 \cdot 4}\frac{x^5}{5} + \cdots\right) \qquad |x| < 1$

22.29. $$\tan^{-1} x = \begin{cases} x - \dfrac{x^3}{3} + \dfrac{x^5}{5} - \dfrac{x^7}{7} + \cdots & |x| < 1 \\ \pm\dfrac{\pi}{2} - \dfrac{1}{x} + \dfrac{1}{3x^3} - \dfrac{1}{5x^5} + \cdots & (+ \text{ if } x \geqq 1, - \text{ if } x \leqq -1) \end{cases}$$

22.30. $$\cot^{-1} x = \frac{\pi}{2} - \tan^{-1} x = \begin{cases} \dfrac{\pi}{2} - \left(x - \dfrac{x^3}{3} + \dfrac{x^5}{5} - \cdots\right) & |x| < 1 \\ p\pi + \dfrac{1}{x} - \dfrac{1}{3x^3} + \dfrac{1}{5x^5} - \cdots & (p = 0 \text{ if } x > 1, p = 1 \text{ if } x < -1) \end{cases}$$

22.31. $$\sec^{-1} x = \cos^{-1}(1/x) = \frac{\pi}{2} - \left(\frac{1}{x} + \frac{1}{2 \bullet 3x^3} + \frac{1 \cdot 3}{2 \bullet 4 \bullet 5x^5} + \cdots\right) \qquad |x| > 1$$

22.32. $$\csc^{-1} x = \sin^{-1}(1/x) = \frac{1}{x} + \frac{1}{2 \bullet 3x^3} + \frac{1 \cdot 3}{2 \bullet 4 \bullet 5x^5} + \cdots \qquad |x| > 1$$

Series for Hyperbolic Functions

22.33. $$\sinh x = x + \frac{x^3}{3!} + \frac{x^5}{5!} + \frac{x^7}{7!} + \cdots \qquad -\infty < x < \infty$$

22.34. $$\cosh x = 1 + \frac{x^2}{2!} + \frac{x^4}{4!} + \frac{x^6}{6!} + \cdots \qquad -\infty < x < \infty$$

22.35. $$\tanh x = x - \frac{x^3}{3} + \frac{2x^5}{15} - \frac{17x^7}{315} + \cdots \frac{(-1)^{n-1} 2^{2n}(2^{2n} - 1) B_n x^{2n-1}}{(2n)!} + \cdots \qquad |x| < \frac{\pi}{2}$$

22.36. $$\coth x = \frac{1}{x} + \frac{x}{3} - \frac{x^3}{45} + \frac{2x^5}{945} + \cdots \frac{(-1)^{n-1} 2^{2n} B_n x^{2n-1}}{(2n)!} + \cdots \qquad 0 < |x| < \pi$$

22.37. $$\operatorname{sech} x = 1 - \frac{x^2}{2} + \frac{5x^4}{24} - \frac{61x^6}{720} + \cdots \frac{(-1)^n E_n x^{2n}}{(2n)!} + \cdots \qquad |x| < \frac{\pi}{2}$$

22.38. $$\operatorname{csch} x = \frac{1}{x} - \frac{x}{6} + \frac{7x^3}{360} - \frac{31x^5}{15{,}120} + \cdots \frac{(-1)^n 2(2^{2n-1} - 1) B_n x^{2n-1}}{(2n)!} + \cdots \qquad 0 < |x| < \pi$$

22.39. $$\sinh^{-1} x = \begin{cases} x - \dfrac{x^3}{2 \bullet 3} + \dfrac{1 \bullet 3x^5}{2 \bullet 4 \bullet 5} - \dfrac{1 \bullet 3 \bullet 5x^7}{2 \bullet 4 \bullet 6 \bullet 7} + \cdots & |x| < 1 \\ \pm\left(\ln|2x| + \dfrac{1}{2 \bullet 2x^2} - \dfrac{1 \bullet 3}{2 \bullet 4 \bullet 4x^4} + \dfrac{1 \bullet 3 \bullet 5}{2 \bullet 4 \bullet 6 \bullet 6x^6} - \cdots\right) & \begin{bmatrix} + \text{ if } x \geqq 1 \\ - \text{ if } x \leqq -1 \end{bmatrix} \end{cases}$$

22.40. $$\cosh^{-1} x = \pm\left\{\ln(2x) - \left(\frac{1}{2 \bullet 2x^2} + \frac{1 \bullet 3}{2 \bullet 4 \bullet 4x^4} + \frac{1 \bullet 3 \bullet 5}{2 \bullet 4 \bullet 6 \bullet 6x^6} + \cdots\right)\right\} \qquad \begin{bmatrix} + \text{ if } \cosh^{-1} x > 0, x \geqq 1 \\ - \text{ if } \cosh^{-1} x < 0, x \geqq 1 \end{bmatrix}$$

22.41. $$\tanh^{-1} x = x + \frac{x^3}{3} + \frac{x^5}{5} + \frac{x^7}{7} + \cdots \qquad |x| < 1$$

22.42. $$\coth^{-1} x = \frac{1}{x} + \frac{1}{3x^3} + \frac{1}{5x^5} + \frac{1}{7x^7} + \cdots \qquad |x| > 1$$

Miscellaneous Series

22.43. $$e^{\sin x} = 1 + x + \frac{x^2}{2} - \frac{x^4}{8} - \frac{x^5}{15} + \cdots \qquad -\infty < x < \infty$$

22.44. $$e^{\cos x} = e\left(1 - \frac{x^2}{2} + \frac{x^4}{6} - \frac{31x^6}{720} + \cdots\right) \qquad -\infty < x < \infty$$

22.45. $e^{\tan x} = 1 + x + \frac{x^2}{2} + \frac{x^3}{2} + \frac{3x^4}{8} + \cdots \qquad |x| < \frac{\pi}{2}$

22.46. $e^x \sin x = x + x^2 + \frac{x^3}{3} - \frac{x^5}{30} - \frac{x^6}{90} + \cdots + \frac{2^{n/2}\sin(n\pi/4)x^n}{n!} + \cdots \qquad -\infty < x < \infty$

22.47. $e^x \cos x = 1 + x - \frac{x^3}{3} - \frac{x^4}{6} + \cdots + \frac{2^{n/2}\cos(n\pi/4)x^n}{n!} + \cdots \qquad -\infty < x < \infty$

22.48. $\ln|\sin x| = \ln|x| - \frac{x^2}{6} - \frac{x^4}{180} - \frac{x^6}{2835} - \cdots - \frac{2^{2n-1}B_n x^{2n}}{n(2n)!} + \cdots \qquad 0 < |x| < \pi$

22.49. $\ln|\cos x| = -\frac{x^2}{2} - \frac{x^4}{12} - \frac{x^6}{45} - \frac{17x^8}{2520} - \cdots - \frac{2^{2n-1}(2^{2n}-1)B_n x^{2n}}{n(2n)!} + \cdots \qquad |x| < \frac{\pi}{2}$

22.50. $\ln|\tan x| = \ln|x| + \frac{x^2}{3} + \frac{7x^4}{90} + \frac{62x^6}{2835} + \cdots + \frac{2^{2n}(2^{2n-1}-1)B_n x^{2n}}{n(2n)!} + \cdots \qquad 0 < |x| < \frac{\pi}{2}$

22.51. $\frac{\ln(1+x)}{1+x} = x - (1+\tfrac{1}{2})x^2 + (1+\tfrac{1}{2}+\tfrac{1}{3})x^3 - \cdots \qquad |x| < 1$

Reversion of Power Series

Suppose

22.52. $y = C_1x + C_2x^2 + C_3x^3 + C_4x^4 + C_5x^5 + C_6x^6 + \cdots$

then

22.53. $x = C_1y + C_2y^2 + C_3y^3 + C_4y^4 + C_5y^5 + C_6y^6 + \cdots$

where

22.54. $c_1C_1 = 1$

22.55. $c_1^3C_2 = -c_2$

22.56. $c_1^5C_3 = 2c_2^2 - c_1c_3$

22.57. $c_1^7C_4 = 5c_1c_2c_3 - 5c_2^3 - c_1^2c_4$

22.58. $c_1^9C_5 = 6c_1^2c_2c_4 + 3c_1^2c_3^2 - c_1^3c_5 + 14c_2^4 - 21c_1c_2^2c_3$

22.59. $c_1^{11}C_6 = 7c_1^3c_2c_5 + 84c_1c_2^3c_3 + 7c_1^3c_3c_4 - 28c_1^2c_2c_3^2 - c_1^4c_6 - 28c_1^2c_2^2c_4 - 42c_2^5$

Taylor Series for Functions of Two Variables

22.60. $f(x,y) = f(a,b) + (x-a)f_x(a,b) + (y-b)f_y(a,b)$

$$+\frac{1}{2!}\{(x-a)^2 f_{xx}(a,b) + 2(x-a)(y-b)f_{xy}(a,b) + (y-b)^2 f_{yy}(a,b)\} + \cdots$$

where $f_x(a,b)$, $f_y(a,b)$,... denote partial derivatives with respect to x, y, ... evaluated at $x = a$, $y = b$.

23 BERNOULLI and EULER NUMBERS

Definition of Bernoulli Numbers

The *Bernoulli numbers* $B_1, B_2, B_3, \ldots$ are defined by the series

23.1. $$\frac{x}{e^x - 1} = 1 - \frac{x}{2} + \frac{B_1 x^2}{2!} - \frac{B_2 x^4}{4!} + \frac{B_3 x^6}{6!} - \cdots \qquad |x| < 2\pi$$

23.2. $$1 - \frac{x}{2}\cot\frac{x}{2} = \frac{B_1 x^2}{2!} + \frac{B_3 x^4}{4!} + \frac{B_3 x^6}{6!} + \cdots \qquad |x| < \pi$$

Definition of Euler Numbers

The *Euler numbers* $E_1, E_2, E_3, \ldots$ are defined by the series

23.3. $$\operatorname{sech} x = 1 - \frac{E_1 x^2}{2!} + \frac{E_2 x^4}{4!} - \frac{E_3 x^6}{6!} + \cdots \qquad |x| < \frac{\pi}{2}$$

23.4. $$\sec x = 1 + \frac{E_1 x^2}{2!} + \frac{E_2 x^4}{4!} - \frac{E_3 x^6}{6!} + \cdots \qquad |x| < \frac{\pi}{2}$$

Table of First Few Bernoulli and Euler Numbers

Bernoulli Numbers	Euler Numbers
$B_1 = 1/6$	$E_1 = 1$
$B_2 = 1/30$	$E_2 = 5$
$B_3 = 1/42$	$E_3 = 61$
$B_4 = 1/30$	$E_4 = 1385$
$B_5 = 5/66$	$E_5 = 50,521$
$B_6 = 691/2730$	$E_6 = 2,702,765$
$B_7 = 7/6$	$E_7 = 199,360,981$
$B_8 = 3617/510$	$E_8 = 19,391,512,145$
$B_9 = 43,867/798$	$E_9 = 2,404,879,675,441$
$B_{10} = 174,611/330$	$E_{10} = 370,371,188,237,525$
$B_{11} = 854,513/138$	$E_{11} = 69,348,874,393,137,901$
$B_{12} = 236,364,091/2730$	$E_{12} = 15,514,534,163,557,086,905$

Relationships of Bernoulli and Euler Numbers

23.5. $\binom{2n+1}{2}2^2B_1 - \binom{2n+1}{4}2^4B_2 + \binom{2n+1}{6}2^6B_3 - \cdots (-1)^{n-1}(2n+1)2^{2n}B_n = 2n$

23.6. $E_n = \binom{2n}{2}E_{n-1} - \binom{2n}{4}E_{n-2} + \binom{2n}{6}E_{n-3} - \cdots (-1)^n$

23.7. $B_n = \dfrac{2n}{2^{2n}(2^{2n}-1)}\left\{\binom{2n-1}{1}E_{n-1} - \binom{2n-1}{3}E_{n-2} + \binom{2n-1}{5}E_{n-3} - \cdots (-1)^{n-1}\right\}$

Series Involving Bernoulli and Euler Numbers

23.8. $B_n = \dfrac{(2n)!}{2^{2n-1}\pi^{2n}}\left\{1 + \dfrac{1}{2^{2n}} + \dfrac{1}{3^{2n}} + \cdots\right\}$

23.9. $B_n = \dfrac{2(2n)!}{(2^{2n}-1)\pi^{2n}}\left\{1 + \dfrac{1}{3^{2n}} + \dfrac{1}{5^{2n}} + \cdots\right\}$

23.10. $B_n = \dfrac{2(2n)!}{(2^{2n-1}-1)\pi^{2n}}\left\{1 - \dfrac{1}{2^{2n}} + \dfrac{1}{3^{2n}} - \cdots\right\}$

23.11. $E_n = \dfrac{2^{2n+2}(2n)!}{\pi^{2n+1}}\left\{1 - \dfrac{1}{3^{2n+1}} + \dfrac{1}{5^{2n+1}} - \cdots\right\}$

Asymptotic Formula for Bernoulli Numbers

23.12. $B_n \sim 4n^{2n}(\pi e)^{-2n}\sqrt{\pi n}$

24 FOURIER SERIES

Definition of a Fourier Series

The Fourier series corresponding to a function $f(x)$ defined in the interval $c \leqq x \leqq c+2L$ where c and $L>0$ are constants, is defined as

24.1. $$\frac{a_0}{2}+\sum_{n=1}^{\infty}\left(a_n\cos\frac{n\pi x}{L}+b_n\sin\frac{n\pi x}{L}\right)$$

where

24.2. $$\begin{cases} a_n = \dfrac{1}{L}\displaystyle\int_c^{c+2L} f(x)\cos\frac{n\pi x}{L}\,dx \\ b_n = \dfrac{1}{L}\displaystyle\int_c^{c+2L} f(x)\sin\frac{n\pi x}{L}\,dx \end{cases}$$

If $f(x)$ and $f'(x)$ are piecewise continuous and $f(x)$ is defined by periodic extension of period $2L$, i.e., $f(x+2L)=f(x)$, then the series converges to $f(x)$ if x is a point of continuity and to $\frac{1}{2}\{f(x+0)+f(x-0)\}$ if x is a point of discontinuity.

Complex Form of Fourier Series

Assuming that the series 24.1 converges to $f(x)$, we have

24.3. $$f(x)=\sum_{n=-\infty}^{\infty} c_n e^{in\pi x/L}$$

where

24.4. $$c_n=\frac{1}{2L}\int_c^{c+2L} f(x)e^{-in\pi x/L}\,dx=\begin{cases}\frac{1}{2}(a_n-ib_n) & n>0\\ \frac{1}{2}(a_{-n}+ib_{-n}) & n<0\\ \frac{1}{2}a_0 & n=0\end{cases}$$

Parseval's Identity

24.5. $$\frac{1}{L}\int_c^{c+2L}\{f(x)\}^2\,dx=\frac{a_0^2}{2}+\sum_{n=1}^{\infty}(a_n^2+b_n^2)$$

Generalized Parseval Identity

24.6. $$\frac{1}{L}\int_c^{c+2L} f(x)g(x)\,dx=\frac{a_0c_0}{2}+\sum_{n=1}^{\infty}(a_nc_n+b_nd_n)$$

where a_n, b_n and c_n, d_n are the Fourier coefficients corresponding to $f(x)$ and $g(x)$, respectively.

Special Fourier Series and Their Graphs

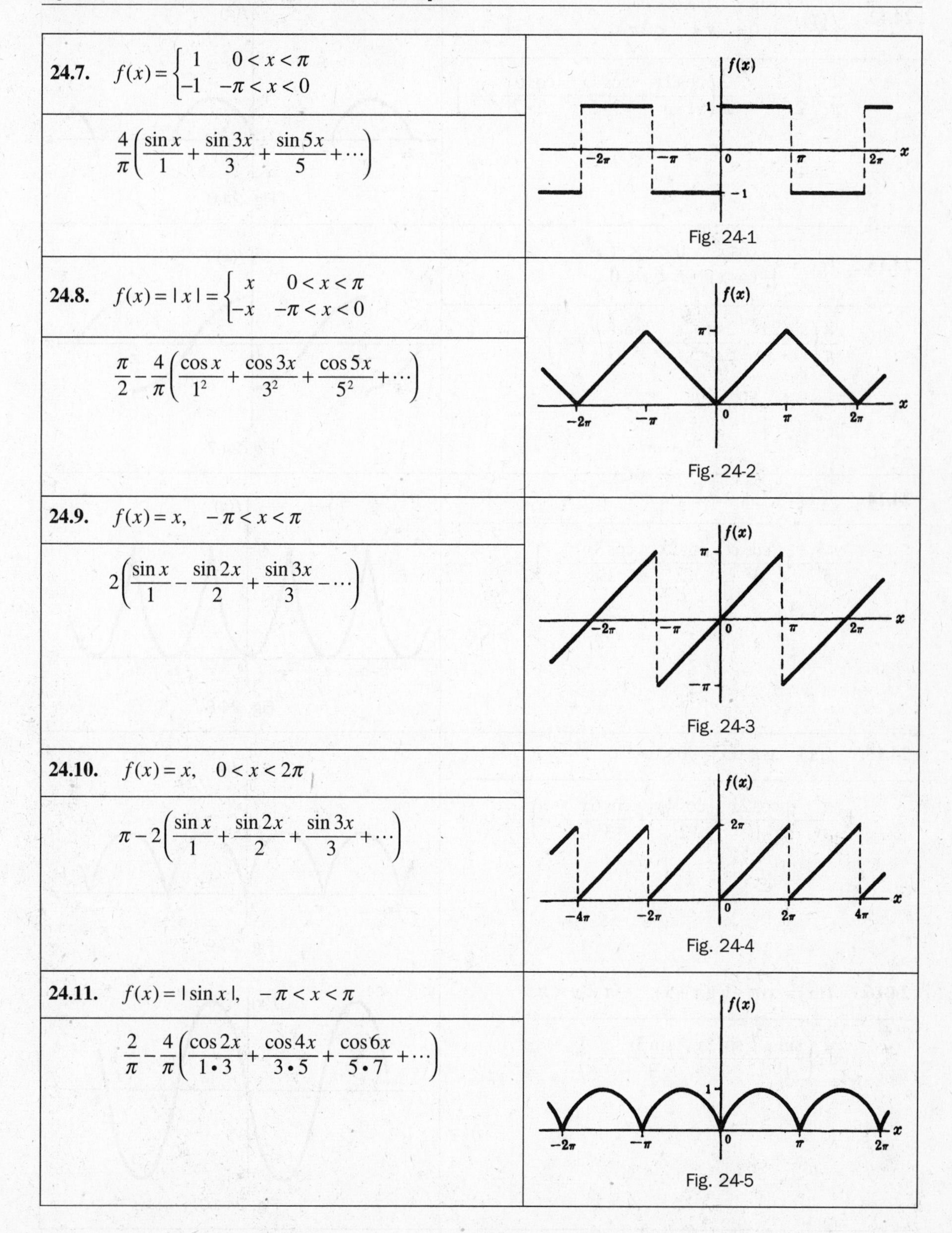

24.7. $f(x) = \begin{cases} 1 & 0 < x < \pi \\ -1 & -\pi < x < 0 \end{cases}$

$$\frac{4}{\pi}\left(\frac{\sin x}{1} + \frac{\sin 3x}{3} + \frac{\sin 5x}{5} + \cdots\right)$$

Fig. 24-1

24.8. $f(x) = |x| = \begin{cases} x & 0 < x < \pi \\ -x & -\pi < x < 0 \end{cases}$

$$\frac{\pi}{2} - \frac{4}{\pi}\left(\frac{\cos x}{1^2} + \frac{\cos 3x}{3^2} + \frac{\cos 5x}{5^2} + \cdots\right)$$

Fig. 24-2

24.9. $f(x) = x, \quad -\pi < x < \pi$

$$2\left(\frac{\sin x}{1} - \frac{\sin 2x}{2} + \frac{\sin 3x}{3} - \cdots\right)$$

Fig. 24-3

24.10. $f(x) = x, \quad 0 < x < 2\pi$

$$\pi - 2\left(\frac{\sin x}{1} + \frac{\sin 2x}{2} + \frac{\sin 3x}{3} + \cdots\right)$$

Fig. 24-4

24.11. $f(x) = |\sin x|, \quad -\pi < x < \pi$

$$\frac{2}{\pi} - \frac{4}{\pi}\left(\frac{\cos 2x}{1 \cdot 3} + \frac{\cos 4x}{3 \cdot 5} + \frac{\cos 6x}{5 \cdot 7} + \cdots\right)$$

Fig. 24-5

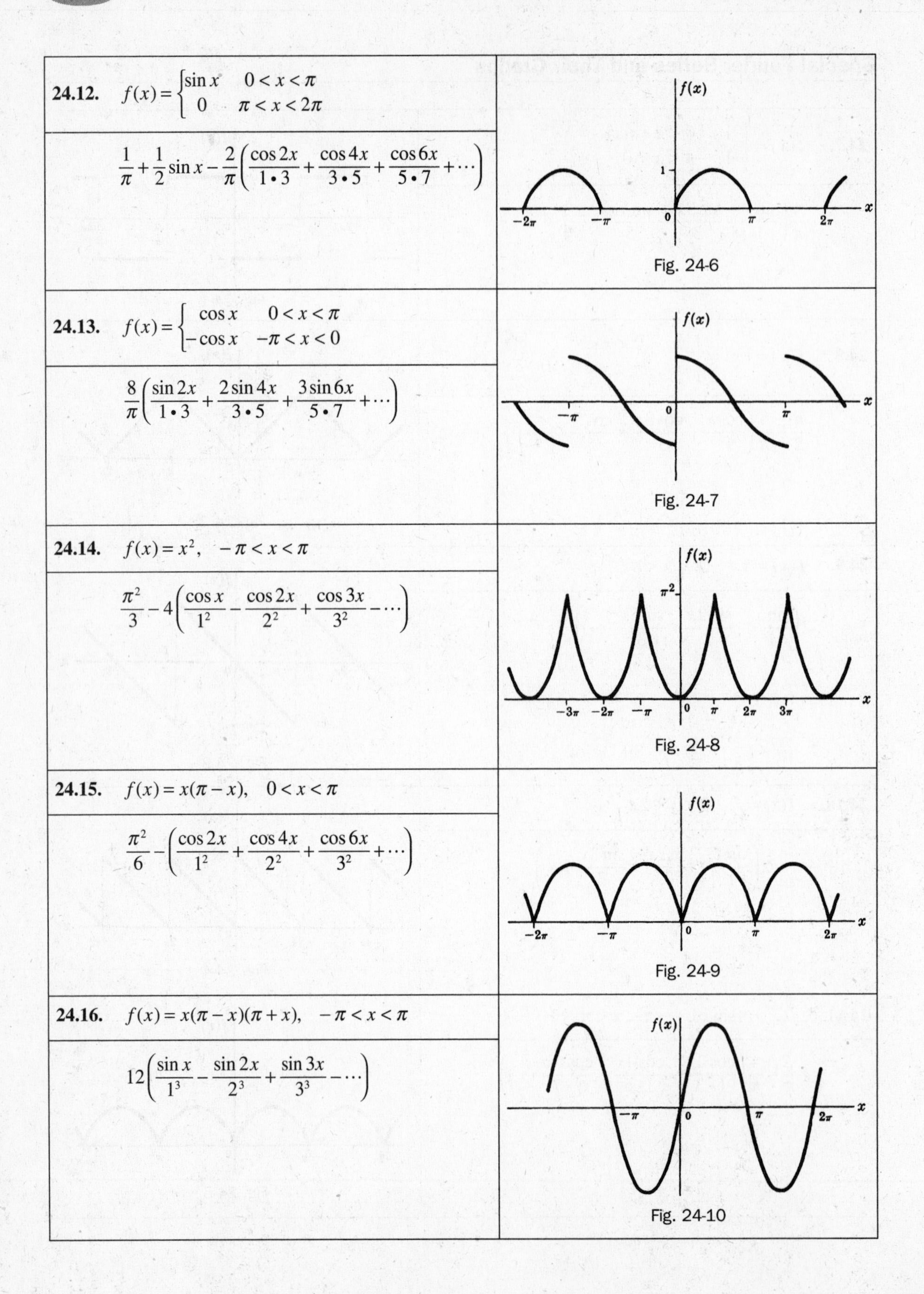

24.12. $f(x)=\begin{cases}\sin x & 0<x<\pi\\ 0 & \pi<x<2\pi\end{cases}$	
$\frac{1}{\pi}+\frac{1}{2}\sin x-\frac{2}{\pi}\left(\frac{\cos 2x}{1\cdot 3}+\frac{\cos 4x}{3\cdot 5}+\frac{\cos 6x}{5\cdot 7}+\cdots\right)$	Fig. 24-6
24.13. $f(x)=\begin{cases}\cos x & 0<x<\pi\\ -\cos x & -\pi<x<0\end{cases}$	
$\frac{8}{\pi}\left(\frac{\sin 2x}{1\cdot 3}+\frac{2\sin 4x}{3\cdot 5}+\frac{3\sin 6x}{5\cdot 7}+\cdots\right)$	Fig. 24-7
24.14. $f(x)=x^2,\quad -\pi<x<\pi$	
$\frac{\pi^2}{3}-4\left(\frac{\cos x}{1^2}-\frac{\cos 2x}{2^2}+\frac{\cos 3x}{3^2}-\cdots\right)$	Fig. 24-8
24.15. $f(x)=x(\pi-x),\quad 0<x<\pi$	
$\frac{\pi^2}{6}-\left(\frac{\cos 2x}{1^2}+\frac{\cos 4x}{2^2}+\frac{\cos 6x}{3^2}+\cdots\right)$	Fig. 24-9
24.16. $f(x)=x(\pi-x)(\pi+x),\quad -\pi<x<\pi$	
$12\left(\frac{\sin x}{1^3}-\frac{\sin 2x}{2^3}+\frac{\sin 3x}{3^3}-\cdots\right)$	Fig. 24-10

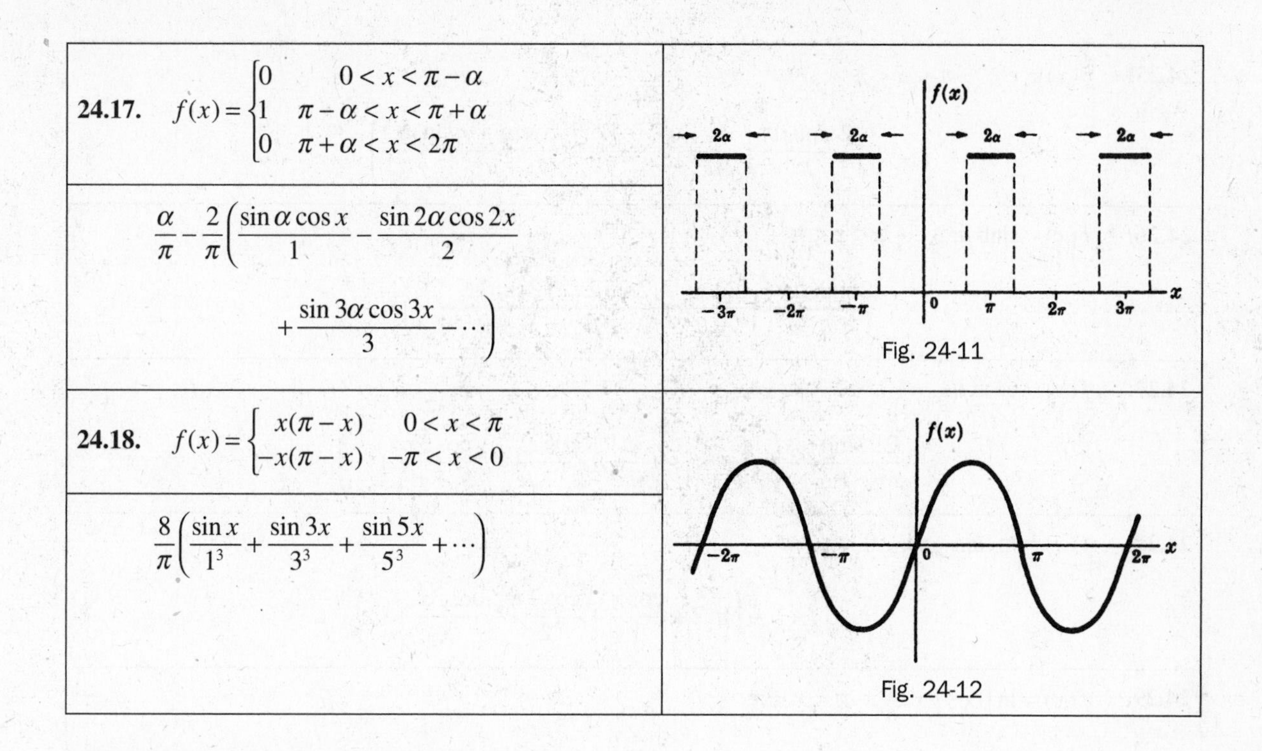

24.17. $f(x)=\begin{cases}0 & 0<x<\pi-\alpha\\ 1 & \pi-\alpha<x<\pi+\alpha\\ 0 & \pi+\alpha<x<2\pi\end{cases}$

$$\frac{\alpha}{\pi}-\frac{2}{\pi}\left(\frac{\sin\alpha\cos x}{1}-\frac{\sin 2\alpha\cos 2x}{2}+\frac{\sin 3\alpha\cos 3x}{3}-\cdots\right)$$

Fig. 24-11

24.18. $f(x)=\begin{cases}x(\pi-x) & 0<x<\pi\\ -x(\pi-x) & -\pi<x<0\end{cases}$

$$\frac{8}{\pi}\left(\frac{\sin x}{1^3}+\frac{\sin 3x}{3^3}+\frac{\sin 5x}{5^3}+\cdots\right)$$

Fig. 24-12

Miscellaneous Fourier Series

24.19. $f(x)=\sin\mu x,\quad -\pi<x<\pi,\quad \mu\neq\text{integer}$

$$\frac{2\sin\mu\pi}{\pi}\left(\frac{\sin x}{1^2-\mu^2}-\frac{2\sin 2x}{2^2-\mu^2}+\frac{3\sin 3x}{3^2-\mu^2}-\cdots\right)$$

24.20. $f(x)=\cos\mu x,\quad -\pi<x<\pi,\quad \mu\neq\text{integer}$

$$\frac{2\mu\sin\mu\pi}{\pi}\left(\frac{1}{2\mu^2}+\frac{\cos x}{1^2-\mu^2}-\frac{\cos 2x}{2^2-\mu^2}+\frac{\cos 3x}{3^2-\mu^2}-\cdots\right)$$

24.21. $f(x)=\tan^{-1}[(a\sin x)/(1-a\cos x)],\quad -\pi<x<\pi,\quad |a|<1$

$$a\sin x+\frac{a^2}{2}\sin 2x+\frac{a^3}{3}\sin 3x+\cdots$$

24.22. $f(x)=\ln(1-2a\cos x+a^2),\quad -\pi<x<\pi,\quad |a|<1$

$$-2\left(a\cos x+\frac{a^2}{2}\cos 2x+\frac{a^3}{3}\cos 3x+\cdots\right)$$

24.23. $f(x)=\frac{1}{2}\tan^{-1}[(2a\sin x)/(1-a^2)],\quad -\pi<x<\pi,\quad |a|<1$

$$a\sin x+\frac{a^3}{3}\sin 3x+\frac{a^5}{5}\sin 5x+\cdots$$

24.24. $f(x)=\frac{1}{2}\tan^{-1}[(2a\cos x)/(1-a^2)],\quad -\pi<x<\pi,\quad |a|<1$

$$a\cos x-\frac{a^3}{3}\cos 3x+\frac{a^5}{5}\cos 5x-\cdots$$

24.25. $f(x) = e^{\mu x}, \ -\pi < x < \pi$	$\dfrac{2\sinh \mu\pi}{\pi}\left(\dfrac{1}{2\mu} + \sum_{n=1}^{\infty} \dfrac{(-1)^n(\mu\cos nx - n\sin nx)}{\mu^2 + n^2}\right)$
24.26. $f(x) = \sinh \mu x, \quad -\pi < x < \pi$	$\dfrac{2\sinh \mu\pi}{\pi}\left(\dfrac{\sin x}{1^2+\mu^2} - \dfrac{2\sin 2x}{2^2+\mu^2} + \dfrac{3\sin 3x}{3^2+\mu^2} - \cdots\right)$
24.27. $f(x) = \cosh \mu x, \quad -\pi < x < \pi$	$\dfrac{2\mu\sinh \mu\pi}{\pi}\left(\dfrac{1}{2\mu^2} - \dfrac{\cos x}{1^2+\mu^2} + \dfrac{\cos 2x}{2^2+\mu^2} - \dfrac{\cos 3x}{3^2+\mu^2} + \cdots\right)$
24.28. $f(x) = \ln\lvert\sin \tfrac{1}{2}x\rvert, \quad 0 < x < \pi$	$-\left(\ln 2 + \dfrac{\cos x}{1} + \dfrac{\cos 2x}{2} + \dfrac{\cos 3x}{3} + \cdots\right)$
24.29. $f(x) = \ln\lvert\cos \tfrac{1}{2}x\rvert, \quad -\pi < x < \pi$	$-\left(\ln 2 - \dfrac{\cos x}{1} + \dfrac{\cos 2x}{2} - \dfrac{\cos 3x}{3} + \cdots\right)$
24.30. $f(x) = \tfrac{1}{6}\pi^2 - \tfrac{1}{2}\pi x + \tfrac{1}{4}x^2, \quad 0 \leqq x \leqq 2\pi$	$\dfrac{\cos x}{1^2} + \dfrac{\cos 2x}{2^2} + \dfrac{\cos 3x}{3^2} + \cdots$
24.31. $f(x) = \tfrac{1}{12}x(x-\pi)(x-2\pi), \quad 0 \leqq x \leqq 2\pi$	$\dfrac{\sin x}{1^3} + \dfrac{\sin 2x}{2^3} + \dfrac{\sin 3x}{3^3} + \cdots$
24.32. $f(x) = \tfrac{1}{90}\pi^4 - \tfrac{1}{12}\pi^2x^2 + \tfrac{1}{12}\pi x^3 - \tfrac{1}{48}x^4, \quad 0 \leqq x \leqq 2\pi$	$\dfrac{\cos x}{1^4} + \dfrac{\cos 2x}{2^4} + \dfrac{\cos 3x}{3^4} + \cdots$

Section VII: Special Functions and Polynomials

25 THE GAMMA FUNCTION

Definition of the Gamma Function $\Gamma(n)$ for $n > 0$

25.1. $\Gamma(n) = \int_0^\infty t^{n-1} e^{-t} dt \qquad n > 0$

Recursion Formula

25.2. $\Gamma(n+1) = n\Gamma(n)$

If $n = 0, 1, 2, \ldots$, a nonnegative integer, we have the following (where $0! = 1$):

25.3. $\Gamma(n+1) = n!$

The Gamma Function for $n < 0$

For $n < 0$ the gamma function can be defined by using 25.2, that is,

25.4. $\Gamma(n) = \dfrac{\Gamma(n+1)}{n}$

Graph of the Gamma Function

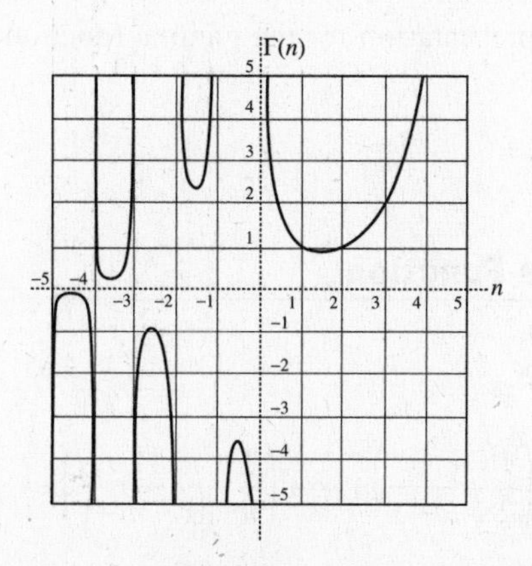

Fig. 25-1

Special Values for the Gamma Function

25.5. $\Gamma(\tfrac{1}{2}) = \sqrt{\pi}$

25.6. $\Gamma(m+\tfrac{1}{2}) = \dfrac{1 \cdot 3 \cdot 5 \cdots (2m-1)}{2^m}\sqrt{\pi} \qquad m = 1, 2, 3, \ldots$

25.7. $\Gamma(-m+\tfrac{1}{2}) = \dfrac{(-1)^m 2^m \sqrt{\pi}}{1 \cdot 3 \cdot 5 \cdots (2m-1)} \qquad m = 1, 2, 3, \ldots$

Relationships Among Gamma Functions

25.8. $\Gamma(p)\Gamma(1-p) = \dfrac{\pi}{\sin p\pi}$

25.9. $2^{2x-1}\Gamma(x)\Gamma(x+\tfrac{1}{2}) = \sqrt{\pi}\,\Gamma(2x)$

This is called the *duplication formula*.

25.10. $\Gamma(x)\Gamma\left(x+\dfrac{1}{m}\right)\Gamma\left(x+\dfrac{2}{m}\right)\cdots\Gamma\left(x+\dfrac{m-1}{m}\right) = m^{1/2-mx}(2\pi)^{(m-1)/2}\Gamma(mx)$

For $m = 2$ this reduces to 25.9.

Other Definitions of the Gamma Function

25.11. $\Gamma(x+1) = \lim\limits_{k\to\infty} \dfrac{1 \cdot 2 \cdot 3 \cdots k}{(x+1)(x+2)\cdots(x+k)} k^x$

25.12. $\dfrac{1}{\Gamma(x)} = xe^{\gamma x} \prod\limits_{m=1}^{\infty}\left\{\left(1+\dfrac{x}{m}\right)e^{-x/m}\right\}$

This is an infinite product representation for the gamma function where γ is Euler's constant defined in 1.3, page 3.

Derivatives of the Gamma Function

25.13. $\Gamma'(1) = \int_0^\infty e^{-x} \ln x \, dx = -\gamma$

25.14. $\dfrac{\Gamma'(x)}{\Gamma(x)} = -\gamma + \left(\dfrac{1}{1} - \dfrac{1}{x}\right) + \left(\dfrac{1}{2} - \dfrac{1}{x+1}\right) + \cdots + \left(\dfrac{1}{n} - \dfrac{1}{x+n-1}\right) + \cdots$

Here again is Euler's constant γ.

Asymptotic Expansions for the Gamma Function

25.15. $\Gamma(x+1) = \sqrt{2\pi x}\, x^x e^{-x} \left\{1 + \frac{1}{12x} + \frac{1}{288x^2} - \frac{139}{51{,}840x^3} + \cdots\right\}$

This is called *Stirling's asymptotic series.*

If we let $x = n$ a positive integer in 25.15, then a useful approximation for $n!$ where n is large (e.g., $n > 10$) is given by *Stirling's formula*

25.16. $n! \sim \sqrt{2\pi n}\, n^n e^{-n}$

where ~ is used to indicate that the ratio of the terms on each side approaches 1 as $n \to \infty$.

Miscellaneous Results

25.17. $|\Gamma(ix)|^2 = \frac{\pi}{x \sinh \pi x}$

26 THE BETA FUNCTION

Definition of the Beta Function $B(m, n)$

26.1. $B(m,n) = \int_0^1 t^{m-1}(1-t)^{n-1}\,dt \qquad m > 0, n > 0$

Relationship of Beta Function to Gamma Function

26.2. $B(m, n) = \dfrac{\Gamma(m)\Gamma(n)}{\Gamma(m+n)}$

Extensions of $B(m, n)$ to $m < 0$, $n < 0$ are provided by using 25.4.

Some Important Results

26.3. $B(m, n) = B(n, m)$

26.4. $B(m, n) = 2\int_0^{\pi/2} \sin^{2m-1}\theta \cos^{2n-1}\theta\, d\theta$

26.5. $B(m, n) = \int_0^{\infty} \dfrac{t^{m-1}}{(1+t)^{m+n}}\,dt$

26.6. $B(m, n) = r^n(r+1)^m \int_0^1 \dfrac{t^{m-1}(1-t)^{n-1}}{(r+t)^{m+n}}\,dt$

27 BESSEL FUNCTIONS

Bessel's Differential Equation

27.1. $x^2y'' + xy' + (x^2 - n^2)y = 0 \quad n \geqq 0$

Solutions of this equation are called *Bessel functions of order n*.

Bessel Functions of the First Kind of Order n

27.2. $$J_n(x) = \frac{x^n}{2^n\Gamma(n+1)}\left\{1 - \frac{x^2}{2(2n+2)} + \frac{x^4}{2\cdot4(2n+2)(2n+4)} - \cdots\right\}$$

$$= \sum_{k=0}^{\infty} \frac{(-1)^k (x/2)^{n+2k}}{k!\,\Gamma(n+k+1)}$$

27.3. $$J_{-n}(x) = \frac{x^{-n}}{2^{-n}\Gamma(1-n)}\left\{1 - \frac{x^2}{2(2-2n)} + \frac{x^4}{2\cdot4(2-2n)(4-2n)} - \cdots\right\}$$

$$= \sum_{k=0}^{\infty} \frac{(-1)^k (x/2)^{2k-n}}{k!\,\Gamma(k+1-n)}$$

27.4. $J_{-n}(x) = (-1)^n J_n(x) \qquad n = 0, 1, 2, \ldots$

If $n \neq 0, 1, 2, \ldots, J_n(x)$ and $J_{-n}(x)$ are linearly independent.
If $n \neq 0, 1, 2, \ldots, J_n(x)$ is bounded at $x = 0$ while $J_{-n}(x)$ is unbounded.
For $n = 0, 1$ we have

27.5. $$J_0(x) = 1 - \frac{x^2}{2^2} + \frac{x^4}{2^2\cdot4^2} - \frac{x^6}{2^2\cdot4^2\cdot6^2} + \cdots$$

27.6. $$J_1(x) = \frac{x}{2} - \frac{x^3}{2^2\cdot4} + \frac{x^5}{2^2\cdot4^2\cdot6^2} - \frac{x^7}{2^2\cdot4^2\cdot6^2\cdot8} + \cdots$$

27.7. $J_0'(x) = -J_1(x)$

Bessel Functions of the Second Kind of Order n

27.8. $$Y_n(x) = \begin{cases} \dfrac{J_n(x)\cos n\pi - J_{-n}(x)}{\sin n\pi} & n \neq 0, 1, 2, \ldots \\[2ex] \lim\limits_{p\to n} \dfrac{J_p(x)\cos p\pi - J_{-p}(x)}{\sin p\pi} & n = 0, 1, 2, \ldots \end{cases}$$

This is also called *Weber's function* or *Neumann's function* [also denoted by $N_n(x)$].

For $n = 0, 1, 2, \ldots$, L' Hospital's rule yields

27.9. $$Y_n(x) = \frac{2}{\pi}\{\ln(x/2) + \gamma\}J_n(x) - \frac{1}{\pi}\sum_{k=0}^{n-1}\frac{(n-k-1)!}{k!}(x/2)^{2k-n}$$

$$-\frac{1}{\pi}\sum_{k=0}^{\infty}(-1)^k\{\Phi(k) + \Phi(n+k)\}\frac{(x/2)^{2k+n}}{k!(n+k)!}$$

where $\gamma = .5772156 \ldots$ is Euler's constant (see 1.20) and

27.10. $$\Phi(p) = 1 + \frac{1}{2} + \frac{1}{3} + \cdots + \frac{1}{P}, \qquad \Phi(0) = 0$$

For $n = 0$,

27.11. $$Y_0(x) = \frac{2}{\pi}\{\ln(x/2) + \gamma\}J_0(x) + \frac{2}{\pi}\left\{\frac{x^2}{2^2} - \frac{x^4}{2^2 4^2}\left(1 + \frac{1}{2}\right) + \frac{x^6}{2^2 4^2 6^2}\left(1 + \frac{1}{2} + \frac{1}{3}\right) - \cdots\right\}$$

27.12. $$Y_{-n}(x) = (-1)^n Y_n(x) \qquad n = 0, 1, 2, \ldots$$

For any value $n \geqq 0$, $J_n(x)$ is bounded at $x = 0$ while $Y_n(x)$ is unbounded.

General Solution of Bessel's Differential Equation

27.13. $$y = AJ_n(x) + BJ_{-n}(x) \qquad n \neq 0, 1, 2, \ldots$$

27.14. $$y = AJ_n(x) + BY_n(x) \qquad \text{all } n$$

27.15. $$y = AJ_n(x) + BJ_n(x)\int\frac{dx}{xJ_n^2(x)} \qquad \text{all } n$$

where A and B are arbitrary constants.

Generating Function for $J_n(x)$

27.16. $$e^{x(t-1/t)/2} = \sum_{n=-\infty}^{\infty} J_n(x)t^n$$

Recurrence Formulas for Bessel Functions

27.17. $$J_{n+1}(x) = \frac{2n}{x}J_n(x) - J_{n-1}(x)$$

27.18. $$J_n'(x) = \tfrac{1}{2}\{J_{n-1}(x) - J_{n+1}(x)\}$$

27.19. $$xJ_n'(x) = xJ_{n-1}(x) - nJ_n(x)$$

27.20. $$xJ_n'(x) = nJ_n(x) - xJ_{n+1}(x)$$

27.21. $\frac{d}{dx}\{x^n J_n(x)\} = x^n J_{n-1}(x)$

27.22. $\frac{d}{dx}\{x^{-n} J_n(x)\} = -x^{-n} J_{n+1}(x)$

The functions $Y_n(x)$ satisfy identical relations.

Bessel Functions of Order Equal to Half an Odd Integer

In this case the functions are expressible in terms of sines and cosines.

27.23. $J_{1/2}(x) = \sqrt{\frac{2}{\pi x}} \sin x$

27.24. $J_{-1/2}(x) = \sqrt{\frac{2}{\pi x}} \cos x$

27.25. $J_{3/2}(x) = \sqrt{\frac{2}{\pi x}} \left(\frac{\sin x}{x} - \cos x\right)$

27.26. $J_{-3/2}(x) = \sqrt{\frac{2}{\pi x}} \left(\frac{\cos x}{x} + \sin x\right)$

27.27. $J_{5/2}(x) = \sqrt{\frac{2}{\pi x}} \left\{\left(\frac{3}{x^2} - 1\right)\sin x - \frac{3}{x}\cos x\right\}$

27.28. $J_{-5/2}(x) = \sqrt{\frac{2}{\pi x}} \left\{\frac{3}{x}\sin x + \left(\frac{3}{x} - 1\right)\cos x\right\}$

For further results use the recurrence formula. Results for $Y_{1/2}(x), Y_{3/2}(x), \ldots$ are obtained from 27.8.

Hankel Functions of First and Second Kinds of Order n

27.29. $H_n^{(1)}(x) = J_n(x) + iY_n(x)$

27.30. $H_n^{(2)}(x) = J_n(x) - iY_n(x)$

Bessel's Modified Differential Equation

27.31. $x^2y'' + xy' - (x^2 + n^2)y = 0 \quad n \geqq 0$

Solutions of this equation are called *modified Bessel functions of order n*.

Modified Bessel Functions of the First Kind of Order n

27.32. $I_n(x) = i^{-n} J_n(ix) = e^{-n\pi i/2} J_n(ix)$

$$= \frac{x^n}{2^n \Gamma(n+1)}\left\{1 + \frac{x^2}{2(2n+2)} + \frac{x^4}{2 \cdot 4(2n+2)(2n+4)} + \cdots\right\} = \sum_{k=0}^{\infty} \frac{(x/2)^{n+2k}}{k!\,\Gamma(n+k+1)}$$

27.33. $I_{-n}(x) = i^n J_{-n}(ix) = e^{n\pi i/2} J_{-n}(ix)$

$$= \frac{x^{-n}}{2^{-n} \Gamma(1-n)}\left\{1 + \frac{x^2}{2(2-2n)} + \frac{x^4}{2 \cdot 4(2-2n)(4-2n)} + \cdots\right\} = \sum_{k=0}^{\infty} \frac{(x/2)^{2k-n}}{k!\,\Gamma(k+1-n)}$$

27.34. $I_{-n}(x) = I_n(x) \quad n = 0, 1, 2, \ldots$

If $n \neq 0, 1, 2, \ldots$, then $I_n(x)$ and $I_{-n}(x)$ are linearly independent.
For $n = 0, 1$, we have

27.35. $$I_0(x) = 1 + \frac{x^2}{2^2} + \frac{x^4}{2^2 \cdot 4^2} + \frac{x^6}{2^2 \cdot 4^2 \cdot 6^2} + \cdots$$

27.36. $$I_1(x) = \frac{x}{2} + \frac{x^3}{2^2 \cdot 4} + \frac{x^5}{2^2 \cdot 4^2 \cdot 6} + \frac{x^7}{2^2 \cdot 4^2 \cdot 6^2 \cdot 8} + \cdots$$

27.37. $$I_0'(x) = I_1(x)$$

Modified Bessel Functions of the Second Kind of Order *n*

27.38. $$K_n(x) = \begin{cases} \dfrac{\pi}{2 \sin n\pi}\{I_{-n}(x) - I_n(x)\} & n \neq 0, 1, 2, \ldots \\ \lim\limits_{p \to n} \dfrac{\pi}{2 \sin p\pi}\{I_{-p}(x) - I_p(x)\} & n = 0, 1, 2, \ldots \end{cases}$$

For $n = 0, 1, 2, \ldots$, L' Hospital's rule yields

27.39. $$K_n(x) = (-1)^{n+1}\{\ln(x/2) + \gamma\}I_n(x) + \frac{1}{2}\sum_{k=0}^{n-1}(-1)^k(n-k-1)!(x/2)^{2k-n}$$
$$+ \frac{(-1)^n}{2}\sum_{k=0}^{\infty}\frac{(x/2)^{n+2k}}{k!(n+k)!}\{\Phi(k) + \Phi(n+k)\}$$

where $\Phi(p)$ is given by 27.10.
For $n = 0$,

27.40. $$K_0(x) = -\{\ln(x/2) + \gamma\}I_0(x) + \frac{x^2}{2^2} + \frac{x^4}{2^2 \cdot 4^2}\left(1 + \frac{1}{2}\right) + \frac{x^6}{2^2 \cdot 4^2 \cdot 6^2}\left(1 + \frac{1}{2} + \frac{1}{3}\right) + \cdots$$

27.41. $$K_{-n}(x) = K_n(x) \qquad n = 0, 1, 2, \ldots$$

General Solution of Bessel's Modified Equation

27.42. $$y = AI_n(x) + BI_{-n}(x) \qquad n \neq 0, 1, 2, \ldots$$

27.43. $$y = AI_n(x) + BK_n(x) \qquad \text{all } n$$

27.44. $$y = AI_n(x) + BI_n(x)\int\frac{dx}{xI_n^2(x)} \qquad \text{all } n$$

where A and B are arbitrary constants.

Generating Function for $I_n(x)$

27.45. $$e^{x(t+1/t)/2} = \sum_{n=-\infty}^{\infty} I_n(x)t^n$$

Recurrence Formulas for Modified Bessel Functions

27.46. $I_{n+1}(x) = I_{n-1}(x) - \frac{2n}{x} I_n(x)$

27.47. $I'_n(x) = \frac{1}{2}\{I_{n-1}(x) + I_{n+1}(x)\}$

27.48. $xI'_n(x) = xI_{n-1}(x) - nI_n(x)$

27.49. $xI'_n(x) = xI_{n+1}(x) + nI_n(x)$

27.50. $\frac{d}{dx}\{x^n I_n(x)\} = x^n I_{n-1}(x)$

27.51. $\frac{d}{dx}\{x^{-n} I_n(x)\} = x^{-n} I_{n+1}(x)$

27.52. $K_{n+1}(x) = K_{n-1}(x) + \frac{2n}{x} K_n(x)$

27.53. $K'_n(x) = -\frac{1}{2}\{K_{n-1}(x) + K_{n+1}(x)\}$

27.54. $xK'_n(x) = -xK_{n-1}(x) - nK_n(x)$

27.55. $xK'_n(x) = nK_n(x) - xK_{n+1}(x)$

27.56. $\frac{d}{dx}\{x^n K_n(x)\} = -x^n K_{n-1}(x)$

27.57. $\frac{d}{dx}\{x^{-n} K_n(x)\} = -x^{-n} K_{n+1}(x)$

Modified Bessel Functions of Order Equal to Half an Odd Integer

In this case the functions are expressible in terms of hyperbolic sines and cosines.

27.58. $I_{1/2}(x) = \sqrt{\frac{2}{\pi x}} \sinh x$

27.59. $I_{-1/2}(x) = \sqrt{\frac{2}{\pi x}} \cosh x$

27.60. $I_{3/2}(x) = \sqrt{\frac{2}{\pi x}} \left(\cosh x - \frac{\sinh x}{x}\right)$

27.61. $I_{-3/2}(x) = \sqrt{\frac{2}{\pi x}} \left(\sinh x - \frac{\cosh x}{x}\right)$

27.62. $I_{5/2}(x) = \sqrt{\frac{2}{\pi x}} \left\{\left(\frac{3}{x^2} + 1\right)\sinh x - \frac{3}{x}\cosh x\right\}$

27.63. $I_{-5/2}(x) = \sqrt{\frac{2}{\pi x}} \left\{\left(\frac{3}{x^2} + 1\right)\cosh x - \frac{3}{x}\sinh x\right\}$

For further results use the recurrence formula 27.46. Results for $K_{1/2}(x)$, $K_{3/2}(x)$, … are obtained from 27.38.

Ber and Bei Functions

The real and imaginary parts of $J_n(xe^{3\pi i/4})$ are denoted by $\text{Ber}_n(x)$ and $\text{Bei}_n(x)$ where

27.64. $\text{Ber}_n(x) = \sum_{k=0}^{\infty} \frac{(x/2)^{2k+n}}{k!\Gamma(n+k+1)} \cos \frac{(3n+2k)\pi}{4}$

27.65. $\text{Bei}_n(x) = \sum_{k=0}^{\infty} \frac{(x/2)^{2k+n}}{k!\Gamma(n+k+1)} \sin \frac{(3n+2k)\pi}{4}$

If $n = 0$.

27.66. $\text{Ber}(x) = 1 - \frac{(x/2)^4}{2!^2} + \frac{(x/2)^8}{4!^2} - \cdots$

27.67. $\text{Bei}(x) = (x/2)^2 - \frac{(x/2)^6}{3!^2} + \frac{(x/2)^{10}}{5!^2} - \cdots$

Ker and Kei Functions

The real and imaginary parts of $e^{-n\pi i/2} K_n(xe^{\pi i/4})$ are denoted by $\text{Ker}_n(x)$ and $\text{Kei}_n(x)$ where

27.68. $$\text{Ker}_n(x) = -\{\ln(x/2)+\gamma\}\text{Ber}_n(x) + \tfrac{1}{4}\pi\,\text{Bei}_n(x)$$
$$+\frac{1}{2}\sum_{k=0}^{n-1}\frac{(n-k-1)!(x/2)^{2k-n}}{k!}\cos\frac{(3n+2k)\pi}{4}$$
$$+\frac{1}{2}\sum_{k=0}^{\infty}\frac{(x/2)^{n+2k}}{k!(n+k)!}\{\Phi(k)+\Phi(n+k)\}\cos\frac{(3n+2k)\pi}{4}$$

27.69. $$\text{Kei}_n(x) = -\{\ln(x/2)+\gamma\}\text{Bei}_n(x) - \tfrac{1}{4}\pi\,\text{Ber}_n(x)$$
$$-\frac{1}{2}\sum_{k=0}^{n-1}\frac{(n-k-1)!(x/2)^{2k-n}}{k!}\sin\frac{(3n+2k)\pi}{4}$$
$$+\frac{1}{2}\sum_{k=0}^{\infty}\frac{(x/2)^{n+2k}}{k!(n+k)!}\{\Phi(k)+\Phi(n+k)\}\sin\frac{(3n+2k)\pi}{4}$$

and Φ is given by 27.10.

If $n = 0$,

27.70. $$\text{Ker}(x) = -\{\ln(x/2)+\gamma\}\text{Ber}(x) + \frac{\pi}{4}\text{Bei}(x) + 1 - \frac{(x/2)^4}{2!^2}(1+\tfrac{1}{2}) + \frac{(x/2)^8}{4!^2}(1+\tfrac{1}{2}+\tfrac{1}{3}+\tfrac{1}{4}) - \cdots$$

27.71. $$\text{Kei}(x) = -\{\ln(x/2)+\gamma\}\text{Bei}(x) - \frac{\pi}{4}\text{Ber}(x) + (x/2)^2 - \frac{(x/2)^6}{3!^2}(1+\tfrac{1}{2}+\tfrac{1}{3}) + \cdots$$

Differential Equation For Ber, Bei, Ker, Kei Functions

27.72. $$x^2y'' + xy' - (ix^2+n^2)y = 0$$

The general solution of this equation is

27.73. $$y = A\{Ber_n(x) + i\,\text{Bei}_n(x)\} + B\{\text{Ker}_n(x) + i\,\text{Kei}_n(x)\}$$

Graphs of Bessel Functions

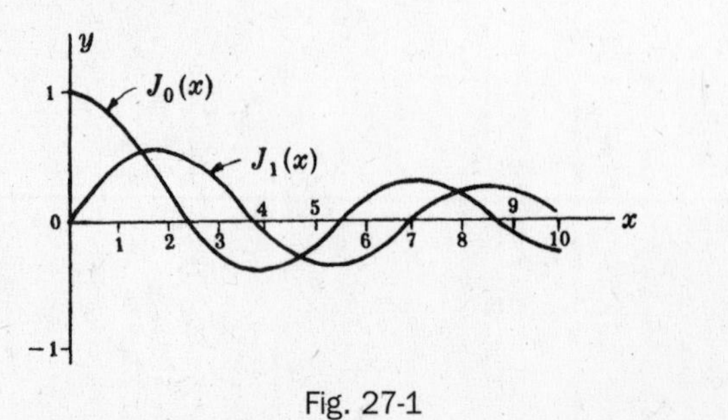

Fig. 27-1

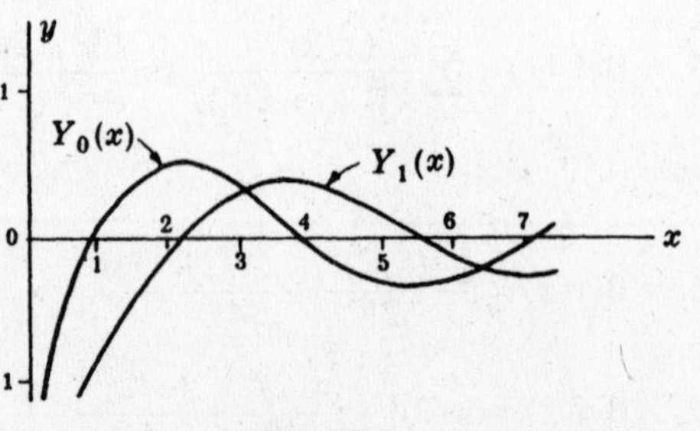

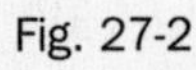

Fig. 27-2

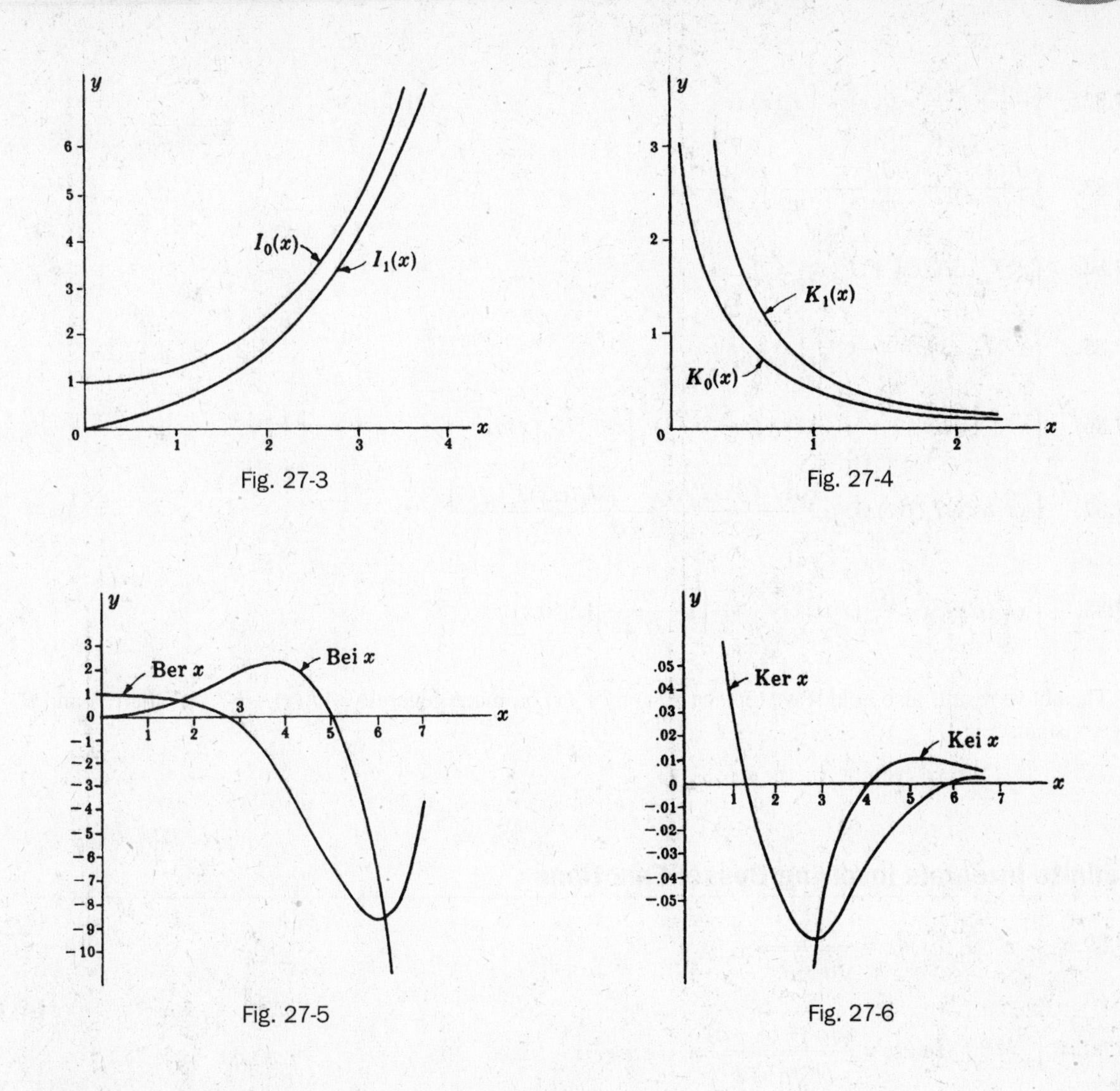

Fig. 27-3

Fig. 27-4

Fig. 27-5

Fig. 27-6

Indefinite Integrals Involving Bessel Functions

27.74. $\int xJ_0(x)dx = xJ_1(x)$

27.75. $\int x^2J_0(x)dx = x^2J_1(x) + xJ_0(x) - \int J_0(x)dx$

27.76. $\int x^mJ_0(x)dx = x^mJ_1(x) + (m-1)x^{m-1}J_0(x) - (m-1)^2 \int x^{m-2}J_0(x)dx$

27.77. $\int \frac{J_0(x)}{x^2}dx = J_1(x) - \frac{J_0(x)}{x} - \int J_0(x)dx$

27.78. $\int \frac{J_0(x)}{x^m}dx = \frac{J_1(x)}{(m-1)^2x^{m-2}} - \frac{J_0(x)}{(m-1)x^{m-1}} - \frac{1}{(m-1)^2} \int \frac{J_0(x)}{x^{m-2}}dx$

27.79. $\int J_1(x)dx = -J_0(x)$

27.80. $\int xJ_1(x)dx = -xJ_0(x) + \int J_0(x)dx$

27.81. $\int x^mJ_1(x)dx = -x^mJ_0(x) + m \int x^{m-1}J_0(x)dx$

27.82. $\int \frac{J_1(x)}{x}dx = -J_1(x) + \int J_0(x)dx$

27.83. $\int \frac{J_1(x)}{x^m}dx = -\frac{J_1(x)}{mx^{m-1}} + \frac{1}{m}\int \frac{J_0(x)}{x^{m-1}}dx$

27.84. $\int x^n J_{n-1}(x)dx = x^n J_n(x)$

27.85. $\int x^{-n} J_{n+1}(x)dx = -x^{-n} J_n(x)$

27.86. $\int x^m J_n(x)dx = -x^m J_{n-1}(x) + (m+n-1)\int x^{m-1} J_{n-1}(x)dx$

27.87. $\int xJ_n(\alpha x)J_n(\beta x)dx = \frac{x\{\alpha J_n(\beta x)J_n'(\alpha x) - \beta J_n(\alpha x)J_n'(\beta x)\}}{\beta^2 - \alpha^2}$

27.88. $\int xJ_n^2(\alpha x)dx = \frac{x^2}{2}\{J_n'(\alpha x)\}^2 + \frac{x^2}{2}\left(1 - \frac{n^2}{\alpha^2 x^2}\right)\{J_n(\alpha x)\}^2$

The above results also hold if we replace $J_n(x)$ by $Y_n(x)$ or, more generally, $AJ_n(x) + BY_n(x)$ where A and B are constants.

Definite Integrals Involving Bessel Functions

27.89. $\int_0^\infty e^{-ax} J_0(bx)dx = \frac{1}{\sqrt{a^2+b^2}}$

27.90. $\int_0^\infty e^{-ax} J_n(bx)dx = \frac{(\sqrt{a^2+b^2} - a)^n}{b^n\sqrt{a^2+b^2}} \qquad n > -1$

27.91. $\int_0^\infty \cos ax\, J_0(bx)dx = \begin{cases} \frac{1}{\sqrt{a^2-b^2}} & a > b \\ 0 & a < b \end{cases}$

27.92. $\int_0^\infty J_n(bx)dx = \frac{1}{b}, \qquad n > -1$

27.93. $\int_0^\infty \frac{J_n(bx)}{x}dx = \frac{1}{n}, \qquad n = 1, 2, 3, \ldots$

27.94. $\int_0^\infty e^{-ax} J_0(b\sqrt{x})dx = \frac{e^{-b^2/4a}}{a}$

27.95. $\int_0^1 xJ_n(\alpha x)J_n(\beta x)dx = \frac{\alpha J_n(\beta)J_n'(\alpha) - \beta J_n(\alpha)J_n'(\beta)}{\beta^2 - \alpha^2}$

27.96. $\int_0^1 xJ_n^2(\alpha x)dx = \frac{1}{2}\{J_n'(\alpha)\}^2 + \frac{1}{2}(1 - n^2/\alpha^2)\{J_n(\alpha)\}^2$

27.97. $\int_0^1 xJ_0(\alpha x)I_0(\beta x)dx = \frac{\beta J_0(\alpha)I_0'(\beta) - \alpha J_0'(\alpha)I_0(\beta)}{\alpha^2 + \beta^2}$

Integral Representations for Bessel Functions

27.98. $J_0(x) = \frac{1}{\pi}\int_0^{\pi} \cos(x\sin\theta)d\theta$

27.99. $J_n(x) = \frac{1}{\pi}\int_0^{\pi} \cos(n\theta - x\sin\theta)d\theta \qquad n = \text{integer}$

27.100. $J_n(x) = \frac{x^n}{2^n\sqrt{\pi}\,\Gamma(n+\frac{1}{2})}\int_0^{\pi} \cos(x\sin\theta)\cos^{2n}\theta d\theta, \qquad n > -\frac{1}{2}$

27.101. $Y_0(x) = -\frac{2}{\pi}\int_0^{\infty} \cos(x\cosh u)du$

27.102. $I_0(x) = \frac{1}{\pi}\int_0^{\pi} \cosh(x\sin\theta)d\theta = \frac{1}{2\pi}\int_0^{2\pi} e^{x\sin\theta}d\theta$

Asymptotic Expansions

27.103. $J_n(x) \sim \sqrt{\frac{2}{\pi x}}\cos\left(x - \frac{n\pi}{2} - \frac{\pi}{4}\right)$ where x is large

27.104. $Y_n(x) \sim \sqrt{\frac{2}{\pi x}}\sin\left(x - \frac{n\pi}{2} - \frac{\pi}{4}\right)$ where x is large

27.105. $J_n(x) \sim \frac{1}{\sqrt{2\pi n}}\left(\frac{ex}{2n}\right)^n$ where n is large

27.106. $Y_n(x) \sim -\sqrt{\frac{2}{\pi n}}\left(\frac{ex}{2n}\right)^{-n}$ where n is large

27.107 $I_n(x) \sim \frac{e^x}{\sqrt{2\pi x}}$ where x is large

27.108 $K_n(x) \sim \frac{e^{-x}}{\sqrt{2\pi x}}$ where x is large

Orthogonal Series of Bessel Functions

Let $\lambda_1, \lambda_2, \lambda_3, \ldots$ be the positive roots of $RJ_n(x) + SxJ_n'(x) = 0,\ n > -1$. Then the following series expansions hold under the conditions indicated.

$S = 0,\ R \neq 0$, i.e., $\lambda_1, \lambda_2, \lambda_3, \ldots$ are positive roots of $J_N(x) = 0$

27.109. $f(x) = A_1J_n(\lambda_1x) + A_2J_n(\lambda_2x) + A_3J_n(\lambda_3x) + \cdots$

where

27.110. $A_k = \frac{2}{J_{n+1}^2(\lambda_k)}\int_0^1 xf(x)J_n(\lambda_kx)dx$

In psarticular if $n = 0$,

27.111. $f(x) = A_1J_0(\lambda_1x) + A_2J_0(\lambda_2x) + A_3J_0(\lambda_3x) + \cdots$

where

27.112. $A_k = \frac{2}{J_1^2(\lambda_k)}\int_0^1 xf(x)J_0(\lambda_kx)dx$

$R/S > -n$

27.113. $f(x) = A_1 J_n(\lambda_1 x) + A_2 J_n(\lambda_2 x) + A_3 J_n(\lambda_3 x) + \cdots$

where

27.114. $A_k = \dfrac{2}{J_n^2(\lambda_k) - J_{n-1}(\lambda_k) J_{n+1}(\lambda_k)} \int_0^1 x f(x) J_n(\lambda_k x)\, dx$

In particular if $n = 0$.

27.115. $f(x) = A_1 J_0(\lambda_1 x) + A_2 J_0(\lambda_2 x) + A_3 J_0(\lambda_3 x) + \cdots$

where

27.116. $A_k = \dfrac{2}{J_0^2(\lambda_k) + J_1^2(\lambda_k)} \int_0^1 x f(x) J_0(\lambda_k x)\, dx$

The next formulas refer to the expansion of Bessel functions where $S \neq 0$.

$R/S = -n$

27.117. $f(x) = A_0 x^n + A_1 J_n(\lambda_1 x) + A_2 J_n(\lambda_2 x) + \cdots$

where

27.118. $\begin{cases} A_0 = 2(n+1) \int_0^1 x^{n+1} f(x)\, dx \\ A_k = \dfrac{2}{J_n^2(\lambda_k) - J_{n-1}(\lambda_k) J_{n+1}(\lambda_k)} \int_0^1 x f(x) J_n(\lambda_k x)\, dx \end{cases}$

In particular if $n = 0$ so that $R = 0$ [i.e., $\lambda_1, \lambda_2, \lambda_3, \ldots$ are the positive roots of $J_1(x) = 0$],

27.119. $f(x) = A_0 + A_1 J_0(\lambda_1 x) + A_2 J_0(\lambda_2 x) + \cdots$

where

27.120. $\begin{cases} A_0 = 2 \int_0^1 x f(x)\, dx \\ A_k = \dfrac{2}{J_0^2(\lambda_k)} \int_0^1 x f(x) J_0(\lambda_k x)\, dx \end{cases}$

$R/S < -N$

In this case there are two pure imaginary roots $\pm i\lambda_0$ as well the positive roots $\lambda_1, \lambda_2, \lambda_3, \ldots$ and we have

27.121. $f(x) = A_0 I_n(\lambda_0 x) + A_1 J_n(\lambda_1 x) + A_2 J_n(\lambda_2 x) + \cdots$

where

27.122. $\begin{cases} A_0 = \dfrac{2}{I_n^2(\lambda_0) + I_{n-1}(\lambda_0) I_{n+1}(\lambda_0)} \int_0^1 x f(x) I_n(\lambda_0 x)\, dx \\ A_k = \dfrac{2}{J_n^2(\lambda_k) - J_{n-1}(\lambda_k) J_{n+1}(\lambda_k)} \int_0^1 x f(x) J_n(\lambda_k x)\, dx \end{cases}$

Miscellaneous Results

27.123. $\cos(x\sin\theta) = J_0(x) + 2J_2(x)\cos 2\theta + 2J_4(x)\cos 4\theta + \cdots$

27.124. $\sin(x\sin\theta) = 2J_1(x)\sin\theta + 2J_3(x)\sin 3\theta + 2J_5(x)\sin 5\theta + \cdots$

27.125. $J_n(x+y) = \sum_{k=-\infty}^{\infty} J_k(x)J_{n-k}(y) \qquad n = 0, \pm 1, \pm 2, \ldots$

This is called the *addition formula* for Bessel functions.

27.126. $1 = J_0(x) + 2J_2(x) + \cdots + 2J_{2n}(x) + \cdots$

27.127. $x = 2\{J_1(x) + 3J_3(x) + 5J_5(x) + \cdots + (2n+1)J_{2n+1}(x) + \cdots\}$

27.128. $x^2 = 2\{4J_2(x) + 16J_4(x) + 36J_6(x) + \cdots + (2n)^2 J_{2n}(x) + \cdots\}$

27.129. $\dfrac{xJ_1(x)}{4} = J_2(x) - 2J_4(x) + 3J_6(x) - \cdots$

27.130. $1 = J_0^2(x) + 2J_1^2(x) + 2J_2^2(x) + 2J_3^2(x) + \cdots$

27.131. $J_n''(x) = \frac{1}{4}\{J_{n-2}(x) - 2J_n(x) + J_{n+2}(x)\}$

27.132. $J_n'''(x) = \frac{1}{8}\{J_{n-3}(x) - 3J_{n-1}(x) + 3J_{n+1}(x) - J_{n+3}(x)\}$

Formulas 27.131 and 27.132 can be generalized.

27.133. $J_n'(x)J_{-n}(x) - J_{-n}'J_n(x) = \dfrac{2\sin n\pi}{\pi x}$

27.134. $J_n(x)J_{-n+1}(x) + J_{-n}(x)J_{n-1}(x) = \dfrac{2\sin n\pi}{\pi x}$

27.135. $J_{n+1}(x)Y_n(x) - J_n(x)Y_{n+1}(x) = J_n(x)Y_n'(x) - J_n'(x)Y_n(x) = \dfrac{2}{\pi x}$

27.136. $\sin x = 2\{J_1(x) - J_3(x) + J_5(x) - \cdots\}$

27.137. $\cos x = J_0(x) - 2J_2(x) + 2J_4(x) - \cdots$

27.138. $\sinh x = 2\{I_1(x) + I_3(x) + I_5(x) + \cdots\}$

27.139. $\cosh x = I_0(x) + 2\{I_2(x) + I_4(x) + I_6(x) + \cdots\}$

28 LEGENDRE and ASSOCIATED LEGENDRE FUNCTIONS

Legendre's Differential Equation

28.1. $(1-x^2)y'' - 2xy' + n(n+1)y = 0$

Solutions of this equation are called *Legendre functions of order n.*

Legendre Polynomials

If $n = 0, 1, 2, \ldots$, a solution of 28.1 is the Legendre polynomial $P_n(x)$ given by *Rodrigues' formula*

28.2. $P_n(x) = \dfrac{1}{2^n n!} \dfrac{d^n}{dx^n}(x^2-1)^n$

Special Legendre Polynomials

28.3. $P_0(x) = 1$

28.4. $P_1(x) = x$

28.5. $P_2(x) = \frac{1}{2}(3x^2 - 1)$

28.6. $P_3(x) = \frac{1}{2}(5x^3 - 3x)$

28.7. $P_4(x) = \frac{1}{8}(35x^4 - 30x^2 + 3)$

28.8. $P_5(x) = \frac{1}{8}(63x^5 - 70x^3 + 15x)$

28.9. $P_6(x) = \frac{1}{16}(231x^6 - 315x^4 + 105x^2 - 5)$

28.10. $P_7(x) = \frac{1}{16}(429x^7 - 693x^5 + 315x^3 - 35x)$

Legendre Polynomials in Terms of θ where $x = \cos\theta$

28.11. $P_0(\cos\theta) = 1$

28.12. $P_1(\cos\theta) = \cos\theta$

28.13. $P_2(\cos\theta) = \frac{1}{4}(1 + 3\cos 2\theta)$

28.14. $P_3(\cos\theta) = \frac{1}{8}(3\cos\theta + 5\cos 3\theta)$

28.15. $P_4(\cos\theta) = \frac{1}{64}(9 + 20\cos 2\theta + 35\cos 4\theta)$

28.16. $P_5(\cos\theta) = \frac{1}{128}(30\cos\theta + 35\cos 3\theta + 63\cos 5\theta)$

28.17. $P_6(\cos\theta) = \frac{1}{512}(50 + 105\cos 2\theta + 126\cos 4\theta + 231\cos 6\theta)$

28.18. $P_7(\cos\theta) = \frac{1}{1024}(175\cos\theta + 189\cos 3\theta + 231\cos 5\theta + 429\cos 7\theta)$

Generating Function for Legendre Polynomials

28.19. $$\frac{1}{\sqrt{1-2tx+t^2}} = \sum_{n=0}^{\infty} P_n(x)t^n$$

Recurrence Formulas for Legendre Polynomials

28.20. $(n+1)P_{n+1}(x) - (2n+1)x\,P_n(x) + nP_{n-1}(x) = 0$

28.21. $P'_{n+1}(x) - xP'_n(x) = (n+1)P_n(x)$

28.22. $xP'_n(x) - P'_{n-1}(x) = nP_n(x)$

28.23. $P'_{n+1}(x) - P'_{n-1}(x) = (2n+1)P_n(x)$

28.24. $(x^2-1)P'_n(x) - nxP_n(x) - nP_{n-1}(x)$

Orthogonality of Legendre Polynomials

28.25. $\int_{-1}^{1} P_m(x)P_n(x)dx = 0 \quad m \neq n$

28.26. $$\int_{-1}^{1} \{P_n(x)\}^2 dx = \frac{2}{2n+1}$$

Because of 28.25, $P_m(x)$ and $P_n(x)$ are called *orthogonal* in $-1 \leqq x \leqq 1$.

Orthogonal Series of Legendre Polynomials

28.27. $f(x) = A_0P_0(x) + A_1P_1(x) + A_2P_2(x) + \cdots$

where

28.28. $$A_k = \frac{2k+1}{2}\int_{-1}^{1} f(x)P_k(x)dx$$

Special Results Involving Legendre Polynomials

28.29. $P_n(1) = 1$

28.30. $P_n(-1) = (-1)^n$

28.31. $P_n(-x) = (-1)^n P_n(x)$

28.32. $$P_n(0) = \begin{cases} 0 & n \text{ odd} \\ (-1)^{n/2} \dfrac{1 \cdot 3 \cdot 5 \cdots (n-1)}{2 \cdot 4 \cdot 6 \cdots n} & n \text{ even} \end{cases}$$

28.33. $$P_n(x) = \frac{1}{\pi} \int_0^{\pi} \left(x + \sqrt{x^2 - 1} \cos \phi\right)^n d\phi$$

28.34. $$\int P_n(x)dx = \frac{P_{n+1}(x) - P_{n-1}(x)}{2n+1}$$

28.35. $|P_n(x)| \leqq 1$

28.36. $$P_n(x) = \frac{1}{2^{n+1}\pi i} \oint_c \frac{(z^2-1)^n}{(z-x)^{n+1}} dz$$

where C is a simple closed curve having x as interior point.

General Solution of Legendre's Equation

The general solution of Legendre's equation is

28.37. $y = AU_n(x) + BV_n(x)$

where

28.38. $$U_n(x) = 1 - \frac{n(n+1)}{2!} x^2 + \frac{n(n-2)(n+1)(n+3)}{4!} x^4 - \cdots$$

28.39. $$V_n(x) = x - \frac{(n-1)(n+2)}{3!} x^3 + \frac{(n-1)(n-3)(n+2)(n+4)}{5!} x^5 - \cdots$$

These series converge for $-1 < x < 1$.

Legendre Functions of the Second Kind

If $n = 0, 1, 2, \ldots$ one of the series 28.38, 28.39 terminates. In such cases,

28.40. $$P_n(x) = \begin{cases} U_n(x)/U_n(1) & n = 0, 2, 4, \ldots \\ V_n(x)/V_n(1) & n = 1, 3, 5, \ldots \end{cases}$$

where

28.41. $$U_n(1) = (-1)^{n/2} 2^n \left[\left(\frac{n}{2}\right)!\right]^2 \Big/ n! \qquad n = 0, 2, 4, \ldots$$

28.42. $V_n(1) = (-1)^{(n-1)/2} 2^{n-1} \left[\left(\frac{n-1}{2} \right)! \right]^2 \Big/ n! \qquad n = 1, 3, 5, \ldots$

The nonterminating series in such a case with a suitable multiplicative constant is denoted by $Q_n(x)$ and is called *Legendre's function of the second kind of order n*. We define

28.43. $Q_n(x) = \begin{cases} U_n(1)V_n(x) & n = 0, 2, 4, \ldots \\ -V_n(1)U_n(x) & n = 1, 3, 5, \ldots \end{cases}$

Special Legendre Functions of the Second Kind

28.44. $Q_0(x) = \frac{1}{2} \ln\left(\frac{1+x}{1-x} \right)$

28.45. $Q_1(x) = \frac{x}{2} \ln\left(\frac{1+x}{1-x} \right) - 1$

28.46. $Q_2(x) = \frac{3x^2 - 1}{4} \ln\left(\frac{1+x}{1-x} \right) - \frac{3x}{2}$

28.47. $Q_3(x) = \frac{5x^3 - 3x}{4} \ln\left(\frac{1+x}{1-x} \right) - \frac{5x^2}{2} + \frac{2}{3}$

The functions $Q_n(x)$ satisfy recurrence formulas exactly analogous to 28.20 through 28.24. Using these, the general solution of Legendre's equation can also be written as

28.48. $y = AP_n(x) + BQ_n(x)$

Legendre's Associated Differential Equation

28.49. $(1 - x^2)y'' - 2xy' + \left\{ n(n+1) - \frac{m^2}{1 - x^2} \right\} y = 0$

Solutions of this equation are called *associated Legendre functions*. We restrict ourselves to the important case where m, n are nonnegative integers.

Associated Legendre Functions of the First Kind

28.50. $P_n^m(x) = (1 - x^2)^{m/2} \frac{d^m}{dx^m} P_n(x) = \frac{(1 - x^2)^{m/2}}{2^n n!} \frac{d^{m+n}}{dx^{m+n}} (x^2 - 1)^n$

where $P_n(x)$ are Legendre polynomials (page 164). We have

28.51. $P_n^0(x) = P_n(x)$

28.52. $P_n^m(x) = 0 \qquad \text{if } m > n$

Special Associated Legendre Functions of the First Kind

28.53. $P_1^1(x) = (1-x^2)^{1/2}$

28.54. $P_2^1(x) = 3x(1-x^2)^{1/2}$

28.55. $P_2^2(x) = 3(1-x^2)$

28.56. $P_3^1(x) = \frac{3}{2}(5x^2-1)(1-x^2)^{1/2}$

28.57. $P_3^2(x) = 15x(1-x^2)$

28.58. $P_3^3(x) = 15(1-x^2)^{3/2}$

Generating Function for $P_n^m(x)$

28.59. $$\frac{(2m)!(1-x^2)^{m/2}t^m}{2^m m!(1-2tx+t^2)^{m+1/2}} = \sum_{n=m}^{\infty} P_n^m(x)t^n$$

Recurrence Formulas

28.60. $$(n+1-m)P_{n+1}^m(x) - (2n+1)xP_n^m(x) + (n+m)P_{n-1}^m(x) = 0$$

28.61. $$P_n^{m+2}(x) - \frac{2(m+1)x}{(1-x^2)^{1/2}}P_n^{m+1}(x) + (n-m)(n+m+1)P_n^m(x) = 0$$

Orthogonality of $P_n^m(x)$

28.62. $$\int_{-1}^{1} P_l^m(x)P_l^m(x)dx = 0 \qquad \text{if } n \neq l$$

28.63. $$\int_{-1}^{1} \{P_n^m(x)\}^2 dx = \frac{2}{2n+1}\frac{(n+m)!}{(n-m)!}$$

Orthogonal Series

28.64. $$f(x) = A_m P_m^m(x) + A_{m+1}P_{m+1}^m(x) + A_{m+2}P_{m+2}^m(x) + \cdots$$

where

28.65. $$A_k = \frac{2k+1}{2}\frac{(k-m)!}{(k+m)!}\int_{-1}^{1} f(x)P_k^m(x)dx$$

Associated Legendre Functions of the Second Kind

28.66. $$Q_n^m(x) = (1-x^2)^{m/2}\frac{d^m}{dx^m}Q_n(x)$$

where $Q_n(x)$ are Legendre functions of the second kind (page 166).

These functions are unbounded at $x = \pm 1$, whereas $P_n^m(x)$ are bounded at $x = \pm 1$.

The functions $Q_n^m(x)$ satisfy the same recurrence relations as $P_n^m(x)$ (see 28.60 and 28.61).

General Solution of Legendre's Associated Equation

28.67. $$y = AP_n^m(x) + BQ_n^m(x)$$

29 HERMITE POLYNOMIALS

Hermite's Differential Equation

29.1. $y'' - 2xy' + 2ny = 0$

Hermite Polynomials

If $n = 0, 1, 2, \ldots$, then a solution of Hermite's equation is the Hermite polynomial $H_n(x)$ given by *Rodrigue's formula.*

29.2. $H_n(x) = (-1)^n e^{x^2} \dfrac{d^n}{dx^n}(e^{-x^2})$

Special Hermite Polynomials

29.3. $H_0(x) = 1$

29.4. $H_1(x) = 2x$

29.5. $H_2(x) = 4x^2 - 2$

29.6. $H_3(x) = 8x^3 - 12x$

29.7. $H_4(x) = 16x^4 - 48x^2 + 12$

29.8. $H_5(x) = 32x^5 - 160x^3 + 120x$

29.9. $H_6(x) = 64x^6 - 480x^4 + 720x^2 - 120$

29.10. $H_7(x) = 128x^7 - 1344x^5 + 3360x^3 - 1680x$

Generating Function

29.11. $e^{2tx - t^2} = \sum_{n=0}^{\infty} \dfrac{H_n(x)t^n}{n!}$

Recurrence Formulas

29.12. $H_{n+1}(x) = 2xH_n(x) - 2nH_{n-1}(x)$

29.13. $H_n'(x) = 2nH_{n-1}(x)$

Orthogonality of Hermite Polynomials

29.14. $\int_{-\infty}^{\infty} e^{-x^2} H_m(x)H_n(x)dx = 0 \qquad m \neq n$

29.15. $\int_{-\infty}^{\infty} e^{-x^2} \{H_n(x)\}^2 dx = 2^n n! \sqrt{\pi}$

Orthogonal Series

29.16. $f(x) = A_0H_0(x) + A_1H_1(x) + A_2H_2(x) + \cdots$

where

29.17. $A_k = \dfrac{1}{2^k k!\sqrt{\pi}} \int_{-\infty}^{\infty} e^{-x^2} f(x) H_k(x)dx$

Special Results

29.18. $H_n(x) = (2x)^n - \dfrac{n(n-1)}{1!}(2x)^{n-2} + \dfrac{n(n-1)(n-2)(n-3)}{2!}(2x)^{n-4} - \cdots$

29.19. $H_n(-x) = (-1)^n H_n(x)$

29.20. $H_{2n-1}(0) = 0$

29.21. $H_{2n}(0) = (-1)^n 2^n \cdot 1 \cdot 3 \cdot 5 \cdots (2n-1)$

29.22. $\int_0^x H_n(t)dt = \dfrac{H_{n+1}(x)}{2(n+1)} - \dfrac{H_{n+1}(0)}{2(n+1)}$

29.23. $\dfrac{d}{dx}\{e^{-x^2} H_n(x)\} = -e^{-x^2} H_{n+1}(x)$

29.24. $\int_0^x e^{-t^2} H_n(t)dt = H_{n-1}(0) - e^{-x^2} H_{n-1}(x)$

29.25. $\int_{-\infty}^{\infty} t^n e^{-t^2} H_n(xt)dt = \sqrt{\pi}\, n!\, P_n(x)$

29.26. $H_n(x+y) = \sum_{k=0}^{n} \dfrac{1}{2^{n/2}} \binom{n}{k} H_k(x\sqrt{2}) H_{n-k}(y\sqrt{2})$

This is called the *addition formula* for Hermite polynomials.

29.27. $\sum_{k=0}^{n} \dfrac{H_k(x)H_k(y)}{2^k k!} = \dfrac{H_{n+1}(x)H_n(y) - H_n(x)H_{n+1}(y)}{2^{n+1} n!(x-y)}$

30 LAGUERRE and ASSOCIATED LAGUERRE POLYNOMIALS

Laguerre's Differential Equation

30.1. $xy'' + (1-x)y' + ny = 0$

Laguerre Polynomials

If $n = 0, 1, 2, \ldots$, then a solution of Laguerre's equation is the Laguerre polynomial $L_n(x)$ given by *Rodrigues' formula*

30.2. $L_n(x) = e^x \dfrac{d^n}{dx^n}(x^n e^{-x})$

Special Laguerre Polynomials

30.3. $L_0(x) = 1$

30.4. $L_1(x) = -x + 1$

30.5. $L_2(x) = x^2 - 4x + 2$

30.6. $L_3(x) = -x^3 + 9x^2 - 18x + 6$

30.7. $L_4(x) = x^4 - 16x^3 + 72x^2 - 96x + 24$

30.8. $L_5(x) = -x^5 + 25x^4 - 200x^3 + 600x^2 - 600x + 120$

30.9. $L_6(x) = x^6 - 36x^5 + 450x^4 - 2400x^3 + 5400x^2 - 4320x + 720$

30.10. $L_7(x) = -x^7 + 49x^6 - 882x^5 + 7350x^4 - 29{,}400x^3 + 52{,}920x^2 - 35{,}280x + 5040$

Generating Function

30.11. $\dfrac{e^{-xt/(1-t)}}{1-t} = \sum_{n=0}^{\infty} \dfrac{L_n(x)t^n}{n!}$

Recurrence Formulas

30.12. $L_{n+1}(x) - (2n+1-x)L_n(x) + n^2 L_{n-1}(x) = 0$

30.13. $L_n'(x) - nL_{n-1}'(x) + nL_{n-1}(x) = 0$

30.14. $xL_n'(x) = nL_n(x) - n^2 L_{n-1}(x)$

Orthogonality of Laguerre Polynomials

30.15. $\int_0^\infty e^{-x} L_m(x) L_n(x)\,dx = 0 \qquad m \neq n$

30.16. $\int_0^\infty e^{-x} \{L_n(x)\}^2\,dx = (n!)^2$

Orthogonal Series

30.17. $f(x) = A_0 L_0(x) + A_1 L_1(x) + A_2 L_2(x) + \cdots$

where

30.18. $A_k = \dfrac{1}{(k!)^2} \int_0^\infty e^{-x} f(x) L_k(x)\,dx$

Special Results

30.19. $L_n(0) = n!$

30.20. $\int_0^x L_n(t)\,dt = L_n(x) - \dfrac{L_{n+1}(x)}{n+1}$

30.21. $L_n(x) = (-1)^n \left\{ x^n - \dfrac{n^2 x^{n-1}}{1!} + \dfrac{n^2(n-1)^2 x^{n-2}}{2!} - \cdots (-1)^n n! \right\}$

30.22. $\int_0^\infty x^p e^{-x} L_n(x)\,dx = \begin{cases} 0 & \text{if } p < n \\ (-1)^n (n!)^2 & \text{if } p = n \end{cases}$

30.23. $\sum_{k=0}^{n} \dfrac{L_k(x) L_k(y)}{(k!)^2} = \dfrac{L_n(x) L_{n+1}(y) - L_{n+1}(x) L_n(y)}{(n!)^2 (x-y)}$

30.24. $\sum_{k=0}^{\infty} \dfrac{t^k L_k(x)}{(k!)^2} = e^t J_0(2\sqrt{xt})$

30.25. $L_n(x) = \int_0^\infty u^n e^{x-u} J_0(2\sqrt{xu})\,du$

Laguerre's Associated Differential Equation

30.26. $xy'' + (m+1-x)y' + (n-m)y = 0$

Associated Laguerre Polynomials

Solutions of 30.26 for nonnegative integers m and n are given by the associated Laguerre polynomials

30.27. $L_n^m(x) = \frac{d^m}{dx^m} L_n(x)$

where $L_n(x)$ are Laguerre polynomials (see page 171).

30.28. $L_n^0(x) = L_n(x)$

30.29. $L_n^m(x) = 0 \qquad \text{if } m > n$

Special Associated Laguerre Polynomials

30.30. $L_1^1(x) = -1$

30.31. $L_2^1(x) = 2x - 4$

30.32. $L_2^2(x) = 2$

30.33. $L_3^1(x) = -3x^2 + 18x - 18$

30.34. $L_3^2(x) = -6x + 18$

30.35. $L_3^3(x) = -6$

30.36. $L_4^1(x) = 4x^3 - 48x^2 + 144x - 96$

30.37. $L_4^2(x) = 12x^2 - 96x + 144$

30.38. $L_4^3(x) = 24x - 96$

30.39. $L_4^4(x) = 24$

Generating Function for $L_n^m(x)$

30.40. $\frac{(-1)^m t^m}{(1-t)^{m+1}} e^{-xt/(1-t)} = \sum_{n=m}^{\infty} \frac{L_n^m(x)}{n!} t^n$

Recurrence Formulas

30.41. $\frac{n-m+1}{n+1} L_{n+1}^m(x) + (x+m-2n-1)L_n^m(x) + n^2 L_{n-1}^m(x) = 0$

30.42. $\frac{d}{dx}\{L_n^m(x)\} = L_n^{m+1}(x)$

30.43. $\frac{d}{dx}\{x^m e^{-x} L_n^m(x)\} = (m-n-1)x^{m-1} e^{-x} L_n^{m-1}(x)$

30.44. $x\frac{d}{dx}\{L_n^m(x)\} = (x-m)L_n^m(x) + (m-n-1)L_n^{m-1}(x)$

Orthogonality

30.45. $\int_0^\infty x^m e^{-x} L_n^m(x) L_p^m(x)dx = 0 \quad p \neq n$

30.46. $\int_0^\infty x^m e^{-x} \{L_n^m(x)\}^2 dx = \frac{(n!)^3}{(n-m)!}$

Orthogonal Series

30.47. $f(x) = A_m L_m^m(x) + A_{m+1} L_{m+1}^m(x) + A_{m+2} L_{m+2}^m(x) + \cdots$

where

30.48. $A_k = \frac{(k-m)!}{(k!)^3} \int_0^\infty x^m e^{-x} L_k^m(x) f(x) dx$

Special Results

30.49. $L_n^m(x) = (-1)^n \frac{n!}{(n-m)!} \left\{ x^{n-m} - \frac{n(n-m)}{1!} x^{n-m-1} + \frac{n(n-1)(n-m)(n-m-1)}{2!} x^{n-m-2} + \cdots \right\}$

30.50. $\int_0^\infty x^{m+1} e^{-x} \{L_n^m(x)\}^2 dx = \frac{(2n-m+1)(n!)^3}{(n-m)!}$

31 CHEBYSHEV POLYNOMIALS

Chebyshev's Differential Equation

31.1. $(1-x^2)y'' - xy' + n^2 y = 0 \qquad n = 0, 1, 2, \ldots$

Chebyshev Polynomials of the First Kind

A solution of 31.1 is given by

31.2. $T_n(x) = \cos(n\cos^{-1} x) = x^n - \binom{n}{2} x^{n-2}(1-x^2) + \binom{n}{4} x^{n-4}(1-x^2)^2 - \cdots$

Special Chebyshev Polynomials of The First Kind

31.3. $T_0(x) = 1$

31.4. $T_1(x) = x$

31.5. $T_2(x) = 2x^2 - 1$

31.6. $T_3(x) = 4x^3 - 3x$

31.7. $T_4(x) = 8x^4 - 8x^2 + 1$

31.8. $T_5(x) = 16x^5 - 20x^3 + 5x$

31.9. $T_6(x) = 32x^6 - 48x^4 + 18x^2 - 1$

31.10. $T_7(x) = 64x^7 - 112x^5 + 56x^3 - 7x$

Generating Function for $T_n(x)$

31.11. $\dfrac{1-tx}{1-2tx+t^2} = \sum_{n=0}^{\infty} T_n(x)t^n$

Special Values

31.12. $T_n(-x) = (-1)^n T_n(x)$

31.13. $T_n(1) = 1$

31.14. $T_n(-1) = (-1)^n$

31.15. $T_{2n}(0) = (-1)^n$

31.16. $T_{2n+1}(0) = 0$

Recursion Formula for $T_n(x)$

31.17. $T_{n+1}(x) - 2xT_n(x) + T_{n-1}(x) = 0$

Orthogonality

31.18. $\int_{-1}^{1} \frac{T_m(x)T_n(x)}{\sqrt{1-x^2}} dx = 0 \qquad m \neq n$

31.19. $\int_{-1}^{1} \frac{\{T_n(x)\}^2}{\sqrt{1-x^2}} dx = \begin{cases} \pi & \text{if } n = 0 \\ \pi/2 & \text{if } n = 1, 2, \ldots \end{cases}$

Orthogonal Series

31.20. $f(x) = \frac{1}{2} A_0 T_0(x) + A_1 T_1(x) + A_2 T_2(x) + \cdots$

where

31.21. $A_k = \frac{2}{\pi} \int_{-1}^{1} \frac{f(x) T_k(x)}{\sqrt{1-x^2}} dx$

Chebyshev Polynomials of The Second Kind

31.22. $U_n(x) = \frac{\sin\{(n+1)\cos^{-1} x\}}{\sin(\cos^{-1} x)}$

$$= \binom{n+1}{1} x^n - \binom{n+1}{3} x^{n-2}(1-x^2) + \binom{n+1}{5} x^{n-4}(1-x^2)^2 - \cdots$$

Special Chebyshev Polynomials of The Second Kind

31.23. $U_0(x) = 1$

31.24. $U_1(x) = 2x$

31.25. $U_2(x) = 4x^2 - 1$

31.26. $U_3(x) = 8x^3 - 4x$

31.27. $U_4(x) = 16x^4 - 12x^2 + 1$

31.28. $U_5(x) = 32x^5 - 32x^3 + 6x$

31.29. $U_6(x) = 64x^6 - 80x^4 + 24x^2 - 1$

31.30. $U_7(x) = 128x^7 - 192x^5 + 80x^3 - 8x$

Generating Function for $U_n(x)$

31.31. $\frac{1}{1 - 2tx + t^2} = \sum_{n=0}^{\infty} U_n(x) t^n$

Special Values

31.32. $U_n(-x) = (-1)^n U_n(x)$

31.33. $U_n(1) = n + 1$

31.34. $U_n(-1) = (-1)^n (n+1)$

31.35. $U_{2n}(0) = (-1)^n$

31.36. $U_{2n+1}(0) = 0$

Recursion Formula for $U_n(x)$

31.37. $U_{n+1}(x) - 2xU_n(x) + U_{n-1}(x) = 0$

Orthogonality

31.38. $\int_{-1}^{1} \sqrt{1-x^2}\, U_m(x) U_n(x) dx = 0 \qquad m \neq n$

31.39. $\int_{-1}^{1} \sqrt{1-x^2}\, \{U_n(x)\}^2 dx = \frac{\pi}{2}$

Orthogonal Series

31.40. $f(x) = A_0 U_0(x) + A_1 U_1(x) + A_2 U_2(x) + \cdots$

where

31.41. $A_k = \frac{2}{\pi} \int_{-1}^{1} \sqrt{1-x^2}\, f(x) U_k(x) dx$

Relationships Between $T_n(x)$ and $U_n(x)$

31.42. $T_n(x) = U_n(x) - xU_{n-1}(x)$

31.43. $(1-x^2)U_{n-1}(x) = xT_n(x) - T_{n+1}(x)$

31.44. $U_n(x) = \frac{1}{\pi} \int_{-1}^{1} \frac{T_{n+1}(v) dv}{(v-x)\sqrt{1-v^2}}$

31.45. $T_n(x) = \frac{1}{\pi} \int_{-1}^{1} \frac{\sqrt{1-v^2}\, U_{n-1}(v)}{x-v} dv$

General Solution of Chebyshev's Differential Equation

31.46. $y = \begin{cases} AT_n(x) + B\sqrt{1-x^2}\, U_{n-1}(x) & \text{if } n = 1, 2, 3, \ldots \\ A + B\sin^{-1} x & \text{if } n = 0 \end{cases}$

32 HYPERGEOMETRIC FUNCTIONS

Hypergeometric Differential Equation

32.1. $x(1-x)y'' + \{c-(a+b+1)x\}y' - aby = 0$

Hypergeometric Functions

A solution of 32.1 is given by

32.2. $F(a,b;c;x) = 1 + \frac{a \cdot b}{1 \cdot c}x + \frac{a(a+1)b(b+1)}{1 \cdot 2 \cdot c(c+1)}x^2 + \frac{a(a+1)(a+2)b(b+1)(b+2)}{1 \cdot 2 \cdot 3 \cdot c(c+1)(c+2)}x^3 + \cdots$

If a, b, c are real, then the series converges for $-1 < x < 1$ provided that $c - (a + b) > -1$.

Special Cases

32.3. $F(-p,1;1;-x) = (1+x)^p$

32.4. $F(1,1;2;-x) = [\ln(1+x)]/x$

32.5. $\lim_{n\to\infty} F(1,n;1;x/n) = e^x$

32.6. $F(\frac{1}{2},-\frac{1}{2};\frac{1}{2};\sin^2 x) = \cos x$

32.7. $F(\frac{1}{2},1;1;\sin^2 x) = \sec x$

32.8. $F(\frac{1}{2},\frac{1}{2};\frac{3}{2};x^2) = (\sin^{-1} x)/x$

32.9. $F(\frac{1}{2},1;\frac{3}{2};-x^2) = (\tan^{-1} x)/x$

32.10. $F(1,p;p;x) = 1/(1-x)$

32.11. $F(n+1,-n;1;(1-x)/2) = P_n(x)$

32.12. $F(n,-n;\frac{1}{2};(1-x)/2) = T_n(x)$

General Solution of The Hypergeometric Equation

If c, $a-b$ and $c-a-b$ are all nonintegers, then the general solution valid for $|x| < 1$ is

32.13. $y = AF(a,b;c;x) + Bx^{1-c}F(a-c+1,b-c+1;2-c;x)$

Miscellaneous Properties

32.14. $F(a,b;c;1) = \dfrac{\Gamma(c)\Gamma(c-a-b)}{\Gamma(c-a)\Gamma(c-b)}$

32.15. $\dfrac{d}{dx}F(a,b;c;x) = \dfrac{ab}{c}F(a+1,b+1;c+1;x)$

32.16. $F(a,b;c;x) = \dfrac{\Gamma(c)}{\Gamma(b)\Gamma(c-b)}\int_0^1 u^{b-1}(1-u)^{c-b-1}(1-ux)^{-a}\,du$

32.17. $F(a,b;c;x) = (1-x)^{c-a-b}F(c-a,c-b;c;x)$

33 LAPLACE TRANSFORMS

Definition of the Laplace Transform of $F(t)$

33.1. $\mathscr{L}\{F(t)\} = \int_0^\infty e^{-st}F(t)dt = f(s)$

In general $f(s)$ will exist for $s > \alpha$ where α is some constant. $\mathscr{L}$ is called the *Laplace transform operator.*

Definition of the Inverse Laplace Transform of $f(s)$

If $\mathscr{L}\{F(t)\} = f(s)$, then we say that $F(t) = \mathscr{L}^{-1}\{f(s)\}$ is the *inverse Laplace transform* of $f(s)$. $\mathscr{L}^{-1}$ is called the *inverse Laplace transform operator.*

Complex Inversion Formula

The inverse Laplace transform of $f(s)$ can be found directly by methods of complex variable theory. The result is

33.2. $F(t) = \dfrac{1}{2\pi i}\displaystyle\int_{c-i\infty}^{c+i\infty} e^{st}f(s)ds = \dfrac{1}{2\pi i}\lim_{T\to\infty}\int_{c-iT}^{c+iT} e^{st}f(s)ds$

where c is chosen so that all the singular points of $f(s)$ lie to the left of the line $\text{Re}\{s\} = c$ in the complex s plane.

Table of General Properties of Laplace Transforms

	$f(s)$	$F(t)$
33.3.	$af_1(s)+bf_2(s)$	$aF_1(t)+bF_2(t)$
33.4.	$f(s/a)$	$aF(at)$
33.5.	$f(s-a)$	$e^{at}F(t)$
33.6.	$e^{-as}f(s)$	$\mathcal{U}(t-a)=\begin{cases} F(t-a) & t>a \\ 0 & t<a \end{cases}$
33.7.	$sf(s)-F(0)$	$F'(t)$
33.8.	$s^2f(s)-sF(0)-F'(0)$	$F''(t)$
33.9.	$s^nf(s)-s^{n-1}F(0)-s^{n-2}F'(0)-\cdots-F^{(n-1)}(0)$	$F^{(n)}(t)$
33.10.	$f'(s)$	$-tF(t)$
33.11.	$f''(s)$	$t^2F(t)$
33.12.	$f^{(n)}(s)$	$(-1)^nt^nF(t)$
33.13.	$\dfrac{f(s)}{s}$	$\int_0^t F(u)du$
33.14.	$\dfrac{f(s)}{s^n}$	$\int_0^t\cdots\int_0^t F(u)du^n = \int_0^t \dfrac{(t-u)^{n-1}}{(n-1)!}F(u)du$
33.15.	$f(s)g(s)$	$\int_0^t F(u)G(t-u)du$

	$f(s)$	$F(t)$
33.16.	$\int_s^\infty f(u)du$	$\frac{F(t)}{t}$
33.17.	$\frac{1}{1-e^{-sT}}\int_0^T e^{-su}F(u)du$	$F(t) = F(t+T)$
33.18.	$\frac{f(\sqrt{s})}{\sqrt{s}}$	$\frac{1}{\sqrt{\pi t}}\int_0^\infty e^{-u^2/4t}F(u)du$
33.19.	$\frac{1}{s}f\left(\frac{1}{s}\right)$	$\int_0^\infty J_0(2\sqrt{ut})F(u)du$
33.20.	$\frac{1}{s^{n+1}}f\left(\frac{1}{s}\right)$	$t^{n/2}\int_0^\infty u^{-n/2}J_n(2\sqrt{ut})F(u)du$
33.21.	$\frac{f(s+1/s)}{s^2+1}$	$\int_0^t J_0(2\sqrt{u(t-u)})F(u)du$
33.22.	$\frac{1}{2\sqrt{\pi}}\int_0^\infty u^{-3/2}e^{-s^2/4u}f(u)du$	$F(t^2)$
33.23.	$\frac{f(\ln s)}{s\ln s}$	$\int_0^\infty \frac{t^u F(u)}{\Gamma(u+1)}du$
33.24.	$\frac{P(s)}{Q(s)}$ $P(s)$ = polynomial of degree less than n, $Q(s) = (s-\alpha_1)(s-\alpha_2)\ldots(s-\alpha_n)$ where $\alpha_1, \alpha_2, \ldots, \alpha_n$ are all distinct.	$\sum_{k=1}^{n}\frac{P(\alpha_k)}{Q'(\alpha_k)}e^{\alpha_k t}$

Table of Special Laplace Transforms

	$f(s)$	$F(t)$
33.25.	$\dfrac{1}{s}$	1
33.26.	$\dfrac{1}{s^2}$	t
33.27.	$\dfrac{1}{s^n} \quad n = 1, 2, 3, \ldots$	$\dfrac{t^{n-1}}{(n-1)!}, \quad 0! = 1$
33.28.	$\dfrac{1}{s^n} \quad n > 0$	$\dfrac{t^{n-1}}{\Gamma(n)}$
33.29.	$\dfrac{1}{s-a}$	e^{at}
33.30.	$\dfrac{1}{(s-a)^n} \quad n = 1, 2, 3, \ldots$	$\dfrac{t^{n-1}e^{at}}{(n-1)!}, \quad 0! = 1$
33.31.	$\dfrac{1}{(s-a)^n} \quad n > 0$	$\dfrac{t^{n-1}e^{at}}{\Gamma(n)}$
33.32.	$\dfrac{1}{s^2+a^2}$	$\dfrac{\sin at}{a}$
33.33.	$\dfrac{s}{s^2+a^2}$	$\cos at$
33.34.	$\dfrac{1}{(s-b)^2+a^2}$	$\dfrac{e^{bt}\sin at}{a}$
33.35.	$\dfrac{s-b}{(s-b)^2+a^2}$	$e^{bt}\cos at$
33.36.	$\dfrac{1}{s^2-a^2}$	$\dfrac{\sinh at}{a}$
33.37.	$\dfrac{s}{s^2-a^2}$	$\cosh at$
33.38.	$\dfrac{1}{(s-b)^2-a^2}$	$\dfrac{e^{bt}\sinh at}{a}$

	$f(s)$	$F(t)$
33.39.	$\dfrac{s-b}{(s-b)^2-a^2}$	$e^{bt}\cosh at$
33.40.	$\dfrac{1}{(s-a)(s-b)} \quad a \neq b$	$\dfrac{e^{bt}-e^{at}}{b-a}$
33.41.	$\dfrac{s}{(s-a)(s-b)} \quad a \neq b$	$\dfrac{be^{bt}-ae^{at}}{b-a}$
33.42.	$\dfrac{1}{(s^2+a^2)^2}$	$\dfrac{\sin at - at\cos at}{2a^3}$
33.43.	$\dfrac{s}{(s^2+a^2)^2}$	$\dfrac{t\sin at}{2a}$
33.44.	$\dfrac{s^2}{(s^2+a^2)^2}$	$\dfrac{\sin at + at\cos at}{2a}$
33.45.	$\dfrac{s^3}{(s^2+a^2)^2}$	$\cos at - \frac{1}{2}at\sin at$
33.46.	$\dfrac{s^2-a^2}{(s^2+a^2)^2}$	$t\cos at$
33.47.	$\dfrac{1}{(s^2-a^2)^2}$	$\dfrac{at\cosh at - \sinh at}{2a^3}$
33.48.	$\dfrac{s}{(s^2-a^2)^2}$	$\dfrac{t\sinh at}{2a}$
33.49.	$\dfrac{s^2}{(s^2-a^2)^2}$	$\dfrac{\sinh at + at\cosh at}{2a}$
33.50.	$\dfrac{s^3}{(s^2-a^2)^2}$	$\cosh at + \frac{1}{2}at\sinh at$
33.51.	$\dfrac{s^2}{(s^2-a^2)^{3/2}}$	$t\cosh at$
33.52.	$\dfrac{1}{(s^2+a^2)^3}$	$\dfrac{(3-a^2t^2)\sin at - 3at\cos at}{8a^5}$
33.53.	$\dfrac{s}{(s^2+a^2)^3}$	$\dfrac{t\sin at - at^2\cos at}{8a^3}$
33.54.	$\dfrac{s^2}{(s^2+a^2)^3}$	$\dfrac{(1+a^2t^2)\sin at - at\cos at}{8a^3}$
33.55.	$\dfrac{s^3}{(s^2+a^2)^3}$	$\dfrac{3t\sin at + at^2\cos at}{8a}$

	$f(s)$	$F(t)$
33.56.	$\dfrac{s^4}{(s^2+a^2)^3}$	$\dfrac{(3-a^2t^2)\sin at + 5at\cos at}{8a}$
33.57.	$\dfrac{s^5}{(s^2+a^2)^3}$	$\dfrac{(8-a^2t^2)\cos at - 7at\sin at}{8}$
33.58.	$\dfrac{3s^2-a^2}{(s^2+a^2)^3}$	$\dfrac{t^2\sin at}{2a}$
33.59.	$\dfrac{s^3-3a^2s}{(s^2+a^2)^3}$	$\frac{1}{2}t^2\cos at$
33.60.	$\dfrac{s^4-6a^2s^2+a^4}{(s^2+a^2)^4}$	$\frac{1}{6}t^3\cos at$
33.61.	$\dfrac{s^3-a^2s}{(s^2+a^2)^4}$	$\dfrac{t^3\sin at}{24a}$
33.62.	$\dfrac{1}{(s^2-a^2)^3}$	$\dfrac{(3+a^2t^2)\sinh at - 3at\cosh at}{8a^5}$
33.63.	$\dfrac{s}{(s^2-a^2)^3}$	$\dfrac{at^2\cosh at - t\sinh at}{8a^3}$
33.64.	$\dfrac{s^2}{(s^2-a^2)^3}$	$\dfrac{at\cosh at + (a^2t^2-1)\sinh at}{8a^3}$
33.65.	$\dfrac{s^3}{(s^2-a^2)^3}$	$\dfrac{3t\sinh at + at^2\cosh at}{8a}$
33.66.	$\dfrac{s^4}{(s^2-a^2)^3}$	$\dfrac{(3+a^2t^2)\sinh at + 5at\cosh at}{8a}$
33.67.	$\dfrac{s^5}{(s^2-a^2)^3}$	$\dfrac{(8+a^2t^2)\cosh at + 7at\sinh at}{8}$
33.68.	$\dfrac{3s^2+a^2}{(s^2-a^2)^3}$	$\dfrac{t^2\sinh at}{2a}$
33.69.	$\dfrac{s^3+3a^2s}{(s^2-a^2)^3}$	$\frac{1}{2}t^2\cosh at$
33.70.	$\dfrac{s^4+6a^2s^2+a^4}{(s^2-a^2)^4}$	$\frac{1}{6}t^3\cosh at$
33.71.	$\dfrac{s^3+a^2s}{(s^2-a^2)^4}$	$\dfrac{t^3\sinh at}{24a}$
33.72.	$\dfrac{1}{s^3+a^3}$	$\dfrac{e^{at/2}}{3a^2}\left\{\sqrt{3}\sin\dfrac{\sqrt{3}at}{2} - \cos\dfrac{\sqrt{3}at}{2} + e^{-3at/2}\right\}$

	$f(s)$	$F(t)$
33.73.	$\dfrac{s}{s^3+a^3}$	$\dfrac{e^{at/2}}{3a}\left\{\cos\dfrac{\sqrt{3}at}{2}+\sqrt{3}\sin\dfrac{\sqrt{3}at}{2}-e^{-3at/2}\right\}$
33.74.	$\dfrac{s^2}{s^3+a^3}$	$\dfrac{1}{3}\left(e^{-at}+2e^{at/2}\cos\dfrac{\sqrt{3}at}{2}\right)$
33.75.	$\dfrac{1}{s^3-a^3}$	$\dfrac{e^{-at/2}}{3a^2}\left\{e^{3at/2}-\cos\dfrac{\sqrt{3}at}{2}-\sqrt{3}\sin\dfrac{\sqrt{3}at}{2}\right\}$
33.76.	$\dfrac{s}{s^3-a^3}$	$\dfrac{e^{-at/2}}{3a}\left\{\sqrt{3}\sin\dfrac{\sqrt{3}at}{2}-\cos\dfrac{\sqrt{3}at}{2}+e^{3at/2}\right\}$
33.77.	$\dfrac{s^2}{s^3-a^3}$	$\dfrac{1}{3}\left(e^{at}+2e^{-at/2}\cos\dfrac{\sqrt{3}at}{2}\right)$
33.78.	$\dfrac{1}{s^4+4a^4}$	$\dfrac{1}{4a^3}(\sin at\cosh at-\cos at\sinh at)$
33.79.	$\dfrac{s}{s^4+4a^4}$	$\dfrac{\sin at\sinh at}{2a^2}$
33.80.	$\dfrac{s^2}{s^4+4a^4}$	$\dfrac{1}{2a}(\sin at\cosh at+\cos at\sinh at)$
33.81.	$\dfrac{s^3}{s^4+4a^4}$	$\cos at\cosh at$
33.82.	$\dfrac{1}{s^4-a^4}$	$\dfrac{1}{2a^3}(\sinh at-\sin at)$
33.83.	$\dfrac{s}{s^4-a^4}$	$\dfrac{1}{2a^2}(\cosh at-\cos at)$
33.84.	$\dfrac{s^2}{s^4-a^4}$	$\dfrac{1}{2a}(\sinh at+\sin at)$
33.85.	$\dfrac{s^3}{s^4-a^4}$	$\frac{1}{2}(\cosh at+\cos at)$
33.86.	$\dfrac{1}{\sqrt{s+a}+\sqrt{s+b}}$	$\dfrac{e^{-bt}-e^{-at}}{2(b-a)\sqrt{\pi t^3}}$
33.87.	$\dfrac{1}{s\sqrt{s+a}}$	$\dfrac{\operatorname{erf}\sqrt{at}}{\sqrt{a}}$
33.88.	$\dfrac{1}{\sqrt{s}(s-a)}$	$\dfrac{e^{at}\operatorname{erf}\sqrt{at}}{\sqrt{a}}$
33.89.	$\dfrac{1}{\sqrt{s-a}+b}$	$e^{at}\left\{\dfrac{1}{\sqrt{\pi t}}-b\,e^{b^2t}\operatorname{erfc}(b\sqrt{t})\right\}$

	$f(s)$	$F(t)$
33.90.	$\dfrac{1}{\sqrt{s^2+a^2}}$	$J_0(at)$
33.91.	$\dfrac{1}{\sqrt{s^2-a^2}}$	$I_0(at)$
33.92.	$\dfrac{(\sqrt{s^2+a^2}-s)^n}{\sqrt{s^2+a^2}} \quad n>-1$	$a^n J_n(at)$
33.93.	$\dfrac{(s-\sqrt{s^2-a^2})^n}{\sqrt{s^2-a^2}} \quad n>-1$	$a^n I_n(at)$
33.94.	$\dfrac{e^{b(s-\sqrt{s^2+a^2})}}{\sqrt{s^2+a^2}}$	$J_0(a\sqrt{t(t+2b)})$
33.95.	$\dfrac{e^{-b\sqrt{s^2+a^2}}}{\sqrt{s^2+a^2}}$	$\begin{cases} J_0(a\sqrt{t^2-b^2}) & t>b \\ 0 & t<b \end{cases}$
33.96.	$\dfrac{1}{(s^2+a^2)^{3/2}}$	$\dfrac{tJ_1(at)}{a}$
33.97.	$\dfrac{s}{(s^2+a^2)^{3/2}}$	$tJ_0(at)$
33.98.	$\dfrac{s^2}{(s^2+a^2)^{3/2}}$	$J_0(at)-atJ_1(at)$
33.99.	$\dfrac{1}{(s^2-a^2)^{3/2}}$	$\dfrac{tI_1(at)}{a}$
33.100.	$\dfrac{s}{(s^2-a^2)^{3/2}}$	$tI_0(at)$
33.101.	$\dfrac{s^2}{(s^2-a^2)^{3/2}}$	$I_0(at)+atI_1(at)$
33.102.	$\dfrac{1}{s(e^s-1)}=\dfrac{e^{-s}}{s(1-e^{-s})}$ See also entry 33.165.	$F(t)=n, n \leqq t<n+1, n=0,1,2,\ldots$
33.103.	$\dfrac{1}{s(e^s-r)}=\dfrac{e^{-s}}{s(1-re^{-s})}$	$F(t)=\sum_{k=1}^{[t]} r^k$ where $[t]$ = greatest integer $\leqq t$
33.104.	$\dfrac{e^s-1}{s(e^s-r)}=\dfrac{1-e^{-s}}{s(1-re^{-s})}$ See also entry 33.167.	$F(t)=r^n, n \leqq t<n+1, n=0,1,2,\ldots$
33.105.	$\dfrac{e^{-a/s}}{\sqrt{s}}$	$\dfrac{\cos 2\sqrt{at}}{\sqrt{\pi t}}$

	$f(s)$	$F(t)$
33.106.	$\dfrac{e^{-a/s}}{s^{3/2}}$	$\dfrac{\sin 2\sqrt{at}}{\sqrt{\pi a}}$
33.107.	$\dfrac{e^{-a/s}}{s^{n+1}} \quad n > -1$	$\left(\dfrac{t}{a}\right)^{n/2} J_n(2\sqrt{at})$
33.108.	$\dfrac{e^{-a\sqrt{s}}}{\sqrt{s}}$	$\dfrac{e^{-a^2/4t}}{\sqrt{\pi t}}$
33.109.	$e^{-a\sqrt{s}}$	$\dfrac{a}{2\sqrt{\pi t^3}} e^{-a^2/4t}$
33.110.	$\dfrac{1-e^{-a\sqrt{s}}}{s}$	$\operatorname{erf}(a/2\sqrt{t})$
33.111.	$\dfrac{e^{-a\sqrt{s}}}{s}$	$\operatorname{erfc}(a/2\sqrt{t})$
33.112.	$\dfrac{e^{-a\sqrt{s}}}{\sqrt{s}(\sqrt{s}+b)}$	$e^{b(bt+a)}\operatorname{erfc}\left(b\sqrt{t} + \dfrac{a}{2\sqrt{t}}\right)$
33.113.	$\dfrac{e^{-a/\sqrt{s}}}{s^{n+1}} \quad n > -1$	$\dfrac{1}{\sqrt{\pi t}\, a^{2n+1}} \int_0^\infty u^n e^{-u^2/4a^2t} J_{2n}(2\sqrt{u})\,du$
33.114.	$\ln\left(\dfrac{s+a}{s+b}\right)$	$\dfrac{e^{-bt} - e^{-at}}{t}$
33.115.	$\dfrac{\ln[(s^2+a^2)/a^2]}{2s}$	$Ci(at)$
33.116.	$\dfrac{\ln[(s+a)/a]}{s}$	$Ei(at)$
33.117.	$-\dfrac{(\gamma + \ln s)}{s}$ γ = Euler's constant = .5772156 …	$\ln t$
33.118.	$\ln\left(\dfrac{s^2+a^2}{s^2+b^2}\right)$	$\dfrac{2(\cos at - \cos bt)}{t}$
33.119.	$\dfrac{\pi^2}{6s} + \dfrac{(\gamma + \ln s)^2}{s}$ γ = Euler's constant = .5772156 …	$\ln^2 t$
33.120.	$\dfrac{\ln s}{s}$	$-(\ln t + \gamma)$ γ = Euler's constant = .5772156 …
33.121.	$\dfrac{\ln^2 s}{s}$	$(\ln t + \gamma)^2 - \frac{1}{6}\pi^2$ γ = Euler's constant = .5772156 …

	$f(s)$	$F(t)$
33.122.	$\dfrac{\Gamma'(n+1)-\Gamma(n+1)\ln s}{s^{n+1}} \quad n>-1$	$t^n \ln t$
33.123.	$\tan^{-1}(a/s)$	$\dfrac{\sin at}{t}$
33.124.	$\dfrac{\tan^{-1}(a/s)}{s}$	$Si(at)$
33.125.	$\dfrac{e^{a/s}}{\sqrt{s}}\operatorname{erfc}(\sqrt{a/s})$	$\dfrac{e^{-2\sqrt{at}}}{\sqrt{\pi t}}$
33.126.	$e^{s^2/4a^2}\operatorname{erfc}(s/2a)$	$\dfrac{2a}{\sqrt{\pi}}e^{-a^2t^2}$
33.127.	$\dfrac{e^{s^2/4a^2}\operatorname{erfc}(s/2a)}{s}$	$\operatorname{erf}(at)$
33.128.	$\dfrac{e^{as}\operatorname{erfc}\sqrt{as}}{\sqrt{s}}$	$\dfrac{1}{\sqrt{\pi(t+a)}}$
33.129.	$e^{as}Ei(as)$	$\dfrac{1}{t+a}$
33.130.	$\dfrac{1}{a}\left[\cos as\left\{\dfrac{\pi}{2}-Si(as)\right\}-\sin as\, Ci(as)\right]$	$\dfrac{1}{t^2+a^2}$
33.131.	$\sin as\left\{\dfrac{\pi}{2}-Si(as)\right\}+\cos as\, Ci(as)$	$\dfrac{t}{t^2+a^2}$
33.132.	$\dfrac{\cos as\left\{\dfrac{\pi}{2}-Si(as)\right\}-\sin as\, Ci(as)}{s}$	$\tan^{-1}(t/a)$
33.133.	$\dfrac{\sin as\left\{\dfrac{\pi}{2}-Si(as)\right\}-\cos as\, Ci(as)}{s}$	$\dfrac{1}{2}\ln\left(\dfrac{t^2+a^2}{a^2}\right)$
33.134.	$\left[\dfrac{\pi}{2}-Si(as)\right]^2+Ci^2(as)$	$\dfrac{1}{t}\ln\left(\dfrac{t^2+a^2}{a^2}\right)$
33.135.	0	$\mathcal{N}(t)$ = null function
33.136.	1	$\delta(t)$ = delta function
33.137.	e^{-as}	$\delta(t-a)$
33.138.	$\dfrac{e^{-as}}{s}$ See also entry 33.163.	$\mathcal{U}(t-a)$

	$f(s)$	$F(t)$
33.139.	$\dfrac{\sinh sx}{s \sinh sa}$	$\dfrac{x}{a}+\dfrac{2}{\pi}\sum_{n=1}^{\infty}\dfrac{(-1)^n}{n}\sin\dfrac{n\pi x}{a}\cos\dfrac{n\pi t}{a}$
33.140.	$\dfrac{\sinh sx}{s \cosh sa}$	$\dfrac{4}{\pi}\sum_{n=1}^{\infty}\dfrac{(-1)^n}{2n-1}\sin\dfrac{(2n-1)\pi x}{2a}\sin\dfrac{(2n-1)\pi t}{2a}$
33.141.	$\dfrac{\cosh sx}{s \sinh as}$	$\dfrac{t}{a}+\dfrac{2}{\pi}\sum_{n=1}^{\infty}\dfrac{(-1)^n}{n}\cos\dfrac{n\pi x}{a}\sin\dfrac{n\pi t}{a}$
33.142.	$\dfrac{\cosh sx}{s \cosh sa}$	$1+\dfrac{4}{\pi}\sum_{n=1}^{\infty}\dfrac{(-1)^n}{2n-1}\cos\dfrac{(2n-1)\pi x}{2a}\cos\dfrac{(2n-1)\pi t}{2a}$
33.143.	$\dfrac{\sinh sx}{s^2 \sinh sa}$	$\dfrac{xt}{a}+\dfrac{2a}{\pi^2}\sum_{n=1}^{\infty}\dfrac{(-1)^n}{n^2}\sin\dfrac{n\pi x}{a}\sin\dfrac{n\pi t}{a}$
33.144.	$\dfrac{\sinh sx}{s^2 \cosh sa}$	$x+\dfrac{8a}{\pi^2}\sum_{n=1}^{\infty}\dfrac{(-1)^n}{(2n-1)^2}\sin\dfrac{(2n-1)\pi x}{2a}\cos\dfrac{(2n-1)\pi t}{2a}$
33.145.	$\dfrac{\cosh sx}{s^2 \sinh sa}$	$\dfrac{t^2}{2a}+\dfrac{2a}{\pi^2}\sum_{n=1}^{\infty}\dfrac{(-1)^n}{n^2}\cos\dfrac{n\pi x}{a}\left(1-\cos\dfrac{n\pi t}{a}\right)$
33.146.	$\dfrac{\cosh sx}{s^2 \cosh sa}$	$t+\dfrac{8a}{\pi^2}\sum_{n=1}^{\infty}\dfrac{(-1)^n}{(2n-1)^2}\cos\dfrac{(2n-1)\pi x}{2a}\sin\dfrac{(2n-1)\pi t}{2a}$
33.147.	$\dfrac{\cosh sx}{s^3 \cosh sa}$	$\dfrac{1}{2}(t^2+x^2-a^2)-\dfrac{16a^2}{\pi^3}\sum_{n=1}^{\infty}\dfrac{(-1)^n}{(2n-1)^3}\cos\dfrac{(2n-1)\pi x}{2a}\cos\dfrac{(2n-1)\pi t}{2a}$
33.148.	$\dfrac{\sinh x\sqrt{s}}{\sinh a\sqrt{s}}$	$\dfrac{2\pi}{a^2}\sum_{n=1}^{\infty}(-1)^n ne^{-n^2\pi^2 t/a^2}\sin\dfrac{n\pi x}{a}$
33.149.	$\dfrac{\cosh x\sqrt{s}}{\cosh a\sqrt{s}}$	$\dfrac{\pi}{a^2}\sum_{n=1}^{\infty}(-1)^{n-1}(2n-1)e^{-(2n-1)^2\pi^2 t/4a^2}\cos\dfrac{(2n-1)\pi x}{2a}$
33.150.	$\dfrac{\sinh x\sqrt{s}}{\sqrt{s}\cosh a\sqrt{s}}$	$\dfrac{2}{a}\sum_{n=1}^{\infty}(-1)^{n-1}e^{-(2n-1)^2\pi^2 t/4a^2}\sin\dfrac{(2n-1)\pi x}{2a}$
33.151.	$\dfrac{\cosh x\sqrt{s}}{\sqrt{s}\sinh a\sqrt{s}}$	$\dfrac{1}{a}+\dfrac{2}{a}\sum_{n=1}^{\infty}(-1)^n e^{-n^2\pi^2 t/a^2}\cos\dfrac{n\pi x}{a}$
33.152.	$\dfrac{\sinh x\sqrt{s}}{s\sinh a\sqrt{s}}$	$\dfrac{x}{a}+\dfrac{2}{\pi}\sum_{n=1}^{\infty}\dfrac{(-1)^n}{n}e^{-n^2\pi^2 t/a^2}\sin\dfrac{n\pi x}{a}$
33.153.	$\dfrac{\cosh x\sqrt{s}}{s\cosh a\sqrt{s}}$	$1+\dfrac{4}{\pi}\sum_{n=1}^{\infty}\dfrac{(-1)^n}{2n-1}e^{-(2n-1)^2\pi^2 t/4a^2}\cos\dfrac{(2n-1)\pi x}{2a}$
33.154.	$\dfrac{\sinh x\sqrt{s}}{s^2\sinh a\sqrt{s}}$	$\dfrac{xt}{a}+\dfrac{2a^2}{\pi^3}\sum_{n=1}^{\infty}\dfrac{(-1)^n}{n^3}(1-e^{-n^2\pi^2 t/a^2})\sin\dfrac{n\pi x}{a}$
33.155.	$\dfrac{\cosh x\sqrt{s}}{s^2\cosh a\sqrt{s}}$	$\dfrac{1}{2}(x^2-a^2)+t-\dfrac{16a^2}{\pi^3}\sum_{n=1}^{\infty}\dfrac{(-1)^n}{(2n-1)^3}e^{-(2n-1)^2\pi^2 t/4a^2}\cos\dfrac{(2n-1)\pi x}{2a}$

	$f(s)$	$F(t)$
33.156.	$\dfrac{J_0(ix\sqrt{s})}{sJ_0(ia\sqrt{s})}$	$1-2\sum_{n=1}^{\infty}\dfrac{e^{-\lambda_n^2 t/a^2}J_0(\lambda_n x/a)}{\lambda_n J_1(\lambda_n)}$ where $\lambda_1, \lambda_2, \ldots$ are the positive roots of $J_0(\lambda)=0$
33.157.	$\dfrac{J_0(ix\sqrt{s})}{s^2 J_0(ia\sqrt{s})}$	$\dfrac{1}{4}(x^2-a^2)+t+2a^2\sum_{n=1}^{\infty}\dfrac{e^{-\lambda_n^2 t/a^2}J_0(\lambda_n x/a)}{\lambda_n^3 J_1(\lambda_n)}$ where $\lambda_1, \lambda_2, \ldots$ are the positive roots of $J_0(\lambda)=0$
33.158.	$\dfrac{1}{as^2}\tanh\left(\dfrac{as}{2}\right)$	Triangular wave function 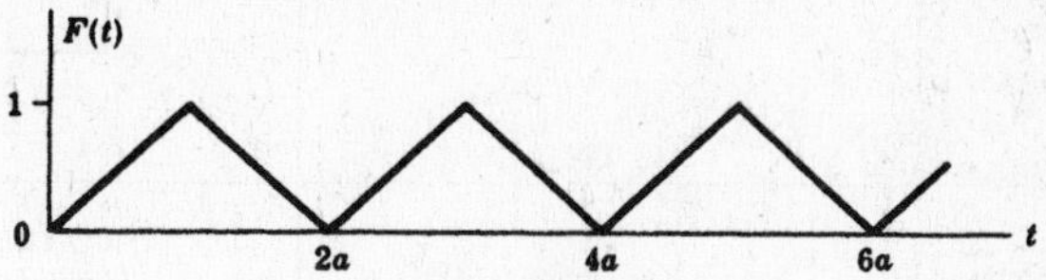 Fig. 33-1
33.159.	$\dfrac{1}{s}\tanh\left(\dfrac{as}{2}\right)$	Square wave function 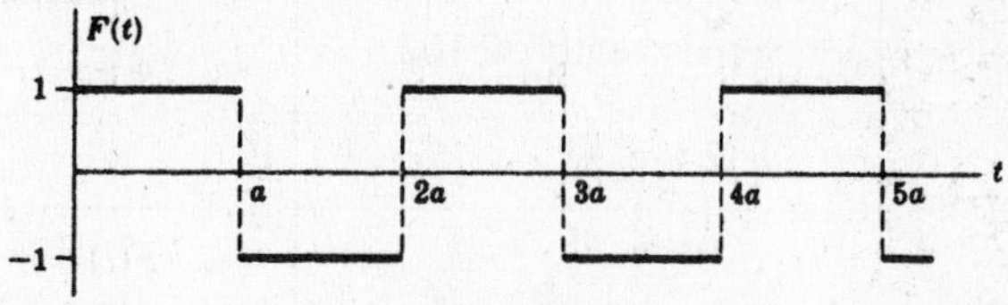 Fig. 33-2
33.160.	$\dfrac{\pi a}{a^2 s^2+\pi^2}\coth\left(\dfrac{as}{2}\right)$	Rectified sine wave function 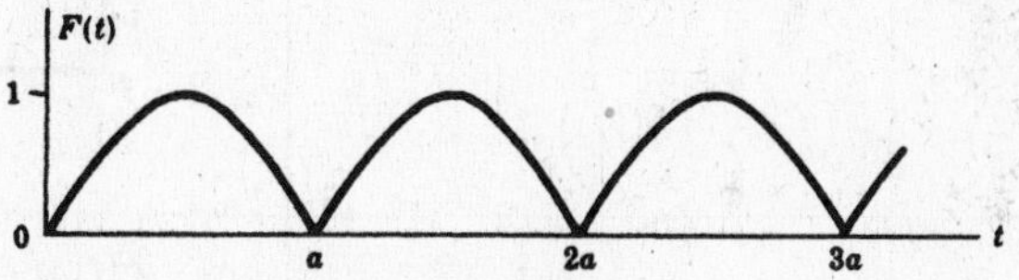 Fig. 33-3
33.161.	$\dfrac{\pi a}{(a^2 s^2+\pi^2)(1-e^{-as})}$	Half-rectified sine wave function 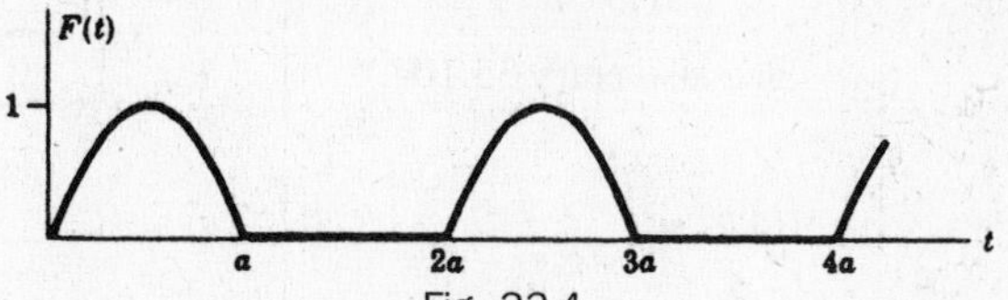 Fig. 33-4
33.162.	$\dfrac{1}{as^2}-\dfrac{e^{-as}}{s(1-e^{-as})}$	Sawtooth wave function 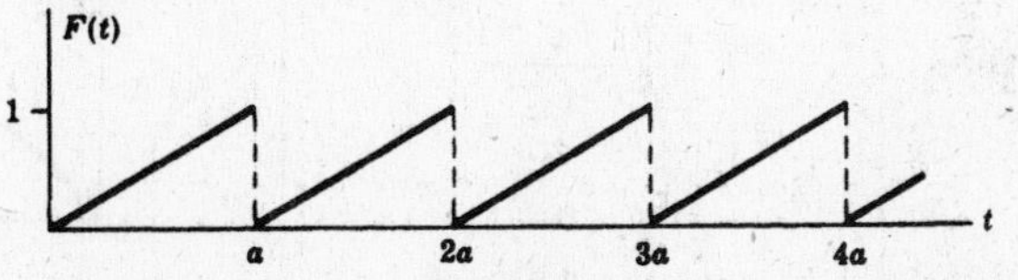 Fig. 33-5

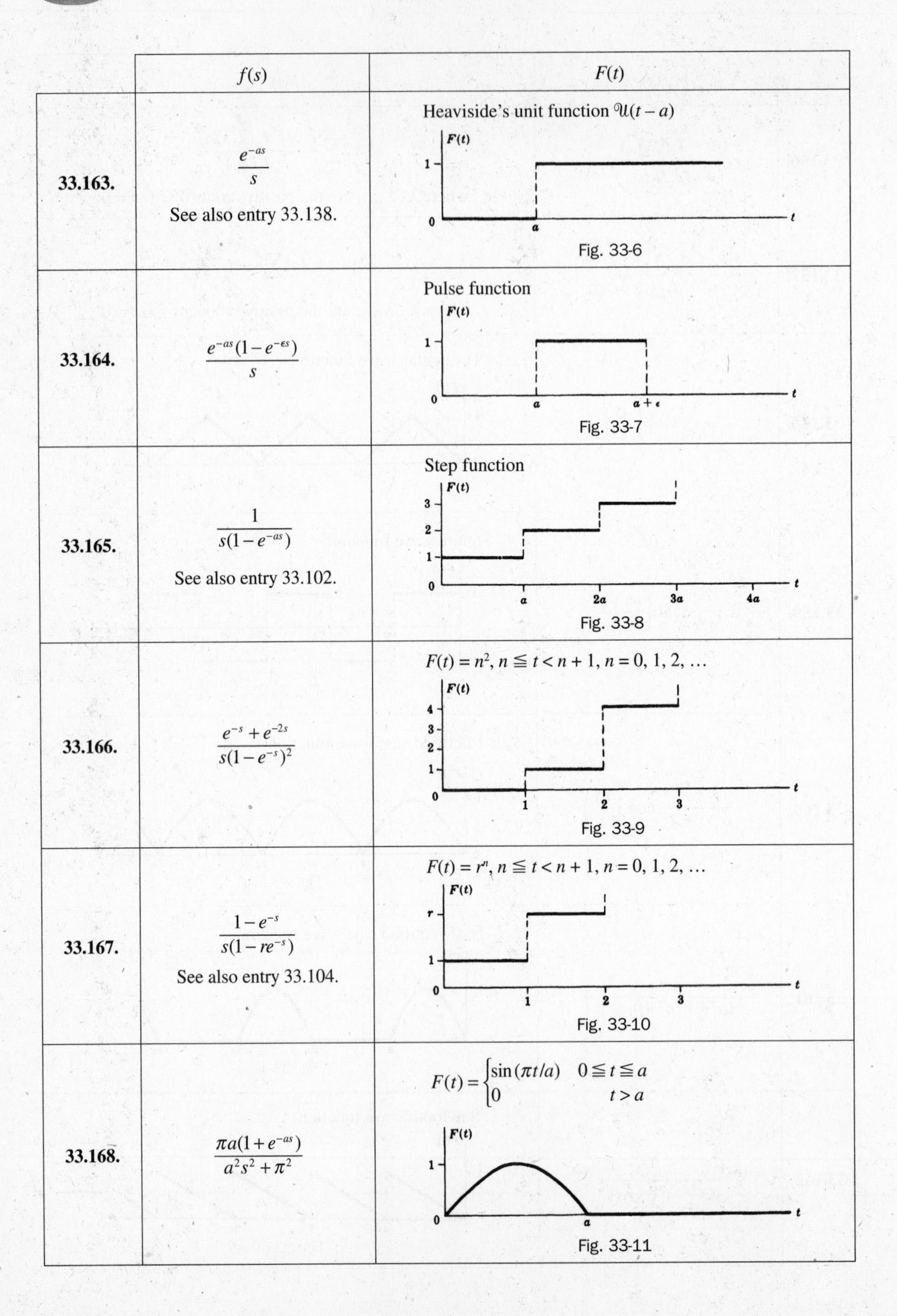

	$f(s)$	$F(t)$
33.163.	$\dfrac{e^{-as}}{s}$ See also entry 33.138.	Heaviside's unit function $\mathscr{U}(t-a)$ Fig. 33-6
33.164.	$\dfrac{e^{-as}(1-e^{-\epsilon s})}{s}$	Pulse function Fig. 33-7
33.165.	$\dfrac{1}{s(1-e^{-as})}$ See also entry 33.102.	Step function Fig. 33-8
33.166.	$\dfrac{e^{-s}+e^{-2s}}{s(1-e^{-s})^2}$	$F(t)=n^2,\ n \leqq t < n+1,\ n=0, 1, 2, \ldots$ Fig. 33-9
33.167.	$\dfrac{1-e^{-s}}{s(1-re^{-s})}$ See also entry 33.104.	$F(t)=r^n,\ n \leqq t < n+1,\ n=0, 1, 2, \ldots$ Fig. 33-10
33.168.	$\dfrac{\pi a(1+e^{-as})}{a^2s^2+\pi^2}$	$F(t)=\begin{cases}\sin(\pi t/a) & 0 \leqq t \leqq a \\ 0 & t > a\end{cases}$ Fig. 33-11

34 FOURIER TRANSFORMS

Fourier's Integral Theorem

34.1. $f(x) = \int_0^\infty \{A(\alpha)\cos\alpha x + B(\alpha)\sin\alpha x\} d\alpha$

where

34.2. $$\begin{cases} A(\alpha) = \dfrac{1}{\pi}\displaystyle\int_{-\infty}^\infty f(x)\cos\alpha x\, dx \\ B(\alpha) = \dfrac{1}{\pi}\displaystyle\int_{-\infty}^\infty f(x)\sin\alpha x\, dx \end{cases}$$

Sufficient conditions under which this theorem holds are:

(i) $f(x)$ and $f'(x)$ are piecewise continuous in every finite interval $-L < x < L$;
(ii) $\int_{-\infty}^\infty |f(x)|\, dx$ converges;
(iii) $f(x)$ is replaced by $\frac{1}{2}\{f(x+0) + f(x-0)\}$ if x is a point of discontinuity.

Equivalent Forms of Fourier's Integral Theorem

34.3. $f(x) = \dfrac{1}{2\pi}\displaystyle\int_{\alpha=-\infty}^\infty \int_{u=-\infty}^\infty f(u)\cos\alpha(x-u)\, du\, d\alpha$

34.4. $$\begin{aligned} f(x) &= \frac{1}{2\pi}\int_{-\infty}^\infty e^{i\alpha x} d\alpha \int_{-\infty}^\infty f(u) e^{-i\alpha u} du \\ &= \frac{1}{2\pi}\int_{-\infty}^\infty \int_{-\infty}^\infty f(u) e^{i\alpha(x-u)} du\, d\alpha \end{aligned}$$

34.5. $f(x) = \dfrac{2}{\pi}\displaystyle\int_0^\infty \sin\alpha x\, d\alpha \int_0^\infty f(u)\sin\alpha u\, du$

where $f(x)$ is an *odd function* $[f(-x) = -f(x)]$.

34.6. $f(x) = \dfrac{2}{\pi}\displaystyle\int_0^\infty \cos\alpha x\, d\alpha \int_0^\infty f(u)\cos\alpha u\, du$

where $f(x)$ is an *even function* $[f(-x) = f(x)]$.

Fourier Transforms

The Fourier transform of $f(x)$ is defined as

34.7. $\mathcal{F}\{f(x)\} = F(\alpha) = \int_{-\infty}^{\infty} f(x)e^{-i\alpha x}\,dx$

Then from 34.7 the inverse Fourier transform of $F(\alpha)$ is

34.8. $\mathcal{F}^{-1}\{F(\alpha)\} = f(x) = \frac{1}{2\pi}\int_{-\infty}^{\infty} F(\alpha)e^{i\alpha x}\,d\alpha$

We call $f(x)$ and $F(\alpha)$ *Fourier transform pairs.*

Convolution Theorem for Fourier Transforms

If $F(\alpha) = \mathcal{F}\{f(x)\}$ and $G(\alpha) = \mathcal{F}\{g(x)\}$, then

34.9. $\frac{1}{2\pi}\int_{-\infty}^{\infty} F(\alpha)G(\alpha)e^{i\alpha x}\,d\alpha = \int_{-\infty}^{\infty} f(u)g(x-u)\,du = f^*g$

where f^*g is called the *convolution* of f and g. Thus,

34.10. $\mathcal{F}\{f^*g\} = \mathcal{F}\{f\}\,\mathcal{F}\{g\}$

Parseval's Identity

If $F(\alpha) = \mathcal{F}\{f(x)\}$, then

34.11. $\int_{-\infty}^{\infty} |f(x)|^2 dx = \frac{1}{2\pi}\int_{-\infty}^{\infty} |F(\alpha)|^2 d\alpha$

More generally if $F(\alpha) = \mathcal{F}\{f(x)\}$ and $G(\alpha) = \mathcal{F}\{g(x)\}$, then

34.12. $\int_{-\infty}^{\infty} f(x)\overline{g(x)}\,dx = \frac{1}{2\pi}\int_{-\infty}^{\infty} F(\alpha)\overline{G(\alpha)}\,d\alpha$

where the bar denotes complex conjugate.

Fourier Sine Transforms

The Fourier sine transform of $f(x)$ is defined as

34.13. $F_S(\alpha) = \mathcal{F}_S\{f(x)\} = \int_0^{\infty} f(x)\sin\alpha x\,dx$

Then from 34.13 the inverse Fourier sine transform of $F_S(\alpha)$ is

34.14. $f(x) = \mathcal{F}_S^{-1}\{F_S(\alpha)\} = \frac{2}{\pi}\int_0^{\infty} F_S(\alpha)\sin\alpha x\,d\alpha$

Fourier Cosine Transforms

The Fourier cosine transform of $f(x)$ is defined as

34.15. $$F_C(\alpha) = \mathcal{F}_C\{f(x)\} = \int_0^\infty f(x)\cos\alpha x\,dx$$

Then from 34.15 the inverse Fourier cosine transform of $F_C(\alpha)$ is

34.16. $$f(x) = \mathcal{F}_C^{-1}\{F_C(\alpha)\} = \frac{2}{\pi}\int_0^\infty F_C(\alpha)\cos\alpha x\,d\alpha$$

Special Fourier Transform Pairs

	$f(x)$	$F(\alpha)$
34.17.	$\begin{cases} 1 & \lvert x\rvert < b \\ 0 & \lvert x\rvert > b \end{cases}$	$\dfrac{2\sin b\alpha}{\alpha}$
34.18.	$\dfrac{1}{x^2+b^2}$	$\dfrac{\pi e^{-b\alpha}}{b}$
34.19.	$\dfrac{x}{x^2+b^2}$	$-i\pi e^{-b\alpha}$
34.20.	$f^{(n)}(x)$	$i^n\alpha^n F(\alpha)$
34.21.	$x^n f(x)$	$i^n\dfrac{d^nF}{d\alpha^n}$
34.22.	$f(bx)e^{itx}$	$\dfrac{1}{b}F\left(\dfrac{\alpha-t}{b}\right)$

Special Fourier Sine Transforms

	$f(x)$	$F_c(\alpha)$
34.23.	$\begin{cases} 1 & 0 < x < b \\ 0 & x > b \end{cases}$	$\dfrac{1 - \cos b\alpha}{\alpha}$
34.24.	x^{-1}	$\dfrac{\pi}{2}$
34.25.	$\dfrac{x}{x^2 + b^2}$	$\dfrac{\pi}{2} e^{-b\alpha}$
34.26.	e^{-bx}	$\dfrac{\alpha}{\alpha^2 + b^2}$
34.27.	$x^{n-1} e^{-bx}$	$\dfrac{\Gamma(n) \sin(n \tan^{-1} \alpha / b)}{(\alpha^2 + b^2)^{n/2}}$
34.28.	xe^{-bx^2}	$\dfrac{\sqrt{\pi}}{4b^{3/2}} \alpha e^{-\alpha^2/4b}$
34.29.	$x^{-1/2}$	$\sqrt{\dfrac{\pi}{2\alpha}}$
34.30.	x^{-n}	$\dfrac{\pi \alpha^{n-1} \csc(n\pi/2)}{2\Gamma(n)} \quad 0 < n < 2$
34.31.	$\dfrac{\sin bx}{x}$	$\dfrac{1}{2} \ln \left(\dfrac{\alpha + b}{\alpha - b} \right)$
34.32.	$\dfrac{\sin bx}{x^2}$	$\begin{cases} \pi\alpha / 2 & \alpha < b \\ \pi b / 2 & \alpha > b \end{cases}$
34.33.	$\dfrac{\cos bx}{x}$	$\begin{cases} 0 & \alpha < b \\ \pi / 4 & \alpha = b \\ \pi / 2 & \alpha > b \end{cases}$
34.34.	$\tan^{-1}(x / b)$	$\dfrac{\pi}{2\alpha} e^{-b\alpha}$
34.35.	$\csc bx$	$\dfrac{\pi}{2b} \tanh \dfrac{\pi\alpha}{2b}$
34.36.	$\dfrac{1}{e^{2x} - 1}$	$\dfrac{\pi}{4} \coth \left(\dfrac{\pi\alpha}{2} \right) - \dfrac{1}{2\alpha}$

Special Fourier Cosine Transforms

	$f(x)$	$F_c(\alpha)$
34.37.	$\begin{cases} 1 & 0<x<b \\ 0 & x>b \end{cases}$	$\dfrac{\sin b\alpha}{\alpha}$
34.38.	$\dfrac{1}{x^2+b^2}$	$\dfrac{\pi e^{-b\alpha}}{2b}$
34.39.	e^{-bx}	$\dfrac{b}{\alpha^2+b^2}$
34.40.	$x^{n-1}e^{-bx}$	$\dfrac{\Gamma(n)\cos(n\tan^{-1}\alpha/b)}{(\alpha^2+b^2)^{n/2}}$
34.41.	e^{-bx^2}	$\dfrac{1}{2}\sqrt{\dfrac{\pi}{b}}e^{-\alpha^2/4b}$
34.42.	$x^{-1/2}$	$\sqrt{\dfrac{\pi}{2\alpha}}$
34.43.	x^{-n}	$\dfrac{\pi\alpha^{n-1}\sec(n\pi/2)}{2\Gamma(n)}, \quad 0<n<1$
34.44.	$\ln\left(\dfrac{x^2+b^2}{x^2+c^2}\right)$	$\dfrac{e^{-c\alpha}-e^{-b\alpha}}{\pi\alpha}$
34.45.	$\dfrac{\sin bx}{x}$	$\begin{cases} \pi/2 & \alpha<b \\ \pi/4 & \alpha=b \\ 0 & \alpha>b \end{cases}$
34.46.	$\sin bx^2$	$\sqrt{\dfrac{\pi}{8b}}\left(\cos\dfrac{\alpha^2}{4b}-\sin\dfrac{\alpha^2}{4b}\right)$
34.47.	$\cos bx^2$	$\sqrt{\dfrac{\pi}{8b}}\left(\cos\dfrac{\alpha^2}{4b}+\sin\dfrac{\alpha^2}{4b}\right)$
34.48.	$\operatorname{sech} bx$	$\dfrac{\pi}{2b}\operatorname{sech}\dfrac{\pi\alpha}{2b}$
34.49.	$\dfrac{\cosh(\sqrt{\pi}x/2)}{\cosh(\sqrt{\pi}x)}$	$\sqrt{\dfrac{\pi}{2}}\dfrac{\cosh(\sqrt{\pi}\alpha/2)}{\cosh(\sqrt{\pi}\alpha)}$
34.50.	$\dfrac{e^{-b\sqrt{x}}}{\sqrt{x}}$	$\sqrt{\dfrac{\pi}{2\alpha}}\{\cos(2b\sqrt{\alpha})-\sin(2b\sqrt{\alpha})\}$

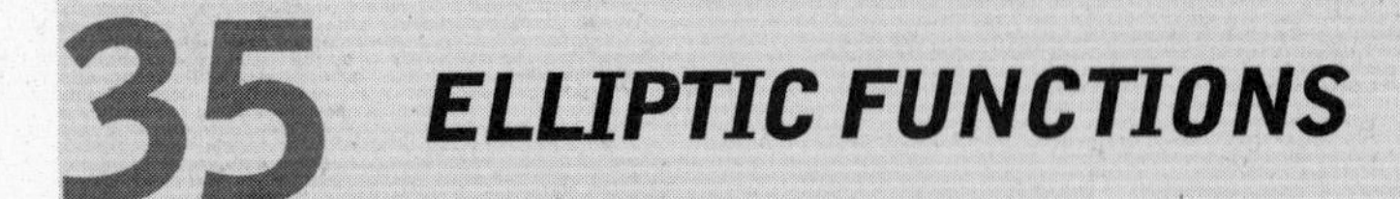

35 ELLIPTIC FUNCTIONS

Incomplete Elliptic Integral of the First Kind

35.1. $$u = F(k,\phi) = \int_0^{\phi} \frac{d\theta}{\sqrt{1-k^2\sin^2\theta}} = \int_0^x \frac{dv}{\sqrt{(1-v^2)(1-k^2v^2)}}$$

where $\phi = \text{am}\, u$ is called the *amplitude* of u and $x = \sin\phi$, and where here and below $0 < k < 1$.

Complete Elliptic Integral of the First Kind

35.2. $$K = F(k,\pi/2) = \int_0^{\pi/2} \frac{d\theta}{\sqrt{1-k^2\sin^2\theta}} = \int_0^1 \frac{dv}{\sqrt{(1-v^2)(1-k^2v^2)}}$$

$$= \frac{\pi}{2}\left\{1 + \left(\frac{1}{2}\right)^2 k^2 + \left(\frac{1\cdot 3}{2\cdot 4}\right)^2 k^4 + \left(\frac{1\cdot 3\cdot 5}{2\cdot 4\cdot 6}\right)^2 k^6 + \cdots\right\}$$

Incomplete Elliptic Integral of the Second Kind

35.3. $$E(k,\phi) = \int_0^{\phi} \sqrt{1-k^2\sin^2\theta}\, d\theta = \int_0^x \frac{\sqrt{1-k^2v^2}}{\sqrt{1-v^2}}\, dv$$

Complete Elliptic Integral of the Second Kind

35.4. $$E = E(k,\pi/2) = \int_0^{\pi/2} \sqrt{1-k^2\sin^2\theta}\, d\theta = \int_0^1 \frac{\sqrt{1-k^2v^2}}{\sqrt{1-v^2}}\, dv$$

$$= \frac{\pi}{2}\left\{1 - \left(\frac{1}{2}\right)^2 k^2 - \left(\frac{1\cdot 3}{2\cdot 4}\right)^2 \frac{k^4}{3} - \left(\frac{1\cdot 3\cdot 5}{2\cdot 4\cdot 6}\right)^2 \frac{k^6}{5} - \cdots\right\}$$

Incomplete Elliptic Integral of the Third Kind

35.5. $$\Pi(k,n,\phi) = \int_0^{\phi} \frac{d\theta}{(1+n\sin^2\theta)\sqrt{1-k^2\sin^2\theta}} = \int_0^x \frac{dv}{(1+nv^2)\sqrt{(1-v^2)(1-k^2v^2)}}$$

Complete Elliptic Integral of the Third Kind

35.6. $\Pi(k,n,\pi/2)=\int_0^{\pi/2}\frac{d\theta}{(1+n\sin^2\theta)\sqrt{1-k^2\sin^2\theta}}=\int_0^1\frac{dv}{(1+nv^2)\sqrt{(1-v^2)(1-k^2v^2)}}$

Landen's Transformation

35.7. $\tan\phi=\frac{\sin 2\phi_1}{k+\cos 2\phi_1}$ or $k\sin\phi=\sin(2\phi_1-\phi)$

This yields

35.8. $F(k,\phi)=\int_0^{\phi}\frac{d\theta}{\sqrt{1-k^2\sin^2\theta}}=\frac{2}{1+k}\int_0^{\phi_1}\frac{d\theta_1}{\sqrt{1-k_1^2\sin^2\theta_1}}$

where $k_1=2\sqrt{k}/(1+k)$. By successive applications, sequences $k_1,k_2,k_3,\ldots$ and ϕ_1, ϕ_2, ϕ_3, … are obtained such that $k<k_1<k_2<k_3<\cdots<1$ where $\lim_{n\to\infty}k_n=1$. It follows that

35.9. $F(k,\Phi)=\sqrt{\frac{k_1k_2k_3\ldots}{k}}\int_0^{\Phi}\frac{d\theta}{\sqrt{1-\sin^2\theta}}=\sqrt{\frac{k_1k_2k_3\ldots}{k}}\ln\tan\left(\frac{\pi}{4}+\frac{\Phi}{2}\right)$

where

35.10. $k_1=\frac{2\sqrt{k}}{1+k},\quad k_2=\frac{2\sqrt{k_1}}{1+k_1},\quad\ldots$ and $\Phi\lim_{n\to\infty}\phi_n$

The result is used in the approximate evaluation of $F(k,\phi)$.

Jacobi's Elliptic Functions

From 35.1 we define the following elliptic functions:

35.11. $x=\sin(\text{am } u)=\text{sn } u$

35.12. $\sqrt{1-x^2}=\cos(\text{am } u)=\text{cn } u$

35.13. $\sqrt{1-k^2x^2}=\sqrt{1-k^2\,\text{sn}^2 u}=\text{dn } u$

We can also define the inverse functions $\text{sn}^{-1}x$, $\text{cn}^{-1}x$, $\text{dn}^{-1}x$ and the following:

35.14. $\text{ns}\,u=\frac{1}{\text{sn}\,u}$

35.15. $\text{nc}\,u=\frac{1}{\text{cn}\,u}$

35.16. $\text{nd}\,u=\frac{1}{\text{dn}\,u}$

35.17. $\text{sc}\,u=\frac{\text{sn}\,u}{\text{cn}\,u}$

35.18. $\text{sd}\,u=\frac{\text{sn}\,u}{\text{dn}\,u}$

35.19. $\text{cd}\,u=\frac{\text{cn}\,u}{\text{dn}\,u}$

35.20. $\text{cs}\,u=\frac{\text{cn}\,u}{\text{sn}\,u}$

35.21. $\text{dc}\,u=\frac{\text{dn}\,u}{\text{cn}\,u}$

35.22. $\text{ds}\,u=\frac{\text{dn}\,u}{\text{dn}\,u}$

Addition Formulas

35.23. $\text{sn}(u+v)=\frac{\text{sn}\,u\ \text{cn}\,v\ \text{dn}\,v+\text{cn}\,u\ \text{sn}\,v\ \text{dn}\,u}{1-k^2\,\text{sn}^2u\,\text{sn}^2v}$

35.24. $\operatorname{cn}(u+v) = \dfrac{\operatorname{cn} u \operatorname{cn} v - \operatorname{sn} u \operatorname{sn} v \operatorname{dn} u \operatorname{dn} v}{1 - k^2 \operatorname{sn}^2 u \operatorname{sn}^2 v}$

35.25. $\operatorname{dn}(u+v) = \dfrac{\operatorname{dn} u \operatorname{dn} v - k^2 \operatorname{sn} u \operatorname{sn} v \operatorname{cn} u \operatorname{cn} v}{1 - k^2 \operatorname{sn}^2 u \operatorname{sn}^2 v}$

Derivatives

35.26. $\dfrac{d}{du} \operatorname{sn} u = \operatorname{cn} u \operatorname{dn} u$

35.27. $\dfrac{d}{du} \operatorname{cn} u = -\operatorname{sn} u \operatorname{dn} u$

35.28. $\dfrac{d}{du} \operatorname{dn} u = -k^2 \operatorname{sn} u \operatorname{cn} u$

35.29. $\dfrac{d}{du} \operatorname{sc} u = \operatorname{dc} u \operatorname{nc} u$

Series Expansions

35.30. $\operatorname{sn} u = u - (1+k^2)\dfrac{u^3}{3!} + (1+14k^2+k^4)\dfrac{u^5}{5!} - (1+135k^2+135k^4+k^6)\dfrac{u^7}{7!} + \cdots$

35.31. $\operatorname{cn} u = 1 - \dfrac{u^2}{2!} + (1+4k^2)\dfrac{u^4}{4!} - (1+44k^2+16k^4)\dfrac{u^6}{6!} + \cdots$

35.32. $\operatorname{dn} u = 1 - k^2 \dfrac{u^2}{2!} + k^2(4+k^2)\dfrac{u^4}{4!} - k^2(16+44k^2+k^4)\dfrac{u^6}{6!} + \cdots$

Catalan's Constant

35.33. $\dfrac{1}{2}\displaystyle\int_0^1 K\,dk = \dfrac{1}{2}\int_{k=0}^1 \int_{\theta=0}^{\pi/2} \dfrac{d\theta\,dk}{\sqrt{1-k^2\sin^2\theta}} = \dfrac{1}{1^2} - \dfrac{1}{3^2} + \dfrac{1}{5^2} - \cdots = .915965594\ldots$

Periods of Elliptic Functions

Let

35.34. $K = \displaystyle\int_0^{\pi/2} \dfrac{d\theta}{\sqrt{1-k^2\sin^2\theta}}, \qquad K' = \int_0^{\pi/2} \dfrac{d\theta}{\sqrt{1-k'^2\sin^2\theta}}$ where $k' = \sqrt{1-k^2}$

Then

35.35. $\operatorname{sn} u$ has periods $4K$ and $2iK'$

35.36. $\operatorname{cn} u$ has periods $4K$ and $2K + 2iK'$

35.37. $\operatorname{dn} u$ has periods $2K$ and $4iK'$

Identities Involving Elliptic Functions

35.38. $\text{sn}^2 u + \text{cn}^2 u = 1$

35.39. $\text{dn}^2 u + k^2\,\text{sn}^2\,u = 1$

35.40. $\text{dn}^2 u - k^2\text{cn}^2 u = k'^2 \quad \text{where } k' = \sqrt{1-k^2}$

35.41. $\text{sn}^2 u = \dfrac{1-\text{cn}\,2u}{1+\text{dn}\,2u}$

35.42. $\text{cn}^2 u = \dfrac{\text{dn}\,2u+\text{cn}\,2u}{1+\text{dn}\,2u}$

35.43. $\text{dn}^2 u = \dfrac{1-k^2+\text{dn}\,2u+k^2\text{cn}\,u}{1+\text{dn}\,2u}$

35.44. $\sqrt{\dfrac{1-\text{cn}\,2u}{1+\text{cn}\,2u}} = \dfrac{\text{sn}\,u\,\text{dn}\,u}{\text{cn}\,u}$

35.45. $\sqrt{\dfrac{1-\text{dn}\,2u}{1+\text{dn}\,2u}} = \dfrac{k\,\text{sn}\,u\,\text{cn}\,u}{\text{dn}\,u}$

Special Values

35.46. $\text{sn}\,0 = 0$ **35.47.** $\text{cn}\,0 = 1$ **35.48.** $\text{dn}\,0 = 1$ **35.49.** $\text{sc}\,0 = 0$ **35.50.** $\text{am}\,0 = 0$

Integrals

35.51. $\int \text{sn}\,u\,du = \dfrac{1}{k}\ln(\text{dn}\,u - k\,\text{cn}\,u)$

35.52. $\int \text{cn}\,u\,du = \dfrac{1}{k}\cos^{-1}(\text{dn}\,u)$

35.53. $\int \text{dn}\,u\,du = \sin^{-1}(\text{sn}\,u)$

35.54. $\int \text{sc}\,u\,du = \dfrac{1}{\sqrt{1-k^2}}\ln\left(\text{dc}\,u + \sqrt{1-k^2}\,\text{nc}\,u\right)$

35.55. $\int \text{cs}\,u\,du = \ln(\text{ns}\,u - \text{ds}\,u)$

35.56. $\int \text{cd}\,u\,du = \dfrac{1}{k}\ln(\text{nd}\,u + k\,\text{sd}\,u)$

35.57. $\int \text{dc}\,u\,du = \ln(\text{nc}\,u + \text{sc}\,u)$

35.58. $\int \text{sd}\,u\,du = \dfrac{-1}{k\sqrt{1-k^2}}\sin^{-1}(k\,\text{cd}\,u)$

35.59. $\int \text{ds}\,u\,du = \ln(\text{ns}\,u - \text{cs}\,u)$

35.60. $\int \text{ns}\,u\,du = \ln(\text{ds}\,u - \text{cs}\,u)$

35.61. $\int \text{nc}\,u\,du = \dfrac{1}{\sqrt{1-k^2}}\ln\left(\text{dc}\,u + \dfrac{\text{sc}\,u}{\sqrt{1-k^2}}\right)$

35.62. $\int \text{nd}\,u\,du = \dfrac{1}{\sqrt{1-k^2}}\cos^{-1}(\text{cd}\,u)$

Legendre's Relation

35.63. $EK' + E'K - KK' = \pi/2$

where

35.64. $E = \int_0^{\pi/2} \sqrt{1 - k^2 \sin^2 \theta}\, d\theta \qquad K = \int_0^{\pi/2} \frac{d\theta}{\sqrt{1 - k^2 \sin^2 \theta}}$

35.65. $E' = \int_0^{\pi/2} \sqrt{1 - k'^2 \sin^2 \theta}\, d\theta \qquad K' = \int_0^{\pi/2} \frac{d\theta}{\sqrt{1 - k'^2 \sin^2 \theta}}$

36 MISCELLANEOUS and RIEMANN ZETA FUNCTIONS

Error Function $\operatorname{erf}(x) = \frac{2}{\sqrt{\pi}} \int_0^x e^{-u^2} du$

36.1. $\operatorname{erf}(x) = \frac{2}{\sqrt{\pi}}\left(x - \frac{x^3}{3 \cdot 1!} + \frac{x^5}{5 \cdot 2!} - \frac{x^7}{7 \cdot 3!} + \cdots\right)$

36.2. $\operatorname{erf}(x) \sim 1 - \frac{e^{-x^2}}{\sqrt{\pi}\, x}\left(1 - \frac{1}{2x^2} + \frac{1 \cdot 3}{(2x^2)^2} - \frac{1 \cdot 3 \cdot 5}{(2x^2)^3} + \cdots\right)$

36.3. $\operatorname{erf}(-x) = -\operatorname{erf}(x), \qquad \operatorname{erf}(0) = 0, \qquad \operatorname{erf}(\infty) = 1$

Complementary Error Function $\operatorname{erfc}(x) = 1 - \operatorname{erf}(x) = \frac{2}{\sqrt{\pi}} \int_x^\infty e^{-u^2} du$

36.4. $\operatorname{erfc}(x) = 1 - \frac{2}{\sqrt{\pi}}\left(x - \frac{x^3}{3 \cdot 1!} + \frac{x^5}{5 \cdot 2!} - \frac{x^7}{7 \cdot 3!} + \cdots\right)$

36.5. $\operatorname{erfc}(x) \sim \frac{e^{-x^2}}{\sqrt{\pi}\, x}\left(1 - \frac{1}{2x^2} + \frac{1 \cdot 3}{(2x^2)^2} - \frac{1 \cdot 3 \cdot 5}{(2x^2)^3} + \cdots\right)$

36.6. $\operatorname{erfc}(0) = 1, \qquad \operatorname{erfc}(\infty) = 0$

Exponential Integral $\operatorname{Ei}(x) = \int_x^\infty \frac{e^{-u}}{u} du$

36.7. $\operatorname{Ei}(x) = -\gamma - \ln x + \int_0^x \frac{1 - e^{-u}}{u} du$

36.8. $\operatorname{Ei}(x) = -\gamma - \ln x + \left(\frac{x}{1 \cdot 1!} - \frac{x^2}{2 \cdot 2!} + \frac{x^3}{3 \cdot 3!} - \cdots\right)$

36.9. $\operatorname{Ei}(x) \sim \frac{e^{-x}}{x}\left(1 - \frac{1!}{x} + \frac{2!}{x^2} - \frac{3!}{x^3} + \cdots\right)$

36.10. $\operatorname{Ei}(\infty) = 0$

Sine Integral $\operatorname{Si}(x) = \int_0^x \frac{\sin u}{u} du$

36.11. $\operatorname{Si}(x) = \frac{x}{1 \cdot 1!} - \frac{x^3}{3 \cdot 3!} + \frac{x^5}{5 \cdot 5!} - \frac{x^7}{7 \cdot 7!} + \cdots$

36.12. $\operatorname{Si}(x) \sim \frac{\pi}{2} - \frac{\sin x}{x}\left(\frac{1}{x} - \frac{3!}{x^3} + \frac{5!}{x^5} - \cdots\right) - \frac{\cos x}{x}\left(1 - \frac{2!}{x^2} + \frac{4!}{x^4} - \cdots\right)$

36.13. $\operatorname{Si}(-x) = -\operatorname{Si}(x), \qquad \operatorname{Si}(0) = 0, \qquad \operatorname{Si}(\infty) = \pi/2$

Cosine Integral $\mathrm{Ci}(x) = \int_x^\infty \frac{\cos u}{u}\, du$

36.14. $\mathrm{Ci}(x) = -\gamma - \ln x + \int_0^x \frac{1 - \cos u}{u}\, du$

36.15. $\mathrm{Ci}(x) = -\gamma - \ln x + \frac{x^2}{2 \cdot 2!} - \frac{x^4}{4 \cdot 4!} + \frac{x^6}{6 \cdot 6!} - \frac{x^8}{8 \cdot 8!} + \cdots$

36.16. $\mathrm{Ci}(x) \sim \frac{\cos x}{x}\left(\frac{1}{x} - \frac{3!}{x^3} + \frac{5!}{x^5} - \cdots\right) - \frac{\sin x}{x}\left(1 - \frac{2!}{x^2} + \frac{4!}{x^4} - \cdots\right)$

36.17. $\mathrm{Ci}(\infty) = 0$

Fresnel Sine Integral $S(x) = \sqrt{\frac{2}{\pi}} \int_0^x \sin u^2\, du$

36.18. $S(x) = \sqrt{\frac{2}{\pi}}\left(\frac{x^3}{3 \cdot 1!} - \frac{x^7}{7 \cdot 3!} + \frac{x^{11}}{11 \cdot 5!} - \frac{x^{15}}{15 \cdot 7!} + \cdots\right)$

36.19. $S(x) \sim \frac{1}{2} - \frac{1}{\sqrt{2\pi}}\left\{(\cos x^2)\left(\frac{1}{x} - \frac{1 \cdot 3}{2^2 x^5} + \frac{1 \cdot 3 \cdot 5 \cdot 7}{2^4 x^9} - \cdots\right) + (\sin x^2)\left(\frac{1}{2x^3} - \frac{1 \cdot 3 \cdot 5}{2^3 x^7} + \cdots\right)\right\}$

36.20. $S(-x) = -S(x), \qquad S(0) = 0, \qquad S(\infty) = \frac{1}{2}$

Fresnel Cosine Integral $C(x) = \sqrt{\frac{2}{\pi}} \int_0^x \cos u^2\, du$

36.21. $C(x) = \sqrt{\frac{2}{\pi}}\left(\frac{x}{1!} - \frac{x^5}{5 \cdot 2!} + \frac{x^9}{9 \cdot 4!} - \frac{x^{13}}{13 \cdot 6!} + \cdots\right)$

36.22. $C(x) \sim \frac{1}{2} + \frac{1}{\sqrt{2\pi}}\left\{(\sin x^2)\left(\frac{1}{x} - \frac{1 \cdot 3}{2^2 x^5} + \frac{1 \cdot 3 \cdot 5 \cdot 7}{2^4 x^9} - \cdots\right) - (\cos x^2)\left(\frac{1}{2x^3} - \frac{1 \cdot 3 \cdot 5}{2^3 x^7} + \cdots\right)\right\}$

36.23. $C(-x) = -C(x), \qquad C(0) = 0, \qquad C(\infty) = \frac{1}{2}$

Riemann Zeta Function $\zeta(x) = \frac{1}{1^x} + \frac{1}{2^x} + \frac{1}{3^x} + \cdots$

36.24. $\zeta(x) = \frac{1}{\Gamma(x)} \int_0^\infty \frac{u^{x-1}}{e^{u-1}}\, du, \quad x > 1$

36.25. $\zeta(1 - x) = 2^{1-x}\pi^{-x}\Gamma(x)\cos(\pi x/2)\zeta(x)$ (extension to other values)

36.26. $\zeta(2k) = \frac{2^{2k-1}\pi^{2k} B_k}{(2k)!} \; k = 1, 2, 3, \ldots$

37 INEQUALITIES

Triangle Inequality

37.1. $\big||a_1| - |a_2|\big| \leqq |a_1 + a_2| \leqq |a_1| + |a_2|$

37.2. $|a_1 + a_2 + \cdots + a_n| \leqq |a_1| + |a_2| + \cdots + |a_n|$

Cauchy-Schwarz Inequality

37.3. $(a_1b_1 + a_2b_2 + \cdots + a_nb_n)^2 \leqq (a_1^2 + a_2^2 + \cdots + a_n^2)(b_1^2 + b_2^2 + \cdots + b_n^2)$

The equality holds if and only if $a_1/b_1 = a_2/b_2 = \cdots = a_n/b_n$.

Inequalities Involving Arithmetic, Geometric, and Harmonic Means

If A, G, and H are the arithmetic, geometric, and harmonic means of the positive numbers $a_1, a_2, \ldots, a_n$, then

37.4. $H \leqq G \leqq A$

where

37.5. $A = \dfrac{a_1 + a_2 + \cdots + a_n}{n}$

37.6. $G = \sqrt[n]{a_1 a_2 \ldots a_n}$

37.7. $\dfrac{1}{H} = \dfrac{1}{n}\left(\dfrac{1}{a_1} + \dfrac{1}{a_2} + \cdots + \dfrac{1}{a_n}\right)$

The equality holds if and only if $a_1 = a_2 = \cdots = a_n$.

Holder's Inequality

37.8. $|a_1b_1 + a_2b_2 + \cdots + a_nb_n| \leqq (|a_1|^p + |a_2|^p + \cdots + |a_n|^p)^{1/p}(|b_1|^q + |b_2|^q + \cdots + |b_n|^q)^{1/q}$

where

37.9. $\dfrac{1}{p} + \dfrac{1}{q} = 1 \qquad p > 1, q > 1$

The equality holds if and only if $|a_1|^{p-1}/|b_1| = |a_2|^{p-1}/|b_2| = \cdots = |a_n|^{p-1}/|b_n|$. For $p = q = 2$ it reduces to 37.3.

Chebyshev's Inequality

If $a_1 \geqq a_2 \geqq \cdots \geqq a_n$ and $b_1 \geqq b_2 \geqq \cdots \geqq b_n$, then

37.10. $$\left(\frac{a_1 + a_2 + \cdots + a_n}{n}\right)\left(\frac{b_1 + b_2 + \cdots + b_n}{n}\right) \leqq \frac{a_1 b_1 + a_2 b_2 + \cdots + a_n b_n}{n}$$

or

37.11. $$(a_1 + a_2 + \cdots + a_n)(b_1 + b_2 + \cdots + b_n) \leqq n(a_1 b_1 + a_2 b_2 + \cdots + a_n b_n)$$

Minkowski's Inequality

If $a_1, a_2, \ldots a_n, b_1, b_2, \ldots b_n$ are all positive and $p > 1$, then

37.12. $$\{(a_1 + b_1)^p + (a_2 + b_2)^p + \cdots + (a_n + b_n)^p\}^{1/p} \leqq (a_1^p + a_2^p + \cdots + a_n^p)^{1/p} + (b_1^p + b_2^p + \cdots + b_n^p)^{1/p}$$

The equality holds if and only if $a_1/b_1 = a_2/b_2 = \cdots = a_n/b_n$.

Cauchy-Schwarz Inequality for Integrals

37.13. $$\left[\int_a^b f(x)g(x)\,dx\right]^2 \leqq \left\{\int_a^b [f(x)]^2\,dx\right\}\left\{\int_a^b [g(x)]^2\,dx\right\}$$

The equality holds if and only if $f(x)/g(x)$ is a constant.

Holder's Inequality for Integrals

37.14. $$\int_a^b |f(x)g(x)|dx \leqq \left\{\int_a^b |f(x)|^p\,dx\right\}^{1/p} \left\{\int_a^b |g(x)|^q\,dx\right\}^{1/q}$$

where $1/p + 1/q = 1$, $p > 1$, $q > 1$. If $p = q = 2$, this reduces to 37.13.

The equality holds if and only if $|f(x)|^{p-1}/|g(x)|$ is a constant.

Minkowski's Inequality for Integrals

If $p > 1$,

37.15. $$\left\{\int_a^b |f(x) + g(x)|^p\,dx\right\}^{1/p} \leqq \left\{\int_a^b |f(x)|^p\,dx\right\}^{1/p} + \left\{\int_a^b |g(x)|^p\,dx\right\}^{1/p}$$

The equality holds if and only if $f(x)/g(x)$ is a constant.

38 INFINITE PRODUCTS

38.1. $\sin x = x\left(1-\frac{x^2}{x^2}\right)\left(1-\frac{x^2}{4\pi^2}\right)\left(1-\frac{x^2}{9\pi^2}\right)\cdots$

38.2. $\cos x = \left(1-\frac{4x^2}{\pi^2}\right)\left(1-\frac{4x^2}{9\pi^2}\right)\left(1-\frac{4x^2}{25\pi^2}\right)\cdots$

38.3. $\sinh x = x\left(1+\frac{x^2}{\pi^2}\right)\left(1+\frac{x^2}{4\pi^2}\right)\left(1+\frac{x^2}{9\pi^2}\right)\cdots$

38.4. $\cosh x = \left(1+\frac{4x^2}{\pi^2}\right)\left(1+\frac{4x^2}{9\pi^2}\right)\left(1+\frac{4x^2}{25\pi^2}\right)\cdots$

38.5. $\frac{1}{\Gamma(x)} = xe^{\gamma x}\left\{\left(1+\frac{x}{1}\right)e^{-x}\right\}\left\{\left(1+\frac{x}{2}\right)e^{-x/2}\right\}\left\{\left(1+\frac{x}{3}\right)e^{-x/3}\right\}\cdots$

See also 25.11.

38.6. $J_0(x) = \left(1-\frac{x^2}{\lambda_1^2}\right)\left(1-\frac{x^2}{\lambda_2^2}\right)\left(1-\frac{x^2}{\lambda_3^2}\right)\cdots$

where $\lambda_1, \lambda_2, \lambda_3, \ldots$ are the positive roots of $J_0(x) = 0$.

38.7. $J_1(x) = x\left(1-\frac{x^2}{\lambda_1^2}\right)\left(1-\frac{x^2}{\lambda_2^2}\right)\left(1-\frac{x^2}{\lambda_3^2}\right)\cdots$

where $\lambda_1, \lambda_2, \lambda_3, \ldots$ are the positive roots of $J_1(x) = 0$.

38.8. $\frac{\sin x}{x} = \cos\frac{x}{2}\cos\frac{x}{4}\cos\frac{x}{8}\cos\frac{x}{16}\cdots$

38.9. $\frac{\pi}{2} = \frac{2}{1}\cdot\frac{2}{3}\cdot\frac{4}{3}\cdot\frac{4}{5}\cdot\frac{6}{5}\cdot\frac{6}{7}\cdots$

This is called Wallis' product.

39 DESCRIPTIVE STATISTICS

The numerical data $x_1, x_2, \ldots$ will either come from a random sample of a larger population or from the larger population itself. We distinguish these two cases using different notation as follows:

n = number of items in a sample,
N = number of items in the population,

$\bar{x}$ = (read: x-bar) = sample mean,	μ (read: mu) = population mean,
s^2 = sample variance,	σ^2 = population variance,
s = sample standard deviation,	σ = population standard deviation

Note that Greek letters are used with the population and are called *parameters,* whereas Latin letters are used with the samples and are called *statistics*. First we give formulas for the data coming from a sample. This is followed by formulas for the population.

Grouped Data

Frequently, the sample data are collected into groups (grouped data). A group refers to a set of numbers all with the same value x_i, or a set (class) of numbers in a given interval with class value x_i. In such a case, we assume there are k groups with f_i denoting the number of elements in the group with value or class value x_i.

Thus, the total number of data items is

39.1. $n = \sum f_i$

As usual, Σ will denote a summation over all the values of the index, unless otherwise specified.

Accordingly, some of the formulas will be designated as (a) or as (b), where (a) indicates ungrouped data and (b) indicates grouped data.

Measures of Central Tendency

Mean (Arithmetic Mean)

The *arithmetic mean* or simply *mean* of a sample $x_1, x_2, \ldots, x_n$, frequently called the "average value," is the sum of the values divided by the number of values. That is:

39.2(a). Sample mean: $$\bar{x} = \frac{x_1 + x_2 + \cdots + x_n}{n} = \frac{\Sigma x_i}{n}$$

39.2(b). Sample mean: $$\bar{x} = \frac{f_1x_1 + f_2x_2 + \cdots + f_kx_k}{f_1 + f_2 + \cdots + f_k} = \frac{\Sigma f_ix_i}{\Sigma f_i}$$

Median

Suppose that the data $x_1, x_2, \ldots, x_n$ are now sorted in increasing order. The *median* of the data, denoted by

M *or* Median

is defined to be the "middle value." That is:

39.3(a). Median $= \begin{cases} x_{k+1} & \text{when } n \text{ is odd and } n = 2k+1, \\ \dfrac{x_k + x_{k+1}}{2} & \text{when } n \text{ is even and } n = 2k. \end{cases}$

The median of grouped data is obtained by first finding the *cumulative frequency* function F_s. Specifically, we define

$$F_s = f_1 + f_2 + \cdots + f_s$$

that is, F_s is the sum of the frequencies up to f_s. Then:

39.3(b.1). Median $= \begin{cases} x_{j+1} & \text{when } n = 2k+1 \text{ (odd) and } F_j < k+1 \le F_{j+1} \\ \dfrac{x_j + x_{j+1}}{2} & \text{when } n = 2k \text{ (even), and } F_j = k. \end{cases}$

Finding the median of data arranged in classes is more complicated. First one finds the median class m, the class with the median value, and then one linearly interpolates in the class using the formula

39.3(b.2). Median $= L_m + c\dfrac{(n/2) - F_{m-1}}{f_m}$

where L_m denotes the lower class boundary of the median class and c denotes its class width (length of the class interval).

Mode

The mode is the value or values which occur most often. Namely:

39.4. Mode x_m = numerical value that occurs the most number of times

The mode is not defined if every x_m occurs the same number of times, and when the mode is defined it may not be unique.

Weighted and grand means

Suppose that each x_i is assigned a weight $w_i \ge 0$. Then:

39.5. Weighted Mean $\bar{x}_w = \dfrac{w_1x_1 + w_2x_2 + \cdots + w_kx_k}{w_1 + w_2 + \cdots + w_k} = \dfrac{\Sigma\, w_ix_i}{\Sigma\, w_i}$

Note that 39.2(b.1) is a special case of 39.4 where the weight w_i of x_i is its frequency.

Suppose that there are k sample sets and that each sample set has n_i elements and a mean $\bar{x}$. Then the *grand mean,* denoted by $\bar{\bar{x}}_i$ is the "mean of the means" where each mean is weighted by the number of elements in its sample. Specifically:

39.6. Grand Mean $\bar{\bar{x}} = \dfrac{n_1\bar{x}_1 + n_2\bar{x}_2 + \cdots + n_k\bar{x}_k}{n_1 + n_2 + \cdots + n_k} = \dfrac{\Sigma\, n_i\bar{x}_i}{\Sigma\, n_i}$

Geometric and Harmonic Means

The *geometric mean* (G.M.) and *harmonic mean* (H.M.) are defined as follows:

39.7(a). G.M. $= \sqrt[n]{x_1x_2\cdots x_n}$

39.7(b). G.M. $= \sqrt[n]{x_1^{f_1}x_2^{f_2}\cdots x_k^{f_k}}$

39.8(a). $\text{H.M.} = \dfrac{n}{1/x_1 + 1/x_2 + \cdots + 1/x_n} = \dfrac{n}{\Sigma\,(1/x_i)}$

39.8(b). $\text{H.M.} = \dfrac{n}{f_1/x_1 + f_2/x_2 + \cdots + f_k/x_k} = \dfrac{n}{\Sigma(f_k/x_i)}$

Relation Between Arithmetic, Geometric, and Harmonic Means

39.9. $\text{H.M.} \le \text{G.M.} \le \bar{x}$

The equality sign holds only when all the sample values are equal.

Midrange

The *midrange* is the average of the smallest value x_1 and the largest value x_n. That is:

39.10. midrange: $\text{mid} = \dfrac{x_1 + x_n}{2}$

Population Mean

The formula for the population mean μ follows:

39.11(a). Population mean: $\mu = \dfrac{x_1 + x_2 + \cdots + x_N}{N} = \dfrac{\Sigma x_i}{N}$

39.11(b). Population mean: $\mu = \dfrac{f_1x_1 + f_2x_2 + \cdots + f_kx_k}{f_1 + f_2 + \cdots + f_k} = \dfrac{\Sigma f_ix_i}{\Sigma f_i}$

(Recall that N denotes the number of elements in a population.)

Observe that the formula for the population mean μ is the same as the formula for the sample mean $\bar{x}$. On the other hand, the formula for the population standard deviation σ is not the same as the formula for the sample standard deviation s. (This is the main reason we give separate formulas for μ and $\bar{x}$.)

Measures of Dispersion

Sample Variance and Standard Deviation

Here the sample set has n elements with mean $\bar{x}$.

39.12(a). Sample variance: $s^2 = \dfrac{\Sigma(x_i - \bar{x})^2}{n-1} = \dfrac{\Sigma x_i^2 - (\Sigma x_i)^2/n}{n-1}$

39.12(b). Sample variance: $s^2 = \dfrac{\Sigma f_i(x_i - \bar{x})^2}{(\Sigma f_i) - 1} = \dfrac{\Sigma f_i x_i^2 - (\Sigma f_i x_i)^2/\Sigma f_i}{(\Sigma f_i) - 1}$

39.13. Sample standard deviation: $s = \sqrt{\text{Variance}} = \sqrt{s^2}$

EXAMPLE 39.1: Consider the following frequency distribution:

x_i	1	2	3	4	5	6
f_i	8	14	7	12	3	1

Then $n = \Sigma f_i = 45$ and $\Sigma f_i x_i = 126$. Hence, by 39.2(b),

$$\text{Mean } \bar{x} = \frac{\Sigma x_i f_i}{\Sigma f_i} = \frac{126}{45} = 2.8$$

Also, $n - 1 = 44$ and $\Sigma f_i x_i^2 = 430$. Hence, by 39.12(b) and 39.13,

$$s^2 = \frac{430 - (126)^2/45}{44} \approx 1.75 \quad \text{and} \quad s = 1.32$$

We find the median M, first finding the cumulative frequencies:

$$F_1 = 8, \quad F_2 = 22, \quad F_3 = 29, \quad F_4 = 41, \quad F_5 = 44, \quad F_6 = 45 = n$$

Here n is odd, and $(n + 1)/2 = 23$. Hence,

$$\text{Median M} = \text{23rd value} = 3$$

The value 2 occurs most often, hence

$$\text{Mode} = 2$$

M.D. and R.M.S.

Here M.D. stands for *mean deviation* and R.M.S. stands for *root mean square.* As previously, $\bar{x}$ is the mean of the data and, for grouped data, $n = \Sigma f_i$.

39.14(a). $\text{M.D.} = \frac{1}{n}|x_i - \bar{x}|$

39.14(b). $\text{M.D.} = \frac{1}{n}|f_i x_i - \bar{x}|$

39.15(a). $\text{R.M.S.} = \sqrt{\frac{1}{n}(\Sigma x_i^2)}$

39.15(b). $\text{R.M.S.} = \sqrt{\frac{1}{n}(\Sigma f_i x_i^2)}$

Measures of Position (Quartiles and Percentiles)

Now we assume that the data $x_1, x_2, \ldots, x_n$ are arranged in increasing order.

39.16. Sample range: $x_n - x_1$.

There are three quartiles: the first or lower quartile, denoted by Q_1 or Q_L; the second quartile or median, denoted by Q_2 or M; and the third or upper quartile, denoted by Q_3 or Q_U. These quartiles (which essentially divide the data into "quarters") are defined as follows, where "half" means $n/2$ when n is even and $(n\text{-}1)/2$ when n is odd:

39.17. $Q_L (= Q_1)$ = median of the first half of the values.
$\text{M} (= Q_2)$ = median of the values.
$Q_U (= Q_3)$ = median of the second half of the values.

39.18. Five-number summary: $[L, Q_L, \text{M}, Q_U, H]$ where $L = x_1$ (lowest value) and $H = x_n$ (highest value).

39.19. Innerquartile range: $Q_U - Q_L$

39.20. Semi-innerquartile range: $Q = \frac{Q_U - Q_L}{2}$

The k*th percentile*, denoted by P_k, is the number for which k percent of the values are at most P_k and $(100\text{–}k)$ percent of the values are greater than P_k. Specifically:

39.21. P_k = largest x_s such that $F_s \le k/100$. Thus, Q_L = 25th percentile, M = 50th percentile, Q_U = 75th percentile.

Higher-Order Statistics

39.22. The rth moment: (a) $m_r = \frac{1}{n}\Sigma x_i^r$, (b) $m_r = \frac{1}{n}\Sigma f_i x_i^r$

39.23. The rth moment about the mean $\bar{x}$:

(a) $\mu_r = \frac{1}{n}\Sigma(x_i - \bar{x})^r$, (b) $\mu_r = \frac{1}{n}\Sigma(f_i x_i - \bar{x})^r$

39.24. The rth absolute moment about mean $\bar{x}$:

(a) $\mu_r = \frac{1}{n}\Sigma|x_i - \bar{x}|^r$, (b) $\mu_r = \frac{1}{n}\Sigma|f_i x_i - \bar{x}|^r$

39.25. The rth moment in standard z units about $z = 0$:

(a) $\alpha_r = \frac{1}{n}\Sigma z_i^r$, (b) $\alpha_r = \frac{1}{n}\Sigma f_i z_i^r$ where $z_i = \frac{x_i - \bar{x}}{\sigma}$

Measures of Skewness and Kurtosis

39.26. Coefficient of skewness: $\gamma_1 = \frac{\mu_3}{\sigma^3} = \alpha_3$

39.27. Momental skewness: $\frac{\mu_3}{2\sigma^3}$

39.28. Coefficient of kurtosis: $\alpha_4 = \frac{\mu_4}{\sigma^4}$

39.29. Coefficient of excess (kurtosis): $\alpha_4 - 3 = \frac{\mu_4}{\sigma^4} - 3$

39.30. Quartile coefficient of skewness: $\frac{Q_U - 2\hat{x} + Q_L}{Q_U - Q_L} = \frac{Q_3 - 2Q_2 + Q_1}{Q_3 - Q_1}$

Population Variance and Standard Deviation

Recall that N denotes the number of values in the population.

39.31. Population variance: $\sigma^2 = \frac{\Sigma(x_i - \bar{x})^2}{N} = \frac{\Sigma x_i^2 - (\Sigma x_i)^2/n}{N}$

39.32. Population standard deviation: $\sigma = \sqrt{\text{Variance}} = \sqrt{\sigma^2}$

Bivariate Data

The following formulas apply to a list of pairs of numerical values:

$$(x_1, y_1), (x_2, y_2), (x_3, y_3), \ldots, (x_n, y_n)$$

where the first values correspond to a variable x and the second to a variable y. The primary objective is to determine whether there is a mathematical relationship, such as a linear relationship, between the data.

The *scatterplot* of the data is simply a picture of the pairs of values as points in a coordinate plane.

Correlation Coefficient

A numerical indicator of a linear relationship between variables x and y is the *sample correlation coefficient* r of x and y, defined as follows:

39.33. Sample correlation coefficient: $$r = \frac{\Sigma(x_i - \bar{x})(y_i - \bar{y})}{\sqrt{\Sigma(x_i - \bar{x})^2 \Sigma(y_i - \bar{y})^2}}$$

We assume that the denominator in Formula 39.33 is not zero. An alternative formula for computing r follows:

39.34. $$r = \frac{\Sigma x_i y_i - (\Sigma x_i)(\Sigma y_i)/n}{\sqrt{\Sigma x_i^2 - (\Sigma x_i)^2/n}\sqrt{\Sigma y_i^2 - (\Sigma y_i)^2/n}}$$

Properties of the correlation coefficient r follow:

39.35. (1) $-1 \leq r \leq 1$ or, equivalently, $|r| \leq 1$.
(2) r is positive or negative according as y tends to increase or decrease as x increases.
(3) The closer $|r|$ is to 1, the stronger the linear relationship between x and y.

The *sample covariance* of x and y is denoted and defined as follows:

39.36. Sample covariance: $$s_{xy} = \frac{\Sigma(x_i - \bar{x})(y_i - \bar{y})}{n-1}$$

Using the sample covariance, Formula 39.33 can be written in the compact form:

39.37. $$r = \frac{s_{xy}}{s_x s_y}$$

where s_x and s_y are the sample standard deviations of x and y, respectively.

EXAMPLE 39.2: Consider the following data:

x	50	45	40	38	32	40	55
y	2.5	5.0	6.2	7.4	8.3	4.7	1.8

The scatterplot of the data appears in Fig. 39-1. The correlation coefficient r for the data may be obtained by first constructing the table in Fig. 39-2. Then, by Formula 39.34 with $n = 7$,

$$r = \frac{1431.8 - (300)(35.9)/7}{\sqrt{13{,}218 - (300)^2/7}\sqrt{218.67 + (35.9)^2/7}} \approx -0.9562$$

Here r is close to -1, and the scatterplot in Fig. 39-1 does indicate a strong negative linear relationship between x and y.

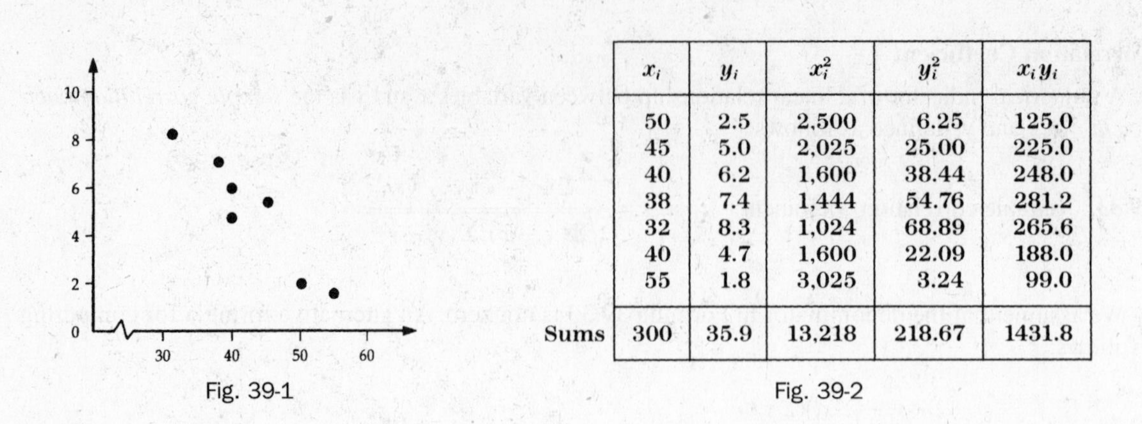

x_i	y_i	x_i^2	y_i^2	$x_i y_i$
50	2.5	2,500	6.25	125.0
45	5.0	2,025	25.00	225.0
40	6.2	1,600	38.44	248.0
38	7.4	1,444	54.76	281.2
32	8.3	1,024	68.89	265.6
40	4.7	1,600	22.09	188.0
55	1.8	3,025	3.24	99.0
Sums 300	35.9	13,218	218.67	1431.8

Fig. 39-1

Fig. 39-2

Regression Line

Consider a given set of n data points P_i (x_i, y_i). Any (nonvertical) line L may be defined by an equation of the form

$$y = a + bx$$

Let y_i^* denote the y value of the point on L corresponding to x_i; that is, let $y_i^* = a + bx_i$. Now let

$$d_i = y_i - y_i^* = y_i - (a + bx_i)$$

that is, d_i is the vertical (directed) distance between the point P_i and the line L. The *squares error* between the line L and the data points is defined by

39.38. $\Sigma d_i^2 = d_1^2 + d_2^2 + \cdots + d_n^2$

The *least-squares line* or the *line of best fit* or the *regression line* of y on x is, by definition, the line L whose squares error is as small as possible. It can be shown that such a line L exists and is unique.

The constants a and b in the equation $y = a + bx$ of the line L of best fit can be obtained from the following two *normal* equations, where a and b are the unknowns and n is the number of points:

39.39. $$\begin{cases} na + (\Sigma x_i)b = \Sigma y_i \\ (\Sigma x_i)a + (\Sigma x_i^2)b = \Sigma x_i y_i \end{cases}$$

The solution of the above normal equations follows:

39.40. $$b = \frac{n\Sigma x_i y_i - (\Sigma x_i)(\Sigma y_i)}{n\Sigma x_i^2 - (\Sigma x_i)^2} = \frac{rs_y}{s_x}; \qquad a = \frac{\Sigma y_i}{n} - b\frac{\Sigma x_i}{n} = \bar{y} - b\bar{x}$$

The second equation tells us that the point $(\bar{x}, \bar{y})$ lies on L, and the first equation tells us that the point $(\bar{x} + s_x, \bar{y} + rs_y)$ also lies on L.

EXAMPLE 39.3: Suppose we want the line L of best fit for the data in Example 39.2. Using the table in Fig. 39-2 and $n = 7$, we obtain the normal equations

$$7a + 300b = 35.9$$

$$300a + 13{,}218b = 1431.8$$

Substitution in 39.40 yields

$$b = \frac{7(1431.8) - (300)(35.9)}{7(13{,}218) - (300)^2} = -0.2959$$

$$a = \frac{35.9}{7} - (-0.2959)\frac{300}{7} = 17.8100$$

Thus, the line L of best fit is

$$y = 17.8100 - 0.2959x$$

The graph of L appears in Fig. 39-3.

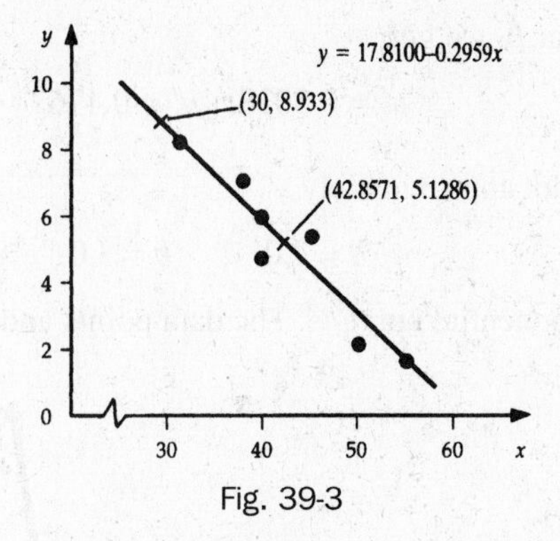

Fig. 39-3

Curve Fitting

Suppose that n data points $P_i\ (x_i, y_i)$ are given, and that the data (using the scatterplot or the correlation coefficient r) do not indicate a linear relationship between the variables x and y, but do indicate that some other standard (well-known) type of curve $y = f(x)$ approximates the data. Then the particular curve C that one uses to approximate that data, called the *best-fitting* or *least-squares* curve, is the curve in the collection which minimizes the squares error sum

$$\Sigma d_i^2 = d_1^2 + d_2^2 + \cdots + d_n^2$$

where $d_i = y_i - f(x_i)$. Three such types of curve are discussed as follows.

Polynomial function of degree m: $y = a_0 + a_1 x + a_2 x^2 + \cdots + a_m x^m$

The coefficients $a_0, a_1, a_2, \ldots, a_m$ of the best-fitting polynomial can be obtained by solving the following system of $m + 1$ normal equations:

39.41.
$$\begin{aligned} na_0 + a_1 \Sigma x_i + a_2 \Sigma x_i^2 + \cdots + a_m \Sigma x_i^m &= \Sigma y_i \\ a_0 \Sigma x_i + a_1 \Sigma x_i^2 + a_2 \Sigma x_i^3 + \cdots + a_m \Sigma x_i^{m+1} &= \Sigma x_i y_i \\ \cdots\cdots\cdots\cdots\cdots\cdots\cdots\cdots\cdots\cdots\cdots\cdots\cdots & \\ a_0 \Sigma x_i^m + a_1 \Sigma x_i^{m+1} + a_2 \Sigma x_i^{m+2} + \cdots + a_m \Sigma x_i^{2m} &= \Sigma x_i^m y_i \end{aligned}$$

Exponential curve: $y = ab^x$ or $\log y = \log a + (\log b)x$

The exponential curve is used if the scatterplot of $\log y$ verses x indicates a linear relationship. Then $\log a$ and $\log b$ are obtained from transformed data points. Namely, the best-fit line L for data points $P'(x_i, \log y_i)$ is

39.42.
$$\begin{cases} na' + (\Sigma x_i) b' = \Sigma (\log y_i) \\ (\Sigma x_i) a' + (\Sigma x_i^2) b' = \Sigma (x_i \log y_i) \end{cases}$$

Then $a = \text{antilog } a'$, $b = \text{antilog } b'$.

EXAMPLE 39.4: Consider the following data which indicates exponential growth:

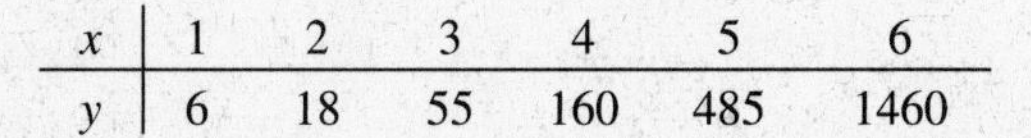

x	1	2	3	4	5	6
y	6	18	55	160	485	1460

Thus, we seek the least-squares line L for the following data:

x	1	2	3	4	5	6
$\log y$	0.7782	1.2553	1.7404	2.2041	2.6857	3.1644

Using the normal equation 39.42 for L, we get

$$a' = 0.3028, \quad b' = 0.4767$$

The antiderivatives of a' and b' yield, approximately,

$$a = 2.0, \qquad b = 3.0$$

Hence, $y = 2(3^x)$ is the required exponential curve C. The data points and C are depicted in Fig. 39-4.

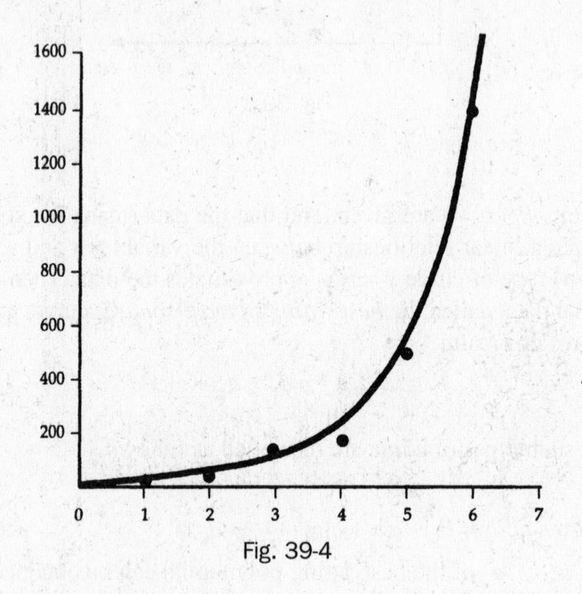

Fig. 39-4

Power function: $y = ax^b$ or $\log y = \log a + b \log x$

The power curve is used if the scatterplot of $\log y$ verses $\log x$ indicates a linear relationship. The $\log a$ and b are obtained from transformed data points. Namely, the best-fit line L for transformed data points $P'(\log x_i, \log y_i)$ is

39.43.
$$\begin{cases} na' + \Sigma(\log x_i)b = \Sigma(\log y_i) \\ \Sigma(\log x_i)a' + \Sigma(\log x_i)^2 b = \Sigma(\log x_i \log y_i) \end{cases}$$

Then $a = \text{antilog } a'$.

40 PROBABILITY

Sample Spaces and Events

Let S be a sample space which consists of the possible outcomes of an experiment where the events are subsets of S. The sample space S itself is called the *certain event*, and the null set $\varnothing$ is called the *impossible event*.

It would be convenient if all subsets of S could be events. Unfortunately, this may lead to contradictions when a probability function is defined on the events. Thus, the events are defined to be a limited collection C of subsets of S as follows.

DEFINITION 40.1: The class C of events of a sample space S form a σ-field. That is, C has the following three properties:

(i) $S \in C$.
(ii) If $A_1, A_2, \ldots$ belong to C, then their union $A_1 \cup A_2 \cup A_3 \cup \ldots$ belongs to C.
(iii) If $A \in C$, then its complement $A^c \in C$.

Although the above definition does not mention intersections, DeMorgan's law (40.3) tells us that the complement of a union is the intersection of the complements. Thus, the events form a collection that is closed under unions, intersections, and complements of denumerable sequences.

If S is finite, then the class of all subsets of S form a σ-field. However, if S is nondenumerable, then only certain subsets of S can be the events. In fact, if B is the collection of all open intervals on the real line **R**, then the smallest σ-field containing B is the collection of Borel sets in **R**.

If Condition (ii) in Definition 40.1 of a σ-field is replaced by finite unions, then the class of subsets of S is called a *field*. Thus a σ-field is a field, but not visa versa.

First, for completeness, we list basic properties of the set operations of union, intersection, and complement.

40.1. Sets satisfy the properties in Table 40-1.

TABLE 40-1 Laws of the Algebra of Sets

Idempotent laws:	(1a)	$A \cup A = A$	(1b)	$A \cap A = A$
Associative laws:	(2a)	$(A \cup B) \cup C = A \cup (B \cup C)$	(2b)	$(A \cap B) \cap C = A \cap (B \cap C)$
Commutative laws:	(3a)	$A \cup B = B \cup A$	(3b)	$A \cap B = B \cap A$
Distributive laws:	(4a)	$A \cup (B \cap C) = (A \cup B) \cap (A \cup C)$	(4b)	$A \cap (B \cup C) = (A \cap B) \cup (A \cap B)$
Identity laws:	(5a)	$A \cup \varnothing = A$	(5b)	$A \cap \mathbf{U} = A$
	(6a)	$A \cup \mathbf{U} = \mathbf{U}$	(6b)	$A \cap \varnothing = \varnothing$
Involution law:	(7)	$(A^C)^C = A$		
Complement laws:	(8a)	$A \cup A^c = \mathbf{U}$	(8b)	$A \cap A^c = \varnothing$
	(9a)	$\mathbf{U}^c = \varnothing$	(9b)	$\varnothing^c = U$
DeMorgan's laws:	(10a)	$(A \cup B)^c = A^c \cap B^c$	(10b)	$(A \cap B)^c = A^c \cup B^c$

40.2. The following are equivalent: (i) $A \subseteq B$, (ii) $A \cap B = A$, (iii) $A \cap B = B$.
Recall that the union and intersection of any collection of sets is defined as follows:

$$\bigcup_j A_j = \{x \mid \text{there exists j such that } x \in A_j\} \quad \text{and} \quad \bigcap_j A_j = \{x \mid \text{for every j we have } x \in A_j\}$$

40.3. (Generalized DeMorgan's Law) (10a)'$(\bigcup_j A_j)^c = \bigcap_j A_j^c$; (10b)'$(\bigcap_j A_j)^c = \bigcup_j A_j^c$

Probability Spaces and Probability Functions

DEFINITION 40.2: Let P be a real-valued function defined on the class C of events of a sample space S. Then P is called a *probability function*, and P(A) is called the *probability* of an event A, when the following axioms hold:

Axiom [P_1] For every event A, $P(A) \geq 0$.

Axiom [P_2] For the certain event S, $P(S) = 1$.

Axiom [P_3] For any sequence of mutually exclusive (disjoint) events $A_1, A_2, \ldots,$

$$P(A_1 \cup A_2 \cup \ldots) = P(A_1) + P(A_2) + \ldots$$

The triple (S, C, P), or simply S when C and P are understood, is called a *probability space*.

Axiom **[P_3]** implies an analogous axiom for any finite number of sets. That is:

Axiom [P_3'] For any finite collection of mutually exclusive events $A_1, A_2, \ldots, A_n$,

$$P(A_1 \cup A_2 \cup \ldots \cup A_n) = P(A_1) + P(A_2) + \ldots + P(A_n)$$

In particular, for two disjoint events A and B, we have $P(A \cup B) = P(A) + P(B)$.

The following properties follow directly from the above axioms.

40.4. (Complement rule) $P(A^c) = 1 - P(A)$. Thus, $P(\emptyset) = 0$.

40.5. (Difference Rule) $P(A \backslash B) = P(A) - P(A \cap B)$.

40.6. (Addition Rule) $P(A \cup B) = P(A) + P(B) - P(A \cap B)$.

40.7. For $n \geq 2$, $P\left(\bigcup_{j=1}^{n} A_j\right) \leq \sum_{j=1}^{n} P(A_j)$

40.8. (Monoticity Rule) If $A \subseteq B$, then $P(A) \leq P(B)$.

Limits of Sequences of Events

40.9. (Continuity) Suppose $A_1, A_2, \ldots$ form a monotonic increasing (decreasing) sequence of events; that is, $A_j \subseteq A_{j+1}$ $(A_j \supseteq A_{j+1})$. Let $A = \cup_j A_j$ $(A = \cap_j A_j)$. Then $\lim P(A_n)$ exists and

$$\lim P(A_n) = P(A)$$

For any sequence of events $A_1, A_2, \ldots,$ we define

$$\liminf A_n = \bigcup_{k=1}^{+\infty} \bigcap_{j=k}^{+\infty} A_j \quad \text{and} \quad \limsup A_n = \bigcap_{k=1}^{+\infty} \bigcup_{j=k}^{+\infty} A_j$$

If $\liminf A_n = \limsup A_n$, then we call this set $\lim A_n$. Note $\lim A_n$ exists when the sequence is monotonic.

40.10. For any sequence A_j of events in a probability space,

$$P(\liminf A_n) \le \liminf P(A_n) \le \limsup P(A_n) \le P(\limsup A_n)$$

Thus, if $\lim A_n$ exists, then $P(\lim A_n) = \lim P(A_n)$.

40.11. For any sequence A_j of events in a probability space, $P(\cup_j A_j) \le \sum_j P(A_j)$.

40.12. (Borel-Cantelli Lemma) Suppose A_j is any sequence of events in a probability space. Furthermore, suppose $\sum_{n=1}^{+\infty} P(A_n) < +\infty$. Then $P(\limsup A_n) = 0$.

40.13. (Extension Theorem) Let F be a field of subsets of S. Let P be a function on F satisfying Axioms $\mathbf{P_1}$, $\mathbf{P_2}$, and $\mathbf{P_3}'$. Then there exists a unique probability function P^* on the smallest σ-field containing F such that P^* is equal to P on F.

Conditional Probability

DEFINITION 40.3: Let E be an event with $P(E) > 0$. The conditional probability of an event A given E is denoted and defined as follows:

$$P(A|E) = \frac{P(A \cap E)}{P(E)}$$

40.14. (Multiplication Theorem for Conditional Probability) $P(A \cap B) = P(A)P(B|A)$. This theorem can be genealized as follows:

40.15. $P(A_1 \cap \cdots \cap A_n) = P(A_1)P(A_2|A_1)P(A_3|A_1 \cap A_2) \ldots P(A_n|A_1 \cap \ldots \cap A_{n-1})$

EXAMPLE 40.1: A lot contains 12 items of which 4 are defective. Three items are drawn at random from the lot one after the other. Find the probabiliy that all three are nondefective.

The probability that the first item is nondefective is 8/12. Assuming the first item is nondefective, the probability that the second item is nondefective is 7/11. Assuming the first and second items are nondefective, the probability that the third item is nondefective is 6/10. Thus,

$$p = \frac{8}{12} \cdot \frac{7}{11} \cdot \frac{6}{10} = \frac{14}{55}$$

Stochastic Processes and Probability Tree Diagrams

A (finite) stochastic process is a finite sequence of experiments where each experiment has a finite number of outcomes with given probabilities. A convenient way of describing such a process is by means of a probability tree diagram, illustrated below, where the multiplication theorem (40.14) is used to compute the probability of an event which is represented by a given path of the tree.

EXAMPLE 40.2: Let X, Y, Z be three coins in a box where X is a fair coin, Y is two-headed, and Z is weighted so the probability of heads is 1/3. A coin is selected at random and is tossed. (a) Find P(H), the probability that heads appears. (b) Find P(X|H), the probability that the fair coin X was picked if heads appears.

The probability tree diagram corresponding to the two-step stochastic process appears in Fig. 40-1a.

(a) Heads appears on three of the paths (from left to right); hence,

$$P(H) = \frac{1}{3}\cdot\frac{1}{2} + \frac{1}{3}\cdot 1 + \frac{1}{3}\cdot\frac{1}{3} = \frac{11}{18}$$

(b) X and heads H appear only along the top path; hence

$$P(X \cap H) = \frac{1}{3}\cdot\frac{1}{2} = \frac{1}{6} \text{ and so } P(X|H) = \frac{P(X \cap H)}{P(H)} = \frac{1/6}{11/18} = \frac{3}{11}$$

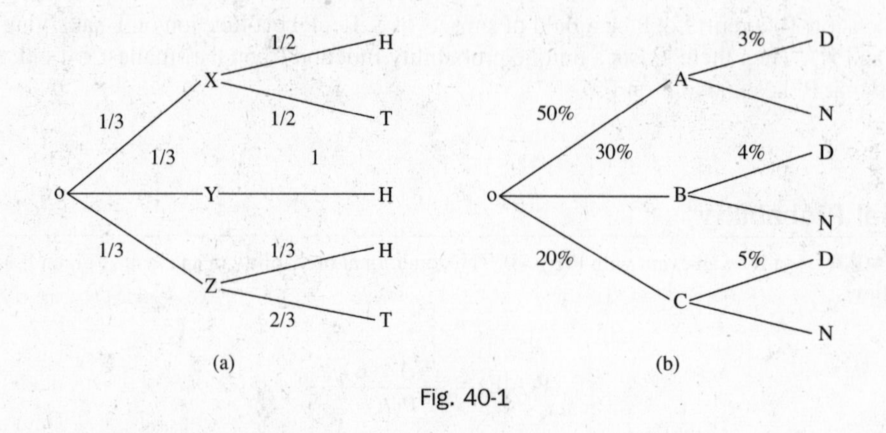

Fig. 40-1

Law of Total Probability and Bayes' Theorem

Here we assume E is an event in a sample space S, and $A_1, A_2, \ldots A_n$ are mutually disjoint events whose union is S; that is, the events $A_1, A_2, \ldots, A_n$ form a partition of S.

40.16. (Law of Total Probability) $P(E) = P(A_1)P(E|A_1) + P(A_2)P(E|A_2) + \ldots + P(A_n)P(E|A_n)$

40.17. (Bayes' Formula) For k = 1, 2, …, n,

$$P(A_k|E) = \frac{P(A_k)P(E\mid A_k)}{P(E)} = \frac{P(A_k)P(E\mid A_k)}{P(A_1)P(E\mid A_1) + P(A_2)P(E\mid A_2) + \cdots + P(A_n)P(E\mid A_n)}$$

EXAMPLE 40.3: Three machines, A, B, C, produce, respectively, 50%, 30%, and 20% of the total number of items in a factory. The percentages of defective output of these machines are, respectively, 3%, 4%, and 5%. An item is randomly selected.

(a) Find P(D), the probability the item is defective.

(b) If the item is defective, find the probability it came from machine: (i) A, (ii) B, (iii) C.

(a) By **40.16** (Total Probability Law),

$$P(D) = P(A)P(D|A) + P(B)P(D|B) + P(C)P(D|C)$$
$$= (0.50)(0.03) + (0.30)(0.04) + (0.20)(0.05) = 3.7\%$$

(b) By **40.17** (Bayes' rule), (i) $P(A|D) = \frac{P(A)P(D\mid A)}{P(D)} = \frac{(0.50)(0.03)}{0.037} = 40.5\%$. Similarly,

(ii) $P(B|D) = \frac{P(B)P(D\mid B)}{P(D)} = 32.5\%$; (iii) $P(C|D) = \frac{P(C)P(D\mid C)}{P(D)} = 27.0\%$

Alternately, we may consider this problem as a two-step stochastic process with a probability tree diagram, as in Fig. 40-1(b). We find P(D) by adding the three probability paths to D:

$$(0.50)(0.03) + (0.30)(0.04) + (0.20)(0.05) = 3.7\%$$

We find P(A|D) by dividing the top path to A and D by the sum of the three paths to D.

$$(0.50)(0.03)/0.037 = 40.5\%$$

Similarly, we find P(B|D) = 32.5% and P(C|D) = 27.0%.

Independent Events

DEFINITION 40.4: Events A and B are independent if $P(A \cap B) = P(A)P(B)$.

40.18. The following are equivalent:
(i) $P(A \cap B) = P(A)P(B)$, (ii) $P(A|B) = P(A)$, (iii) $P(B|A) = P(B)$.

That is, events A and B are independent if the occurrence of one of them does not influence the occurrence of the other.

EXAMPLE 40.4: Consider the following events for a family with children where we assume the sample space S is an equiprobable space:

$$E = \{\text{children of both sexes}\}, \qquad F = \{\text{at most one boy}\}$$

(a) Show that E and F are independent events if a family has three children.

(b) Show that E and F are dependent events if a family has two children.

(a) Here S = {bbb, bbg, bgb, bgg, gbb, gbg, ggb, ggg}. So:

$$E = \{bbg, bgb, bgg, gbb, gbg, ggb\},\ P(E) = 6/8 = 3/4,$$
$$F = \{bgg, gbg, ggb, ggg\},\ P(F) = 4/8 = 1/2$$
$$E \cap F = \{bgg, gbg, ggb\},\ P(E \cap F) = 3/8$$

Therefore, $P(E)P(F) = (3/4)(1/2) = 3/8 = P(E \cap F)$. Hence, E and F are independent.

(b) Here S = {bb, bg, gb, gg}. So:

$$E = \{bg, gb\},\ P(E) = 2/4 = 1/2,$$
$$F = \{bg, gb, gg\},\ P(F) = 3/4$$
$$E \cap F = \{bg, gb\},\ P(E \cap F) = 2/4 = 1/2$$

Therefore, $P(E)P(F) = (1/2)(3/4) = 3/8 \neq P(E \cap F)$. Hence, E and F are dependent.

DEFINITION 40.5: For $n > 2$, the events $A_1, A_2, \ldots, A_n$ are independent if any proper subset of them is independent and

$$P(A_1 \cap A_2 \cap \ldots \cap A_n) = P(A_1)P(A_2) \ldots P(A_n)$$

Observe that induction is used in this definition.

DEFINITION 40.6: A collection $\{A_j \mid j \in J\}$ of events is independent if, for any $n > 0$, the sets $A_{j_1}, A_{j_2}, \ldots, A_{j_n}$ are independent.

The concept of independent repeated trials, when S is a finite set, is formalized as follows.

DEFINITION 40.7: Let S be a finite probability space. The probability space of n independent trials or repeated trials, denoted by S_n, consists of ordered n-tuples $(s_1, s_2, \ldots, s_n)$ of elements of S with the probability of an n-tuple defined by

$$P((s_1, s_2, \ldots, s_n)) = P(s_1)P(s_2) \ldots P(s_n)$$

EXAMPLE 40.5: Suppose whenever horses a, b, c race together, their respective probabilities of winning are 20%, 30%, and 50%. That is, S = {a, b, c} with P(a) = 0.2, P(b) = 0.3, and P(c) = 0.5.

They race three times. Find the probability that

(a) the same horse wins all three times
(b) each horse wins once

(a) Writing xyz for (x, y, z), we seek the probability of the event A = {aaa, bbb, ccc}. Here,

$$P(aaa) = (0.2)^3 = 0.008,\ P(bbb) = (0.3)^3 = 0.027,\ P(ccc) = (0.5)^3 = 0.125$$

Thus, P(A) = 0.008 + 0.027 + 0.125 = 0.160.

(b) We seek the probability of the event B = {abc, acb, bac, bca, cab, cba}. Each element in B has the same probability (0.2)(0.3)(0.5) = 0.03. Thus, P(B) = 6(0.03) = 0.18.

41 RANDOM VARIABLES

Consider a probability space (S, C, P).

DEFINITION 41.1. A random variable X on the sample space S is a function from S into the set **R** of real numbers such that the preimage of every interval of **R** is an event of S.

If S is a discrete sample space in which every subset of S is an event, then every real-valued function on S is a random variable. On the other hand, if S is uncountable, then certain real-valued functions on S may not be random variables.

Let X be a random variable on S, where we let R_X denote the range of X; that is,

$$R_X = \{x \mid \text{there exists } s \in S \text{ for which } X(s) = x\}$$

There are two cases that we treat separately. (i) X is a discrete random variable; that is, R_X is finite or countable. (ii) X is a continuous random variable; that is, R_X is a continuum of numbers such as an interval or a union of intervals.

Let X and Y be random variables on the same sample space S. Then, as usual, X + Y, X + k, kX, and XY (where k is a real number) are the functions on S defined as follows (where s is any point in S):

$$(X + Y)(s) = X(s) + Y(s), \qquad (kX)(s) = kX(s),$$
$$(X + k)(s) = X(s) + k, \qquad (XY)(s) = X(s)Y(s).$$

More generally, for any polynomial, exponential, or continuous function h(t), we define h(X) to be the function on S defined by

$$[h(X)](s) = h[X(s)]$$

One can show that these are also random variables on S.

The following short notation is used:

$P(X = x_i)$	denotes the probability that $X = x_i$.
$P(a \le X \le b$	denotes the probability that X lies in the closed interval [a, b].
μ_X or E(X) or simply μ	denotes the mean or expectation of X.
σ_{X^2} or Var(X) or simply σ^2	denotes the variance of X.
σ_X or simply σ	denotes the standard deviation of X.

Sometimes we let Y be a random variable such that Y = g(X), that is, where Y is some function of X.

Discrete Random Variables

Here X is a random variable with only a finite or countable number of values, say

$R_X = \{x_1, x_2, x_3, \ldots\}$ where, say, $x_1 < x_2, < x_3 < \ldots$. Then X induces a function f(x) on R_X as follows:

$$f(x_i) = P(X = x_i) = P(\{s \in S \mid X(s) = x_i\})$$

The function f(x) has the following properties:

$$\text{(i) } f(x_i) \ge 0 \quad \text{and} \quad \text{(ii) } \Sigma_i\, f(x_i) = 1$$

Thus, f defines a probability function on the range R_X of X. The pair $(x_i, f(x_i))$, usually given by a table, is called the *probability distribution* or *probability mass function* of X.

Mean

41.1. $\mu_X = E(X) = \Sigma x_i f(x_i)$

Here, Y = g(X).

41.2. $\mu_Y = E(Y) = \Sigma g(x_i) f(x_i)$

Variance and Standard Deviation

41.3. $\sigma_X^2 = Var(X) = \Sigma(x_i - \mu)^2 f(x_i) = E((X - \mu)^2)$

Alternately, $Var(X) = \sigma^2$ may be obtained as follows:

41.4. $Var(X) = \Sigma x_i^2 f(x_i) - \mu^2 = E(X^2) - \mu^2$

41.5. $\sigma_X = \sqrt{Var(X)} = \sqrt{E(X^2) - \mu^2}$

REMARK: Both the variance $Var(X) = \sigma^2$ and the standard deviation σ measure the weighted spread of the values x_i about the mean μ; however, the standard deviation has the same units as μ.

EXAMPLE 41.1: Suppose X has the following probability distribution:

x	2	4	6	8
f(x)	0.1	0.2	0.3	0.4

Then:

$$\mu = E(X) = \Sigma x_i f(x_i) = 2(0.1) + 4(0.2) + 6(0.3) + 8(0.4) = 6$$
$$E(X^2) = \Sigma x_i^2 f(x_i) = 2^2(0.1) + 4^2(0.2) + 6^2(0.3) + 8^2(0.4) = 40$$
$$\sigma^2 = Var(X) = E(X^2) - \mu^2 = 40 - 36 = 4$$

$$\sigma = \sqrt{Var(X)} = \sqrt{4} = 2$$

Continuous Random Variable

Here X is a random variable with a continuum number of values. Then X determines a function f(x), called the *density function* of X, such that

$$\text{(i) } f(x) \geq 0 \quad \text{and} \quad \text{(ii) } \int_{-\infty}^{\infty} f(x)\,dx = \int_R f(x)\,dx = 1$$

Furthermore,

$$P(a \leq X \leq b) = \int_a^b f(x)\,dx$$

Mean

41.6. $\mu_X = E(X) = \int_{-\infty}^{\infty} x f(x)\,dx$

Here, Y = g(X).

41.7. $\mu_Y = E(Y) = \int_{-\infty}^{\infty} g(x) f(x)\,dx$

Variance and Standard Deviation

41.8. $\sigma_X{}^2 = \text{Var}(X) = \int_{-\infty}^{\infty} (x - \mu)^2 f(x)dx = E((X - \mu)^2)$

Alternately, $\text{Var}(X) = \sigma^2$ may be obtained as follows:

41.9. $\text{Var}(X) = \int_{-\infty}^{\infty} x^2 f(x)dx - \mu^2 = E(X^2) - \mu^2$

41.10. $\sigma_X = \sqrt{Var(X)} = \sqrt{E(X^2) - \mu^2}$

EXAMPLE 41.2: Let X be the continuous random variable with the following density function:

$$f(x) = \begin{cases} (1/2)x & \text{if } 0 \le x \le 2 \\ 0 & \text{elsewhere} \end{cases}$$

Then:

$$E(X) = \int_{-\infty}^{\infty} xf(x)\,dx = \int_0^2 \frac{1}{2} x^2\,dx = \left[\frac{x^3}{6}\right]_0^2 = \frac{4}{3}$$

$$E(X^2) = \int_{-\infty}^{\infty} x^2 f(x)\,dx = \int_0^2 \frac{1}{2} x^3\,dx = \left[\frac{x^4}{8}\right]_0^2 = 2$$

$$\sigma^2 = \text{Var}(X) = E(X^2) - \mu^2 = 2 - \frac{16}{9} = \frac{2}{9}$$

$$\sigma = \sqrt{Var(X)} = \sqrt{\frac{2}{9}} = \frac{1}{3}\sqrt{2}$$

Cumulative Distribution Function

The *cumulative distribution function* F(x) of a random variable X is the function $F: \mathbf{R} \to \mathbf{R}$ defined by

41.11. $F(a) = P(X \le a)$

The function F is well-defined since the inverse of the interval $(-\infty, a]$ is an event.

The function F(x) has the following properties:

41.12. $F(a) \le F(b)$ whenever $a \le b$.

41.13. $\lim_{x \to -\infty} F(x) = 0$ and $\lim_{x \to +\infty} F(x) = 1$

That is, F(x) is monotonic, and the limit of F to the left is 0 and to the right is 1.

If X is the discrete random variable with distribution f(x), then F(x) is the following step function:

41.14. $F(x) = \sum_{x_i \le x} f(x_i)$

If X is a continuous random variable, then the density funcion f(x) of X can be obtained from the cumulative distribution function F(x) by differentiation. That is,

41.15. $f(x) = \frac{d}{dx} F(x) = F'(x)$

Accordingly, for a continuous random variable X,

41.16. $F(x) = \int_{-\infty}^{x} f(t)\,dt$

Standardized Random Variable

The *standardized random variable* Z of a random variable X with mean μ and standard deviation $\sigma > 0$ is defined by

41.17. $Z = \dfrac{X - \mu}{\sigma}$

Properties of such a standardized random variable Z follow:

$$\mu_Z = E(Z) = 0 \quad \text{and} \quad \sigma_Z = 1$$

EXAMPLE 41.3: Consider the random variable X in Example 41.1 where $\mu_X = 6$ and $\sigma_X = 2$. The distribution of $Z = (X - 6)/2$ where $f(z) = f(x)$ follows:

Z	−2	−1	0	1
f(Z)	0.1	0.2	0.3	0.4

Then:

$$E(Z) = \Sigma z_i f(z_i) = (-2)(0.1) + (-1)(0.2) + 0(0.3) + 1(0.4) = 0$$
$$E(Z^2) = \Sigma z_i^2 f(z_i) = (-2)^2(0.1) + (-1)^2(0.2) + 0^2(0.3) + 1^2(0.4) = 1$$
$$Var(Z) = 1 - 0^2 = 1 \quad \text{and} \quad \sigma_Z = \sqrt{Var(X)} = 1$$

Probability Distributions

41.18. Binomial Distribution: $\Phi(x) = \sum_{t \le x} \binom{n}{t} p^t q^{n-t} \quad p > 0, q > 0, p + q = 1$

41.19. Poisson Distribution: $\Phi(x) = \sum_{t \le x} \dfrac{\lambda^t e^{-\lambda}}{t!}$

41.20. Hypergeometric Distribution: $\Phi(x) = \sum_{t \le x} z \dfrac{\binom{r}{t}\binom{s}{n-t}}{\binom{r+s}{n}}$

41.21. Normal Distribution: $\Phi(x) = \dfrac{1}{\sqrt{2\pi}} \int_{-\infty}^{x} e^{-t^2/2} dt$

41.22. Student's *t* Distribution: $\Phi(x) = \dfrac{1}{\sqrt{n\pi}} \dfrac{\Gamma\left(\frac{n+1}{2}\right)}{\Gamma(n/2)} \int_{-\infty}^{x} \left(1 + \dfrac{t^2}{n}\right)^{-(n+1)/2} dt$

41.23. χ^2 (Chi Square) Distribution: $\Phi(x) = \dfrac{1}{2^{n/2}\Gamma(n/2)} \int_0^x t^{(n-2)/2} e^{-t/2} dt$

41.24. *F* Distribution: $\Phi(x) = \dfrac{\Gamma\left(\frac{n_1 + n_2}{2}\right) n_1^{n_1/2} n_2^{n_2/2}}{\Gamma(n_1/2)\Gamma(n_2/2)} \int_0^x t^{(n_1/2)-1} (n_2 + n_1 t)^{-(n_1+n_2)/2} dt$

Section XII: Numerical Methods

42 INTERPOLATION

Lagrange Interpolation

Two-point formula

42.1. $p(x) = f(x_0)\dfrac{x - x_1}{x_0 - x_1} + f(x_1)\dfrac{x - x_0}{x_1 - x_0}$

where $p(x)$ is a linear polynomial interpolating two points

$$(x_0, f(x_0)), \quad (x_1, f(x_1)), \quad x_0 \neq x_1$$

General formula

42.2. $p(x) = f(x_0)L_{n,0}(x) + f(x_1)L_{n,1}(x) + \cdots + f(x_n)L_{n,n}(x)$

where

$$L_{n,k} = \prod_{i=0, i\neq k}^{n} \frac{x - x_i}{x_k - x_i}$$

and where $p(x)$ is an nth-order polynomial interpolating $n + 1$ points

$$(x_k, f(x_k)), \quad k = 0, 1, \ldots, n; \quad \text{and} \quad x_i \neq x_j \text{ for } i \neq j$$

Remainder formula

Suppose $f(x) \in C^{n+1}[a, b]$. Then there is a $\xi(x) \in (a, b)$ such that:

42.3. $f(x) = p(x) + \dfrac{f^{n+1}(\xi(x))}{(n+1)!}(x - x_0)(x - x_1)\cdots(x - x_n)$

Newton's Interpolation

First-order divided-difference formula

42.4. $f[x_0, x_1] = \dfrac{f(x_1) - f(x_0)}{x_1 - x_0}$

Two-point interpolatory formula

42.5. $p(x) = f(x_0) + f[x_0, x_1](x - x_0)$

where $p(x)$ is a linear polynomial interpolating two points

$$(x_0, f(x_0)), \quad (x_1, f(x_1)), \quad x_0 \neq x_1$$

Second-order divided-difference formula

42.6. $f[x_0,x_1,x_2]=\dfrac{f[x_1,x_2]-f[x_0,x_1]}{x_2-x_0}$

Three-point interpolatory formula

42.7. $p(x)=f(x_0)+f[x_0,x_1](x-x_0)+f[x_0,x_1,x_2](x-x_0)(x-x_1)$

where $p(x)$ is a quadrant polynomial interpolating three points

$$(x_0,f(x_0)),\quad (x_1,f(x_1)),\quad (x_2,f(x_3))$$

General *k*th-order divided-difference formula

42.8. $f[x_0,x_1,\ldots,x_k\}=\dfrac{f[x_1,x_2,\ldots,x_k]-f[x_0,x_1,\ldots,x_{k-1}]}{x_k-x_0}$

General interpolatory formula

42.9. $p(x)=f(x_0)+f[x_0,x_1](x-x_0)+\cdots+f[x_0,x_1,\ldots,x_n](x-x_0)(x-x_1)\cdots(x-x_{n-1})$

where $p(x)$ is an *n*th-order polynomial interpolating $n+1$ points

$$(x_k,f(x_k)),\quad k=0,1,\ldots,n;\quad \text{and}\quad x_i\neq x_j \text{ for } i\neq j$$

Remainder formula

Suppose $f(x)\in C^{n+1}[a,b]$. Then there is a $\xi(x)\in(a,b)$ such that

42.10. $f(x)=p(x)+\dfrac{f^{n+1}(\xi(x))}{(n+1)!}(x-x_0)(x-x_1)\cdots(x-x_n)$

Newton's Forward-Difference Formula

First-order forward-difference at x_0

42.11. $\Delta f(x_0)=f(x_1)-f(x_0)$

Second-order forward difference at x_0

42.12. $\Delta^2 f(x_0)=\Delta f(x_1)-\Delta f(x_0)$

General *k*th-order forward difference at x_0

42.13. $\Delta^k f(x_0)=\Delta^{k-1}f(x_1)-\Delta^{k-1}f(x_0)$

Binomial coefficient

42.14. $\dbinom{s}{k}=\dfrac{s(s-1)\cdots(s-k+1)}{k!}$

Newton's forward-difference formula

42.15. $p(x)=\sum_{k=0}^{n}\dbinom{n}{k}\Delta^k f(x_0)$

where $p(x)$ is an *n*th-order polynomial interpolating $n+1$ equal spaced points

$$(x_k,f(x_k)),\ x_k=x_0+kh\quad k=0,1,\ldots,n$$

Newton's Backward-Difference Formula

First-order backward difference at x_n

42.16. $\nabla f(x_n) = f(x_n) - f(x_{n-1})$

Second-order backward difference at x_n

42.17. $\nabla^2 f(x_n) = \nabla f(x_n) - \nabla f(x_{n-1})$

General *k*th-order backward difference at x_n

42.18. $\nabla^k f(x_n) = \nabla^{k-1} f(x_n) - \nabla^{k-1} f(x_{n-1})$

Newton's backward-difference formula

42.19. $p(x) = \sum_{k=0}^{n} (-1)^k \binom{-n}{k} \nabla^k f(x_n)$

where $p(x)$ is an nth-order polynomial interpolating $n + 1$ equal spaced points

$$(x_k, f(x_k)),\ x_k = x_0 + kh \qquad k = 0, 1, \ldots, n$$

Hermite Interpolation

Two-point basis polynomials

42.20. $H_{1,0} = \left(1 - 2\frac{x - x_0}{x_0 - x_1}\right)\frac{(x - x_1)^2}{(x_0 - x_1)^2}, \quad H_{1,1} = \left(1 - 2\frac{x - x_1}{x_1 - x_0}\right)\frac{(x - x_0)^2}{(x_1 - x_0)^2}$

$$\hat{H}_{1,0} = (x - x_0)\frac{(x - x_1)^2}{(x_0 - x_1)^2}, \quad \hat{H}_{1,1} = (x - x_1)\frac{(x - x_0)^2}{(x_1 - x_0)^2}$$

Two-point interpolatory formula

42.21. $H_3(x) = f(x_0)H_{1,0} + f(x_1)H_{1,1} + f'(x_0)\hat{H}_{1,0} + f'(x_1)\hat{H}_{1,1}$

where $H_3(x)$ is a third-order polynomial, agrees with $f(x)$ and its first-order derivatives at two points, i.e.,

$$H_3(x_0) = f(x_0), \quad H_3'(x_0) = f'(x_0), \qquad H_3(x_1) = f(x_1), \quad H_3'(x_1) = f'(x_1)$$

General basis polynomials

42.22. $H_{n,j} = \left(1 - 2\frac{x - x_j}{L'_{n,j}(x_j)}\right) L^2_{n,j}(x), \quad \hat{H}_{n,j} = (x - x_j)\, L^2_{n,j}(x)$

where

$$L_{n,j} = \prod_{i=0, i \neq j}^{n} \frac{x - x_i}{x_j - x_i}$$

General interpolatory formula

42.23. $$H_{2n+1}(x) = \sum_{j=0}^{n} f(x_j)H_{n,j}(x) + \sum_{j=0}^{n} f'(x_j)\hat{H}_{n,j}(x)$$

where $H_{2n+1}(x)$ is a $(2n + 1)$th-order polynomial, agrees with $f(x)$ and its first order derivatives at $n + 1$ points, i.e.,

$$H_{2n+1}(x_k) = f(x_k),\ H'_{2n+1}(x_k) = f'(x_k) \qquad k = 0, 1, \ldots, n$$

Remainder formula

Suppose $f(x) \in C^{2n+2}[a,b]$. Then there is a $\xi(x) \in (a,b)$ such that

42.24. $$f(x) = H_{2n+1}(x) + \frac{f^{2n+2}(\xi(x))}{(2n+2)!}(x - x_0)^2(x - x_1)^2 \cdots (x - x_n)^2$$

43 QUADRATURE

Trapezoidal Rule

Trapezoidal rule

43.1. $\int_a^b f(x)\,dx \sim \frac{b-a}{2}[f(a)+f(b)]$

Composite trapezoidal rule

43.2. $\int_a^b f(x)\,dx \sim \frac{h}{2}\left(f(a)+2\sum_{i=1}^{n-1} f(a+ih)+f(b)\right)$

where $h=(b-a)/n$ is the grid size.

Simpson's Rule

Simpson's rule

43.3. $\int_a^b f(x)\,dx \sim \frac{b-a}{6}\left[f(a)+4f\left(\frac{a+b}{2}\right)+f(b)\right]$

Composite Simpson's rule

43.4. $\int_a^b f(x)\,dx \sim \frac{h}{3}\left(f(x_0)+2\sum_{i=2}^{n/2} f(x_{2i-2})+4\sum_{i=1}^{n/2} f(x_{2i-1})+f(x_n)\right)$

where n even, $h=(b-a)/n$, $x_i=a+ih$, $i=0, 1, \ldots, n$.

Midpoint Rule

Midpoint rule

43.5. $\int_a^b f(x)\,dx \sim (b-a)f\left(\frac{a+b}{2}\right)$

Composite midpoint rule

43.6. $\int_a^b f(x)\,dx \sim 2h\sum_{i=0}^{n/2} f(x_{2i})$

where n even, $h=(b-a)/(n+2)$, $x_i=a+(i-1)h$, $i=-1, 0, \ldots, n+1$.

Gaussian Quadrature Formula

Legendre polynomial

43.7. $P_n(x) = \frac{1}{2^n n!} \frac{d^n}{dx^n}[(x^2-1)^n]$

Abscissa points and weight formulas

The abscissa points $x_k^{(n)}$ and weight coefficient $\omega_k^{(n)}$ are defined as follows:

43.8. $x_k^{(n)}$ = the kth zero of the Legendre polynomial $P_n(x)$

43.9. $\omega_k^{(n)} = \frac{2P_n'(x_k^{(n)})^2}{1-x_k^{(n)^2}}$

Tables for Gauss-Legendre abscissas and weights appear in Fig. 43-1.

Gauss-Legendre formula in interval (–1, 1)

43.10. $\int_{-1}^{1} f(x)\,dx = \sum_{k=1}^{n} \omega_k^{(n)} f(x_k^{(n)}) + R_n$

Gauss-Legendre formula in general interval (a, b)

43.11. $\int_a^b f(x)\,dx = \frac{b-a}{2}\sum_{k=1}^{n} \omega_k^{(n)} f\left(\frac{a+b}{2} + x_k^{(n)}\frac{b-a}{2}\right) + R_n$

Remainder formula

43.12. $R_n = \frac{(b-a)^{2n+1}(n!)^4}{(2n+1)[(2n)!]^3} f^{(2n)}(\xi)$

for some $a < \xi < b$.

n	$x_k^{(n)}$	$\omega_k^{(n)}$
2	0.5773502692	1.0000000000
	−0.5773502692	1.0000000000
3	0.7745966692	0.5555555556
	0.0000000000	0.8888888889
	−0.7745966692	0.5555555556
4	0.8611363116	0.3478548451
	0.3399810436	0.6521451549
	−0.3399810436	0.6521451549
	−0.8611363116	0.3478548451
5	0.9061798459	0.2369268850
	0.5384693101	0.4786286705
	−0.0000000000	0.5688888889
	−0.5384693101	0.4786286705
	−0.9061798459	0.2369268850

Fig. 43-1

44 SOLUTION of NONLINEAR EQUATIONS

Here we give methods to solve nonlinear equations which come in two forms:

44.1. Nonlinear equation: $f(x) = 0$

44.2. Fixed point nonlinear equation: $x = g(x)$

One can change from 44.1 to 44.2 or from 44.2 to 44.1 by settting:

$$g(x) = f(x) + x \quad \text{or} \quad f(x) = g(x) - x$$

Since the methods are iterative, there are two types of error estimates:

44.3. $|f(x_n)| < \epsilon \quad \text{or} \quad |x_{n+1} - x_n| < \epsilon$

for some preassigned $\epsilon > 0$.

Bisection Method

The following theorem applies:

Intermediate Value Theorem: Suppose f is continuous on an interval $[a, b]$ and $f(a)\,f(b) < 0$. Then there is a root x^* to $f(x) = 0$ in (a, b).

The bisection method approximates one such solution x^*.

44.4. Bisection method:

Initial step: Set $a_0 = a$ and $b_0 = b$.

Repetitive step:

(a) Set $c_n = (a_n + b_n)/2$.

(b) If $f(a_n)f(c_n) < 0$, then set $a_{n+1} = a_n$ and $b_{n+1} = c_n$; else set $a_{n+1} = c_n$ and $b_{n+1} = b_n$.

Newton's Method

Newton method

44.5. $x_{n+1} = x_n - \dfrac{f(x_n)}{f'(x_n)}$

Quadratic convergence

44.6. $\displaystyle \lim_{n\to\infty} \frac{|x_{n+1} - x^*|}{|x_n - x^*|^2} = \frac{f''(x^*)}{2(f'(x^*))^2}$

where x^* is a root of the nonlinear equation 44.1.

Secant Method

Secant method

44.7. $x_{n+1} = x_n - \dfrac{(x_n - x_{n-1})f(x_n)}{f(x_n) - f(x_{n-1})}$

Rate of convergence

44.8. $\lim_{n\to\infty} \dfrac{|x_{n+1} - x^*|}{|x_n - x^*||x_{n-1} - x^*|} = \dfrac{f''(x^*)}{2(f'(x^*))^2}$

where x^* is a root of the nonlinear equation 44.1.

Fixed-Point Iteration

The following definition and theorem apply:

Definition: A function g from (a, b) to (a, b) is called a *contraction mapping* if

$$|g(x) - g(y)| \leq L|x - y| \qquad \text{for any } x,\ y \in (a, b)$$

where $L < 1$ is a positive constant.

Fixed-point theorem: Suppose that g is a contraction mapping on (a, b). Then g has a unique fixed point in (a, b).

Given such a contraction mapping g, the following method may be used.

Fixed-point iteration

44.9. $x_{n+1} = g(x_n)$

45 NUMERICAL METHODS for ORDINARY DIFFERENTIAL EQUATIONS

Here we give methods to solve the following initial-value problem of an ordinary differential equation:

45.1. $$\begin{cases} \dfrac{dx}{dt} = f(x,t) \\ x(t_0) = x_0 \end{cases}$$

The methods will use a computational grid:

45.2. $$t_n = t_0 + nh$$

where h is the grid size.

First-Order Methods

Forward Euler method (first-order explicit method)

45.3. $$x(t+h) = x(t) + hf(x(t),\ t)$$

Backward Euler method (first-order implicit method)

45.4. $$x(t+h) = x(t) + hf(x(t+h),\ t+h)$$

Second-Order Methods

Mid-point rule (second-order explicit method)

45.5. $$\begin{cases} x^* = x(t) + \dfrac{h}{2} f(x(t),\ t) \\ x(t+h) = x(t) + hf\left(x^*,\ t + \dfrac{h}{2}\right) \end{cases}$$

Trapezoidal rule (second-order implicit method)

45.6. $$x(t+h) = x(t) + \frac{h}{2}\{f(x(t),\ t) + f(x(t+h),\ t+h)\}$$

Heun's method (second-order explicit method)

45.7. $$\begin{cases} x^* = x(t) + hf(x(t),\ t) \\ x(t+h) = x(t) + \dfrac{h}{2}\{f(x(t),\ t) + f(x^*,\ t+h)\} \end{cases}$$

Single-Stage High-Order Methods

Fourth-order Runge–Kutta method (fourth-order explicit method)

45.8. $x(t+h) = x(t) + \frac{1}{6}(F_1 + 2F_2 + 2F_3 + F_4)$

where

$$F_1 = hf(x, t), \quad F_2 = hf\left(x + \frac{F_1}{2}, t + \frac{h}{2}\right), \quad F_3 = hf\left(x + \frac{F_2}{2}, t + \frac{h}{2}\right), \quad F_4 = hf(x + F_3, t + h) \, ,$$

Multi-Step High-Order Methods

Adams-Bashforth two-step method

45.9. $x(t+h) = x(t) + h\left(\frac{3}{2} f(x(t), t) - \frac{1}{2} f(x(t-h), t-h)\right)$

Adams-Bashforth three-step method

45.10. $x(t+h) = x(t) + h\left(\frac{23}{12} f(x(t), t) - \frac{4}{3} f(x(t-h), t-h) + \frac{5}{12} f(x(t-2h), t-2h)\right)$

Adams-Bashforth four-step method

45.11. $x(t+h) = x(t)+h\left(\frac{55}{24} f(x(t),t) - \frac{59}{24} f(x(t-h),t-h) + \frac{37}{24} f(x(t-2h),t-2h) - \frac{9}{24} f(x(t-3h),t-3h)\right)$

Milne's method

45.12. $x(t+h) = x(t-3h) + h\left(\frac{8}{3} f(x(t), t) - \frac{4}{3} f(x(t-h), t-h) + \frac{8}{3} f(x(t-2h), t-2h)\right)$

Adams-Moulton two-step method

45.13. $x(t+h) = x(t) + h\left(\frac{5}{12} f(x(t+h), t+h) + \frac{2}{3} f(x(t), t) - \frac{1}{12} f(x(t-h), t-h)\right)$

Adams-Moulton three-step method

45.14. $x(t+h) = x(t) + h\left(\frac{3}{8} f(x(t+h), t+h) + \frac{19}{24} f(x(t), t) - \frac{5}{24} f(x(t-h), t-h) + \frac{1}{24} f(x(t-2h), t-3h)\right)$

46 NUMERICAL METHODS for PARTIAL DIFFERENTIAL EQUATIONS

Finite-Difference Method for Poisson Equation

The following is the Poisson equation in a domain $(a, b) \times (c, d)$:

46.1. $\nabla^2 u = f, \quad \nabla^2 = \dfrac{\partial^2}{\partial x^2} + \dfrac{\partial^2}{\partial y^2}$

Boundary condition:

46.2. $u(x, y) = g(x, y) \qquad \text{for } x = a, b \qquad \text{or} \qquad y = c, d$

Computation grid:

46.3. $x_i = a + i\Delta x \qquad \text{for } i = 0, 1, \ldots, n$

$y_j = c + j\Delta y \qquad \text{for } j = 0, 1, \ldots, m$

where $\Delta x = (b - a)/n$ and $\Delta y = (d - c)/m$ are grid sizes for x and y variables, respectively.

Second-order difference approximation

46.4. $(D_x^2 + D_y^2)u(x_i, y_j) = f(x_i, y_j)$

where

$$D_x^2 u(x_i, y_j) = \frac{u(x_{i+1}, y_j) - 2u(x_i, y_j) + u(x_{i-1}, y_j)}{\Delta x^2}$$

$$D_y^2 u(x_i, y_j) = \frac{u(x_i, y_{j+1}) - 2u(x_i, y_j) + u(x_i, y_{j-1})}{\Delta y^2}$$

Computational boundary condition

46.5. $u(x_0, y_j) = g(a, y_j), \qquad u(x_n, y_j) = g(b, y_j) \qquad \text{for } j = 1, 2, \ldots, m$

$u(x_i, y_0) = g(x_i, c), \qquad u(x_i, y_m) = g(x_i, d) \qquad \text{for } i = 1, 2, \ldots, n$

Finite-Difference Method for Heat Equation

The following is the heat equation in a domain $(a, b) \times (c, d) \times (0, T)$:

46.6. $\dfrac{\partial u}{\partial t} = \nabla^2 u$

Boundary condition:

46.7. $u(x, y, t) = g(x, y)$ for $x = a, b$ or $y = c, d$

Initial condition:

46.8. $u(x, y, 0) = u_0(x, y)$

Computational grid:

46.9. $x_i = a + i\Delta x$ for $i = 0, 1, \ldots, n$

$y_j = c + j\Delta y$ for $j = 0, 1, \ldots, m$

$t_k = k\Delta t$ for $k = 0, 1, \ldots,$

where $\Delta x = (b - a)/n$, $\Delta y = (d - c)/m$, and Δt are grid sizes for x, y and t variables, respectively.

Computational boundary condition

46.10. $u(x_0, y_j) = g(a, y_j), u(x_n, y_j) = g(b, y_j)$ for $j = 1, 2, \ldots, m$

$u(x_i, y_0) = g(x_i, c), u(x_i, y_m) = g(x_i, d)$ for $i = 1, 2, \ldots, n$

Computational initial condition

46.11. $u(x_i, y_j, 0) = u_0(x_i, y_j)$ for $i = 1, 2, \ldots, n; j = 0, 1, \ldots, m$

Forward Euler method with stability condition

46.12. $u(x_i, y_j, t_{k+1}) = u(x_i, y_j, t_k) + \Delta t(D_x^2 + D_y^2)u(x_i, y_j, t_k)$

46.13. $\frac{2\Delta t}{\Delta x^2} + \frac{2\Delta t}{\Delta y^2} \leq 1$

Backward Euler method (unconditional stable)

46.14. $u(x_i, y_j, t_{k+1}) = u(x_i, y_j, t_k) + \Delta t(D_x^2 + D_y^2)u(x_i, y_j, t_{k+1})$

Crank-Nicholson method (unconditional stable)

46.15. $u(x_i, y_j, t_{k+1}) = u(x_i, y_j, t_k) + \Delta t(D_x^2 + D_y^2)\{u(x_i, y_j, t_k) + u(x_i, y_j, t_{k+1})\}/2$

Finite-Difference Method for Wave Equation

The following is a wave equation in a domain $(a, b) \times (c, d) \times (0, T)$:

46.16. $\frac{\partial^2 u}{\partial t^2} = A^2 \nabla^2 u$

where A is a constant representing the speed of the wave.

Boundary condition:

46.17. $u(x, y, t) = g(x, y) \quad \text{for } x = a, b \text{ or } y = c, d$

Initial condition:

46.18. $u(x, y, 0) = u_0(x, y), \quad \frac{\partial u}{\partial t} u(x, y, 0) = u_1(x, y)$

Computational grids:

46.19.
$$x_i = a + i\Delta x \quad \text{for } i = 0, 1, \ldots, n$$
$$y_j = c + j\Delta y \quad \text{for } j = 0, 1, \ldots, m$$
$$t_k = k\Delta t \quad \text{for } k = -1, 0, 1, \ldots$$

where $\Delta x = (b - a)/n$, $\Delta y = (d - c)/m$, and Δt are the grid sizes for x, y, and t variables, respectively.

A second-order finite-difference approximation

46.20. $u(x_i, y_j, t_{k+1}) = 2u(x_i, y_j, t_k) - u(x_i, y_j, t_{k-1}) + \Delta t^2 A^2 (D_x^2 + D_y^2) u(x_i, y_j, t_k)$

Computational boundary condition

46.21.
$$u(x_0, y_j) = g(a, y_j), u(x_n, y_j) = g(b, y_j) \quad \text{for } j = 1, 2, \ldots, m$$
$$u(x_i, y_0) = g(x_i, c), u(x_i, y_m) = g(x_i, d) \quad \text{for } i = 1, 2, \ldots, n$$

Computational initial condition

46.22.
$$u(x_i, y_j, t_0) = u_0(x_i, y_j) \quad \text{for } i = 1, 2, \ldots, n; j = 0, 1, \ldots, m$$
$$u(x_i, y_j, t_{-1}) = u_0(x_i, y_j) + \Delta t^2 u_1(x_i, y_j) \quad \text{for } i = 1, 2, \ldots, n; j = 0, 1, \ldots, m$$

Stability condition

46.23. $\Delta t \leq A \min(\Delta x, \Delta x)$

47 ITERATION METHODS for LINEAR SYSTEMS

Iteration Methods for Poisson Equation

The finite-difference approximation to the Poisson equation follows:

47.1. $$\begin{cases} u_{i+1,j} + u_{i-1,j} + u_{i,j+1} + u_{i,j-1} - 4u_{i,j} = f_{i,j} & \text{for } i, j = 1, 2, \ldots, n-1 \\ u_{0,j} = u_{n,j} = 0 & \text{for } j = 1, 2, \ldots, n-1 \\ u_{i,0} = u_{i,n} = 0 & \text{for } i = 1, 2, \ldots, n-1 \end{cases}$$

Three iteration methods for solving the system follow:

Jacobi method

47.2. $$u_{i,j}^{k+1} = \frac{1}{4}(u_{i+1,j}^{k} + u_{i-1,j}^{k} + u_{i,j+1}^{k} + u_{i,j-1}^{k} - f_{i,j})$$

Gauss-Seidel method

47.3. $$u_{i,j}^{k+1} = \frac{1}{4}(u_{i+1,j}^{k} + u_{i-1,j}^{k+1} + u_{i,j+1}^{k} + u_{i,j-1}^{k+1} - f_{i,j})$$

Successive-overrelaxation (SOR) method

47.4. $$\begin{cases} u_{i,j}^{*} = \frac{1}{4}(u_{i+1,j}^{k} + u_{i-1,j}^{*} + u_{i,j+1}^{k} + u_{i,j-1}^{*} - f_{i,j}) \\ u_{i,j}^{k+1} = (1-\omega)u_{i,j}^{k} + \omega u_{i,j}^{*} \end{cases}$$

Iteration Methods for General Linear Systems

Consider the linear system

47.5. $Ax = b$

where $\boldsymbol{A}$ is an $n \times n$ matrix and $\boldsymbol{x}$ and $\boldsymbol{b}$ are n-vectors. We assume the coefficient matrix $\boldsymbol{A}$ is partitioned as follows:

47.6. $A = D - L - U$

where $D = \text{diag}\,(\boldsymbol{A})$, $\boldsymbol{L}$ is the negative of the strictly lower triangular part of $\boldsymbol{A}$, and $\boldsymbol{U}$ is the negative of the strictly upper triangular part of $\boldsymbol{A}$.

Four iteration methods for solving the system follow:

Richardson method

47.7. $x^{k+1} = (I - A)x^k + b$

Jacobi method

47.8. $Dx^{k+1} = (L + U)x^k + b$

Gauss-Seidel method

47.9. $(D - L)x^{k+1} = Ux^k + b$

Successive-overrelaxation (SOR) method

47.10. $(D - \omega L)x^{k+1} = \omega(Ux^k + b) + (1 - \omega)Dx^k$

TABLES

1 FOUR PLACE COMMON LOGARITHMS

$\log_{10} N$ or $\log N$

N	0	1	2	3	4	5	6	7	8	9	Proportional Parts								
											1	2	3	4	5	6	7	8	9
10	0000	0043	0086	0128	0170	0212	0253	0294	0334	0374	4	8	12	17	21	25	29	33	37
11	0414	0453	0492	0531	0569	0607	0645	0682	0719	0755	4	8	11	15	19	23	26	30	34
12	0792	0828	0864	0899	0934	0969	1004	1038	1072	1106	3	7	10	14	17	21	24	28	31
13	1139	1173	1206	1239	1271	1303	1335	1367	1399	1430	3	6	10	13	16	19	23	26	29
14	1461	1492	1523	1553	1584	1614	1644	1673	1703	1732	3	6	9	12	15	18	21	24	27
15	1761	1790	1818	1847	1875	1903	1931	1959	1987	2014	3	6	8	11	14	17	20	22	25
16	2041	2068	2095	2122	2148	2175	2201	2227	2253	2279	3	5	8	11	13	16	18	21	24
17	2304	2330	2355	2380	2405	2430	2455	2480	2504	2529	2	5	7	10	12	15	17	20	22
18	2553	2577	2601	2625	2648	2672	2695	2718	2742	2765	2	5	7	9	12	14	16	19	21
19	2788	2810	2833	2856	2878	2900	2923	2945	2967	2989	2	4	7	9	11	13	16	18	20
20	3010	3032	3054	3075	3096	3118	3139	3160	3181	3201	2	4	6	8	11	13	15	17	19
21	3222	3243	3263	3284	3304	3324	3345	3365	3385	3404	2	4	6	8	10	12	14	16	18
22	3424	3444	3464	3483	3502	3522	3541	3560	3579	3598	2	4	6	8	10	12	14	15	17
23	3617	3636	3655	3674	3692	3711	3729	3747	3766	3784	2	4	6	7	9	11	13	15	17
24	3802	3820	3838	3856	3874	3892	3909	3927	3945	3962	2	4	5	7	9	11	12	14	16
25	3979	3997	4014	4031	4048	4065	4082	4099	4116	4133	2	3	5	7	9	10	12	14	15
26	4150	4166	4183	4200	4216	4232	4249	4265	4281	4298	2	3	5	7	8	10	11	13	15
27	4314	4330	4346	4362	4378	4393	4409	4425	4440	4456	2	3	5	6	8	9	11	13	14
28	4472	4487	4502	4518	4533	4548	4564	4579	4594	4609	2	3	5	6	8	9	11	12	14
29	4624	4639	4654	4669	4683	4698	4713	4728	4742	4757	1	3	4	6	7	9	10	12	13
30	4771	4786	4800	4814	4829	4843	4857	4871	4886	4900	1	3	4	6	7	9	10	11	13
31	4914	4928	4942	4955	4969	4983	4997	5011	5024	5038	1	3	4	6	7	8	10	11	12
32	5051	5065	5079	5092	5105	5119	5132	5145	5159	5172	1	3	4	5	7	8	9	11	12
33	5185	5198	5211	5224	5237	5250	5263	5276	5289	5302	1	3	4	5	6	8	9	10	12
34	5315	5328	5340	5353	5366	5378	5391	5403	5416	5428	1	3	4	5	6	8	9	10	11
35	5441	5453	5465	5478	5490	5502	5514	5527	5539	5551	1	2	4	5	6	7	9	10	11
36	5563	5575	5587	5599	5611	5623	5635	5647	5658	5670	1	2	4	5	6	7	8	10	11
37	5682	5694	5705	5717	5729	5740	5752	5763	5775	5786	1	2	3	5	6	7	8	9	10
38	5798	5809	5821	5832	5843	5855	5866	5877	5888	5899	1	2	3	5	6	7	8	9	10
39	5911	5922	5933	5944	5955	5966	5977	5988	5999	6010	1	2	3	4	5	7	8	9	10
40	6021	6031	6042	6053	6064	6075	6085	6096	6107	6117	1	2	3	4	5	6	8	9	10
41	6128	6138	6149	6160	6170	6180	6191	6201	6212	6222	1	2	3	4	5	6	7	8	9
42	6232	6243	6253	6263	6274	6284	6294	6304	6314	6325	1	2	3	4	5	6	7	8	9
43	6335	6345	6355	6365	6375	6385	6395	6405	6415	6425	1	2	3	4	5	6	7	8	9
44	6435	6444	6454	6464	6474	6484	6493	6503	6513	6522	1	2	3	4	5	6	7	8	9
45	6532	6542	6551	6561	6571	6580	6590	6599	6609	6618	1	2	3	4	5	6	7	8	9
46	6628	6637	6646	6656	6665	6675	6684	6693	6702	6712	1	2	3	4	5	6	7	7	8
47	6721	6730	6739	6749	6758	6767	6776	6785	6794	6803	1	2	3	4	5	5	6	7	8
48	6812	6821	6830	6839	6848	6857	6866	6875	6884	6893	1	2	3	4	4	5	6	7	8
49	6902	6911	6920	6928	6937	6946	6955	6964	6972	6981	1	2	3	4	4	5	6	7	8
50	6990	6998	7007	7016	7024	7033	7042	7050	7059	7067	1	2	3	3	4	5	6	7	8
51	7076	7084	7093	7101	7110	7118	7126	7135	7143	7152	1	2	3	3	4	5	6	7	8
52	7160	7168	7177	7185	7193	7202	7210	7218	7226	7235	1	2	2	3	4	5	6	7	7
53	7243	7251	7259	7267	7275	7284	7292	7300	7308	7316	1	2	2	3	4	5	6	6	7
54	7324	7332	7340	7348	7356	7364	7372	7380	7388	7396	1	2	2	3	4	5	6	6	7
N	0	1	2	3	4	5	6	7	8	9	1	2	3	4	5	6	7	8	9

1

FOUR PLACE COMMON LOGARITHMS

$\log_{10} N$ or $\log N$ (*Continued*)

N	0	1	2	3	4	5	6	7	8	9	Proportional Parts								
											1	2	3	4	5	6	7	8	9
55	7404	7412	7419	7427	7435	7443	7451	7459	7466	7474	1	2	2	3	4	5	5	6	7
56	7482	7490	7497	7505	7513	7520	7528	7536	7543	7551	1	2	2	3	4	5	5	6	7
57	7559	7566	7574	7582	7589	7597	7604	7612	7619	7627	1	2	2	3	4	5	5	6	7
58	7634	7642	7649	7657	7664	7672	7679	7686	7694	7701	1	1	2	3	4	4	5	6	7
59	7709	7716	7723	7731	7738	7745	7752	7760	7767	7774	1	1	2	3	4	4	5	6	7
60	7782	7789	7796	7803	7810	7818	7825	7832	7839	7846	1	1	2	3	4	4	5	6	6
61	7853	7860	7868	7875	7882	7889	7896	7903	7910	7917	1	1	2	3	4	4	5	6	6
62	7924	7931	7938	7945	7952	7959	7966	7973	7980	7987	1	1	2	3	3	4	5	6	6
63	7993	8000	8007	8014	8021	8028	8035	8041	8048	8055	1	1	2	3	3	4	5	5	6
64	8062	8069	8075	8082	8089	8096	8102	8109	8116	8122	1	1	2	3	3	4	5	5	6
65	8129	8136	8142	8149	8156	8162	8169	8176	8182	8189	1	1	2	3	3	4	5	5	6
66	8195	8202	8209	8215	8222	8228	8235	8241	8248	8254	1	1	2	3	3	4	5	5	6
67	8261	8267	8274	8280	8287	8293	8299	8306	8312	8319	1	1	2	3	3	4	5	5	6
68	8325	8331	8338	8344	8351	8357	8363	8370	8376	8382	1	1	2	3	3	4	4	5	6
69	8388	8395	8401	8407	8414	8420	8426	8432	8439	8445	1	1	2	2	3	4	4	5	6
70	8451	8457	8463	8470	8476	8482	8488	8494	8500	8506	1	1	2	2	3	4	4	5	6
71	8513	8519	8525	8531	8537	8543	8549	8555	8561	8567	1	1	2	2	3	4	4	5	5
72	8573	8579	8585	8591	8597	8603	8609	8615	8621	8627	1	1	2	2	3	4	4	5	5
73	8633	8639	8645	8651	8657	8663	8669	8675	8681	8686	1	1	2	2	3	4	4	5	5
74	8692	8698	8704	8710	8716	8722	8727	8733	8739	8745	1	1	2	2	3	4	4	5	5
75	8751	8756	8762	8768	8774	8779	8785	8791	8797	8802	1	1	2	2	3	3	4	5	5
76	8808	8814	8820	8825	8831	8837	8842	8848	8854	8859	1	1	2	2	3	3	4	5	5
77	8865	8871	8876	8882	8887	8893	8899	8904	8910	8915	1	1	2	2	3	3	4	4	5
78	8921	8927	8932	8938	8943	8949	8954	8960	8965	8971	1	1	2	2	3	3	4	4	5
79	8976	8982	8987	8993	8998	9004	9009	9015	9020	9025	1	1	2	2	3	3	4	4	5
80	9031	9036	9042	9047	9053	9058	9063	9069	9074	9079	1	1	2	2	3	3	4	4	5
81	9085	9090	9096	9101	9106	9112	9117	9122	9128	9133	1	1	2	2	3	3	4	4	5
82	9138	9143	9149	9154	9159	9165	9170	9175	9180	9186	1	1	2	2	3	3	4	4	5
83	9191	9196	9201	9206	9212	9217	9222	9227	9232	9238	1	1	2	2	3	3	4	4	5
84	9243	9248	9253	9258	9263	9269	9274	9279	9284	9289	1	1	2	2	3	3	4	4	5
85	9294	9299	9304	9309	9315	9320	9325	9330	9335	9340	1	1	2	2	3	3	4	4	5
86	9345	9350	9355	9360	9365	9370	9375	9380	9385	9390	1	1	2	2	3	3	4	4	5
87	9395	9400	9405	9410	9415	9420	9425	9430	9435	9440	0	1	1	2	2	3	3	4	4
88	9445	9450	9455	9460	9465	9469	9474	9479	9484	9489	0	1	1	2	2	3	3	4	4
89	9494	9499	9504	9509	9513	9518	9523	9528	9533	9538	0	1	1	2	2	3	3	4	4
90	9542	9547	9552	9557	9562	9566	9571	9576	9581	9586	0	1	1	2	2	3	3	4	4
91	9590	9595	9600	9605	9609	9614	9619	9624	9628	9633	0	1	1	2	2	3	3	4	4
92	9638	9643	9647	9652	9657	9661	9666	9671	9675	9680	0	1	1	2	2	3	3	4	4
93	9685	9689	9694	9699	9703	9708	9713	9717	9722	9727	0	1	1	2	2	3	3	4	4
94	9731	9736	9741	9745	9750	9754	9759	9763	9768	9773	0	1	1	2	2	3	3	4	4
95	9777	9782	9786	9791	9795	9800	9805	9809	9814	9818	0	1	1	2	2	3	3	4	4
96	9823	9827	9832	9836	9841	9845	9850	9854	9859	9863	0	1	1	2	2	3	3	4	4
97	9868	9872	9877	9881	9886	9890	9894	9899	9903	9908	0	1	1	2	2	3	3	4	4
98	9912	9917	9921	9926	9930	9934	9939	9943	9948	9952	0	1	1	2	2	3	3	4	4
99	9956	9961	9965	9969	9974	9978	9983	9987	9991	9996	0	1	1	2	2	3	3	3	4
N	0	1	2	3	4	5	6	7	8	9	1	2	3	4	5	6	7	8	9

Sin x

(x in degrees and minutes)

x	0′	10′	20′	30′	40′	50′
0°	.0000	.0029	.0058	.0087	.0116	.0145
1	.0175	.0204	.0233	.0262	.0291	.0320
2	.0349	.0378	.0407	.0436	.0465	.0494
3	.0523	.0552	.0581	.0610	.0640	.0669
4	.0698	.0727	.0756	.0785	.0814	.0843
5°	.0872	.0901	.0929	.0958	.0987	.1016
6	.1045	.1074	.1103	.1132	.1161	.1190
7	.1219	.1248	.1276	.1305	.1334	.1363
8	.1392	.1421	.1449	.1478	.1507	.1536
9	.1564	.1593	.1622	.1650	.1679	.1708
10°	.1736	.1765	.1794	.1822	.1851	.1880
11	.1908	.1937	.1965	.1994	.2022	.2051
12	.2079	.2108	.2136	.2164	.2193	.2221
13	.2250	.2278	.2306	.2334	.2363	.2391
14	.2419	.2447	.2476	.2504	.2532	.2560
15°	.2588	.2616	.2644	.2672	.2700	.2728
16	.2756	.2784	.2812	.2840	.2868	.2896
17	.2924	.2952	.2979	.3007	.3035	.3062
18	.3090	.3118	.3145	.3173	.3201	.3228
19	.3256	.3283	.3311	.3338	.3365	.3393
20°	.3420	.3448	.3475	.3502	.3529	.3557
21	.3584	.3611	.3638	.3665	.3692	.3719
22	.3746	.3773	.3800	.3827	.3854	.3881
23	.3907	.3934	.3961	.3987	.4014	.4041
24	.4067	.4094	.4120	.4147	.4173	.4200
25°	.4226	.4253	.4279	.4305	.4331	.4358
26	.4384	.4410	.4436	.4462	.4488	.4514
27	.4540	.4566	.4592	.4617	.4643	.4669
28	.4695	.4720	.4746	.4772	.4797	.4823
29	.4848	.4874	.4899	.4924	.4950	.4975
30°	.5000	.5025	.5050	.5075	.5100	.5125
31	.5150	.5175	.5200	.5225	.5250	.5275
32	.5299	.5324	.5348	.5373	.5398	.5422
33	.5446	.5471	.5495	.5519	.5544	.5568
34	.5592	.5616	.5640	.5664	.5688	.5712
35°	.5736	.5760	.5783	.5807	.5831	.5854
36	.5878	.5901	.5925	.5948	.5972	.5995
37	.6018	.6041	.6065	.6088	.6111	.6134
38	.6157	.6180	.6202	.6225	.6248	.6271
39	.6293	.6316	.6338	.6361	.6383	.6406
40°	.6428	.6450	.6472	.6494	.6517	.6539
41	.6561	.6583	.6604	.6626	.6648	.6670
42	.6691	.6713	.6734	.6756	.6777	.6799
43	.6820	.6841	.6862	.6884	.6905	.6926
44	.6947	.6967	.6988	.7009	.7030	.7050
45°	.7071	.7092	.7112	.7133	.7153	.7173

x	0′	10′	20′	30′	40′	50′
45°	.7071	.7092	.7112	.7133	.7153	.7173
46	.7193	.7214	.7234	.7254	.7274	.7294
47	.7314	.7333	.7353	.7373	.7392	.7412
48	.7431	.7451	.7470	.7490	.7509	.7528
49	.7547	.7566	.7585	.7604	.7623	.7642
50°	.7660	.7679	.7698	.7716	.7735	.7753
51	.7771	.7790	.7808	.7826	.7844	.7862
52	.7880	.7898	.7916	.7934	.7951	.7969
53	.7986	.8004	.8021	.8039	.8056	.8073
54	.8090	.8107	.8124	.8141	.8158	.8175
55°	.8192	.8208	.8225	.8241	.8258	.8274
56	.8290	.8307	.8323	.8339	.8355	.8371
57	.8387	.8403	.8418	.8434	.8450	.8465
58	.8480	.8496	.8511	.8526	.8542	.8557
59	.8572	.8587	.8601	.8616	.8631	.8646
60°	.8660	.8675	.8689	.8704	.8718	.8732
61	.8746	.8760	.8774	.8788	.8802	.8816
62	.8829	.8843	.8857	.8870	.8884	.8897
63	.8910	.8923	.8936	.8949	.8962	.8975
64	.8988	.9001	.9013	.9026	.9038	.9051
65°	.9063	.9075	.9088	.9100	.9112	.9124
66	.9135	.9147	.9159	.9171	.9182	.9194
67	.9205	.9216	.9228	.9239	.9250	.9261
68	.9272	.9283	.9293	.9304	.9315	.9325
69	.9336	.9346	.9356	.9367	.9377	.9387
70°	.9397	.9407	.9417	.9426	.9436	.9446
71	.9455	.9465	.9474	.9483	.9492	.9502
72	.9511	.9520	.9528	.9537	.9546	.9555
73	.9563	.9572	.9580	.9588	.9596	.9605
74	.9613	.9621	.9628	.9636	.9644	.9652
75°	.9659	.9667	.9674	.9681	.9689	.9696
76	.9703	.9710	.9717	.9724	.9730	.9737
77	.9744	.9750	.9757	.9763	.9769	.9775
78	.9781	.9787	.9793	.9799	.9805	.9811
79	.9816	.9822	.9827	.9833	.9838	.9843
80°	.9848	.9853	.9858	.9863	.9868	.9872
81	.9877	.9881	.9886	.9890	.9894	.9899
82	.9903	.9907	.9911	.9914	.9918	.9922
83	.9925	.9929	.9932	.9936	.9939	.9942
84	.9945	.9948	.9951	.9954	.9957	.9959
85°	.9962	.9964	.9967	.9969	.9971	.9974
86	.9976	.9978	.9980	.9981	.9983	.9985
87	.9986	.9988	.9989	.9990	.9992	.9993
88	.9994	.9995	.9996	.9997	.9997	.9998
89	.9998	.9999	.9999	1.0000	1.0000	1.0000
90°	1.0000					

3 Cos x

(x in degrees and minutes)

x	0′	10′	20′	30′	40′	50′
0°	1.0000	1.0000	1.0000	1.0000	.9999	.9999
1	.9998	.9998	.9997	.9997	.9996	.9995
2	.9994	.9993	.9992	.9990	.9989	.9988
3	.9986	.9985	.9983	.9981	.9980	.9978
4	.9976	.9974	.9971	.9969	.9967	.9964
5°	.9962	.9959	.9957	.9954	.9951	.9948
6	.9945	.9942	.9939	.9936	.9932	.9929
7	.9925	.9922	.9918	.9914	.9911	.9907
8	.9903	.9899	.9894	.9890	.9886	.9881
9	.9877	.9872	.9868	.9863	.9858	.9853
10°	.9848	.9843	.9838	.9833	.9827	.9822
11	.9816	.9811	.9805	.9799	.9793	.9787
12	.9781	.9775	.9769	.9763	.9757	.9750
13	.9744	.9737	.9730	.9724	.9717	.9710
14	.9703	.9696	.9689	.9681	.9674	.9667
15°	.9659	.9652	.9644	.9636	.9628	.9621
16	.9613	.9605	.9596	.9588	.9580	.9572
17	.9563	.9555	.9546	.9537	.9528	.9520
18	.9511	.9502	.9492	.9483	.9474	.9465
19	.9455	.9446	.9436	.9426	.9417	.9407
20°	.9397	.9387	.9377	.9367	.9356	.9346
21	.9336	.9325	.9315	.9304	.9293	.9283
22	.9272	.9261	.9250	.9239	.9228	.9216
23	.9205	.9194	.9182	.9171	.9159	.9147
24	.9135	.9124	.9112	.9100	.9088	.9075
25°	.9063	.9051	.9038	.9026	.9013	.9001
26	.8988	.8975	.8962	.8949	.8936	.8923
27	.8910	.8897	.8884	.8870	.8857	.8843
28	.8829	.8816	.8802	.8788	.8774	.8760
29	.8746	.8732	.8718	.8704	.8689	.8675
30°	.8660	.8646	.8631	.8616	.8601	.8587
31	.8572	.8557	.8542	.8526	.8511	.8496
32	.8480	.8465	.8450	.8434	.8418	.8403
33	.8387	.8371	.8355	.8339	.8323	.8307
34	.8290	.8274	.8258	.8241	.8225	.8208
35°	.8192	.8175	.8158	.8141	.8124	.8107
36	.8090	.8073	.8056	.8039	.8021	.8004
37	.7986	.7969	.7951	.7934	.7916	.7898
38	.7880	.7862	.7844	.7826	.7808	.7790
39	.7771	.7753	.7735	.7716	.7698	.7679
40°	.7660	.7642	.7623	.7604	.7585	.7566
41	.7547	.7528	.7509	.7490	.7470	.7451
42	.7431	.7412	.7392	.7373	.7353	.7333
43	.7314	.7294	.7274	.7254	.7234	.7214
44	.7193	.7173	.7153	.7133	.7112	.7092
45°	.7071	.7050	.7030	.7009	.6988	.6967

x	0′	10′	20′	30′	40′	50′
45°	.7071	.7050	.7030	.7009	.6988	.6967
46	.6947	.6926	.6905	.6884	.6862	.6841
47	.6820	.6799	.6777	.6756	.6734	.6713
48	.6691	.6670	.6648	.6626	.6604	.6583
49	.6561	.6539	.6517	.6494	.6472	.6450
50°	.6428	.6406	.6383	.6361	.6338	.6316
51	.6293	.6271	.6248	.6225	.6202	.6180
52	.6157	.6134	.6111	.6088	.6065	.6041
53	.6018	.5995	.5972	.5948	.5925	.5901
54	.5878	.5854	.5831	.5807	.5783	.5760
55°	.5736	.5712	.5688	.5664	.5640	.5616
56	.5592	.5568	.5544	.5519	.5495	.5471
57	.5446	.5422	.5398	.5373	.5348	.5324
58	.5299	.5275	.5250	.5225	.5200	.5175
59	.5150	.5125	.5100	.5075	.5050	.5025
60°	.5000	.4975	.4950	.4924	.4899	.4874
61	.4848	.4823	.4797	.4772	.4746	.4720
62	.4695	.4669	.4643	.4617	.4592	.4566
63	.4540	.4514	.4488	.4462	.4436	.4410
64	.4384	.4358	.4331	.4305	.4279	.4253
65°	.4226	.4200	.4173	.4147	.4120	.4094
66	.4067	.4041	.4014	.3987	.3961	.3934
67	.3907	.3881	.3854	.3827	.3800	.3773
68	.3746	.3719	.3692	.3665	.3638	.3611
69	.3584	.3557	.3529	.3502	.3475	.3448
70°	.3420	.3393	.3365	.3338	.3311	.3283
71	.3256	.3228	.3201	.3173	.3145	.3118
72	.3090	.3062	.3035	.3007	.2979	.2952
73	.2924	.2896	.2868	.2840	.2812	.2784
74	.2756	.2728	.2700	.2672	.2644	.2616
75°	.2588	.2560	.2532	.2504	.2476	.2447
76	.2419	.2391	.2363	.2334	.2306	.2278
77	.2250	.2221	.2193	.2164	.2136	.2108
78	.2079	.2051	.2022	.1994	.1965	.1937
79	.1908	.1880	.1851	.1822	.1794	.1765
80°	.1736	.1708	.1679	.1650	.1622	.1593
81	.1564	.1536	.1507	.1478	.1449	.1421
82	.1392	.1363	.1334	.1305	.1276	.1248
83	.1219	.1190	.1161	.1132	.1103	.1074
84	.1045	.1016	.0987	.0958	.0929	.0901
85°	.0872	.0843	.0814	.0785	.0756	.0727
86	.0698	.0669	.0640	.0610	.0581	.0552
87	.0523	.0494	.0465	.0436	.0407	.0378
88	.0349	.0320	.0291	.0262	.0233	.0204
89	.0175	.0145	.0116	.0087	.0058	.0029
90°	.0000					

Tan x

(x in degrees and minutes)

x	0′	10′	20′	30′	40′	50′
0°	.0000	.0029	.0058	.0087	.0116	.0145
1	.0175	.0204	.0233	.0262	.0291	.0320
2	.0349	.0378	.0407	.0437	.0466	.0495
3	.0524	.0553	.0582	.0612	.0641	.0670
4	.0699	.0729	.0758	.0787	.0816	.0846
5°	.0875	.0904	.0934	.0963	.0992	.1022
6	.1051	.1080	.1110	.1139	.1169	.1198
7	.1228	.1257	.1287	.1317	.1346	.1376
8	.1405	.1435	.1465	.1495	.1524	.1554
9	.1584	.1614	.1644	.1673	.1703	.1733
10°	.1763	.1793	.1823	.1853	.1883	.1914
11	.1944	.1974	.2004	.2035	.2065	.2095
12	.2126	.2156	.2186	.2217	.2247	.2278
13	.2309	.2339	.2370	.2401	.2432	.2462
14	.2493	.2524	.2555	.2586	.2617	.2648
15°	.2679	.2711	.2742	.2773	.2805	.2836
16	.2867	.2899	.2931	.2962	.2994	.3026
17	.3057	.3089	.3121	.3153	.3185	.3217
18	.3249	.3281	.3314	.3346	.3378	.3411
19	.3443	.3476	.3508	.3541	.3574	.3607
20°	.3640	.3673	.3706	.3739	.3772	.3805
21	.3839	.3872	.3906	.3939	.3973	.4006
22	.4040	.4074	.4108	.4142	.4176	.4210
23	.4245	.4279	.4314	.4348	.4383	.4417
24	.4452	.4487	.4522	.4557	.4592	.4628
25°	.4663	.4699	.4734	.4770	.4806	.4841
26	.4877	.4913	.4950	.4986	.5022	.5059
27	.5095	.5132	.5169	.5206	.5243	.5280
28	.5317	.5354	.5392	.5430	.5467	.5505
29	.5543	.5581	.5619	.5658	.5696	.5735
30°	.5774	.5812	.5851	.5890	.5930	.5969
31	.6009	.6048	.6088	.6128	.6168	.6208
32	.6249	.6289	.6330	.6371	.6412	.6453
33	.6494	.6536	.6577	.6619	.6661	.6703
34	.6745	.6787	.6830	.6873	.6916	.6959
35°	.7002	.7046	.7089	.7133	.7177	.7221
36	.7265	.7310	.7355	.7400	.7445	.7490
37	.7536	.7581	.7627	.7673	.7720	.7766
38	.7813	.7860	.7907	.7954	.8002	.8050
39	.8098	.8146	.8195	.8243	.8292	.8342
40°	.8391	.8441	.8491	.8541	.8591	.8642
41	.8693	.8744	.8796	.8847	.8899	.8952
42	.9004	.9057	.9110	.9163	.9217	.9271
43	.9325	.9380	.9435	.9490	.9545	.9601
44	.9657	.9713	.9770	.9827	.9884	.9942
45°	1.0000	1.0058	1.0117	1.0176	1.0235	1.0295

x	0′	10′	20′	30′	40′	50′
45°	1.0000	1.0058	1.0117	1.0176	1.0235	1.0295
46	1.0355	1.0416	1.0477	1.0538	1.0599	1.0661
47	1.0724	1.0786	1.0850	1.0913	1.0977	1.1041
48	1.1106	1.1171	1.1237	1.1303	1.1369	1.1436
49	1.1504	1.1571	1.1640	1.1708	1.1778	1.1847
50°	1.1918	1.1988	1.2059	1.2131	1.2203	1.2276
51	1.2349	1.2423	1.2497	1.2572	1.2647	1.2723
52	1.2799	1.2876	1.2954	1.3032	1.3111	1.3190
53	1.3270	1.3351	1.3432	1.3514	1.3597	1.3680
54	1.3764	1.3848	1.3934	1.4019	1.4106	1.4193
55°	1.4281	1.4370	1.4460	1.4550	1.4641	1.4733
56	1.4826	1.4919	1.5013	1.5108	1.5204	1.5301
57	1.5399	1.5497	1.5597	1.5697	1.5798	1.5900
58	1.6003	1.6107	1.6212	1.6319	1.6426	1.6534
59	1.6643	1.6753	1.6864	1.6977	1.7090	1.7205
60°	1.7321	1.7437	1.7556	1.7675	1.7796	1.7917
61	1.8040	1.8165	1.8291	1.8418	1.8546	1.8676
62	1.8807	1.8940	1.9074	1.9210	1.9347	1.9486
63	1.9626	1.9768	1.9912	2.0057	2.0204	2.0353
64	2.0503	2.0655	2.0809	2.0965	2.1123	2.1283
65°	2.1445	2.1609	2.1775	2.1943	2.2113	2.2286
66	2.2460	2.2637	2.2817	2.2998	2.3183	2.3369
67	2.3559	2.3750	2.3945	2.4142	2.4342	2.4545
68	2.4751	2.4960	2.5172	2.5386	2.5605	2.5826
69	2.6051	2.6279	2.6511	2.6746	2.6985	2.7228
70°	2.7475	2.7725	2.7980	2.8239	2.8502	2.8770
71	2.9042	2.9319	2.9600	2.9887	3.0178	3.0475
72	3.0777	3.1084	3.1397	3.1716	3.2041	3.2371
73	3.2709	3.3052	3.3402	3.3759	3.4124	3.4495
74	3.4874	3.5261	3.5656	3.6059	3.6470	3.6891
75°	3.7321	3.7760	3.8208	3.8667	3.9136	3.9617
76	4.0108	4.0611	4.1126	4.1653	4.2193	4.2747
77	4.3315	4.3897	4.4494	4.5107	4.5736	4.6382
78	4.7046	4.7729	4.8430	4.9152	4.9894	5.0658
79	5.1446	5.2257	5.3093	5.3955	5.4845	5.5764
80°	5.6713	5.7694	5.8708	5.9758	6.0844	6.1970
81	6.3138	6.4348	6.5606	6.6912	6.8269	6.9682
82	7.1154	7.2687	7.4287	7.5958	7.7704	7.9530
83	8.1443	8.3450	8.5555	8.7769	9.0098	9.2553
84	9.5144	9.7882	10.078	10.385	10.712	11.059
85°	11.430	11.826	12.251	12.706	13.197	13.727
86	14.301	14.924	15.605	16.350	17.169	18.075
87	19.081	20.206	21.470	22.904	24.542	26.432
88	28.636	31.242	34.368	38.188	42.964	49.104
89	57.290	68.750	85.940	114.59	171.89	343.77
90°	∞					

CONVERSION OF RADIANS TO DEGREES, MINUTES, AND SECONDS OR FRACTIONS OF DEGREES

Radians	Deg.	Min.	Sec.	Fractions of Degrees
1	57°	17′	44.8″	57.2958°
2	114°	35′	29.6″	114.5916°
3	171°	53′	14.4″	171.8873°
4	229°	10′	59.2″	229.1831°
5	286°	28′	44.0″	286.4789°
6	343°	46′	28.8″	343.7747°
7	401°	4′	13.6″	401.0705°
8	458°	21′	58.4″	458.3662°
9	515°	39′	43.3″	515.6620°
10	572°	57′	28.1″	572.9578°
.1	5°	43′	46.5″	
.2	11°	27′	33.0″	
.3	17°	11′	19.4″	
.4	22°	55′	5.9″	
.5	28°	38′	52.4″	
.6	34°	22′	38.9″	
.7	40°	6′	25.4″	
.8	45°	50′	11.8″	
.9	51°	33′	58.3″	
.01	0°	34′	22.6″	
.02	1°	8′	45.3″	
.03	1°	43′	7.9″	
.04	2°	17′	30.6″	
.05	2°	51′	53.2″	
.06	3°	26′	15.9″	
.07	4°	0′	38.5″	
.08	4°	35′	1.2″	
.09	5°	9′	23.8″	
.001	0°	3′	26.3″	
.002	0°	6′	52.5″	
.003	0°	10′	18.8″	
.004	0°	13′	45.1″	
.005	0°	17′	11.3″	
.006	0°	20′	37.6″	
.007	0°	24′	3.9″	
.008	0°	27′	30.1″	
.009	0°	30′	56.4″	
.0001	0°	0′	20.6″	
.0002	0°	0′	41.3″	
.0003	0°	1′	1.9″	
.0004	0°	1′	22.5″	
.0005	0°	1′	43.1″	
.0006	0°	2′	3.8″	
.0007	0°	2′	24.4″	
.0008	0°	2′	45.0″	
.0009	0°	3′	5.6″	

CONVERSION OF DEGREES, MINUTES, AND SECONDS TO RADIANS

Degrees	Radians
1°	.0174533
2°	.0349066
3°	.0523599
4°	.0698132
5°	.0872665
6°	.1047198
7°	.1221730
8°	.1396263
9°	.1570796
10°	.1745329

Minutes	Radians
1′	.00029089
2′	.00058178
3′	.00087266
4′	.00116355
5′	.00145444
6′	.00174533
7′	.00203622
8′	.00232711
9′	.00261800
10′	.00290888

Seconds	Radians
1″	.0000048481
2″	.0000096963
3″	.0000145444
4″	.0000193925
5″	.0000242407
6″	.0000290888
7″	.0000339370
8″	.0000387851
9″	.0000436332
10″	.0000484814

7 NATURAL OR NAPIERIAN LOGARITHMS

$\log_e x$ or $\ln x$

x	0	1	2	3	4	5	6	7	8	9
1.0	.00000	.00995	.01980	.02956	.03922	.04879	.05827	.06766	.07696	.08618
1.1	.09531	.10436	.11333	.12222	.13103	.13976	.14842	.15700	.16551	.17395
1.2	.18232	.19062	.19885	.20701	.21511	.22314	.23111	.23902	.24686	.25464
1.3	.26236	.27003	.27763	.28518	.29267	.30010	.30748	.31481	.32208	.32930
1.4	.33647	.34359	.35066	.35767	.36464	.37156	.37844	.38526	.39204	.39878
1.5	.40547	.41211	.41871	.42527	.43178	.43825	.44469	.45108	.45742	.46373
1.6	.47000	.47623	.48243	.48858	.49470	.50078	.50682	.51282	.51879	.52473
1.7	.53063	.53649	.54232	.54812	.55389	.55962	.56531	.57098	.57661	.58222
1.8	.58779	.59333	.59884	.60432	.60977	.61519	.62058	.62594	.63127	.63658
1.9	.64185	.64710	.65233	.65752	.66269	.66783	.67294	.67803	.68310	.68813
2.0	.69315	.69813	.70310	.70804	.71295	.71784	.72271	.72755	.73237	.73716
2.1	.74194	.74669	.75142	.75612	.76081	.76547	.77011	.77473	.77932	.78390
2.2	.78846	.79299	.79751	.80200	.80648	.81093	.81536	.81978	.82418	.82855
2.3	.83291	.83725	.84157	.84587	.85015	.85442	.85866	.86289	.86710	.87129
2.4	.87547	.87963	.88377	.88789	.89200	.89609	.90016	.90422	.90826	.91228
2.5	.91629	.92028	.92426	.92822	.93216	.93609	.94001	.94391	.94779	.95166
2.6	.95551	.95935	.96317	.96698	.97078	.97456	.97833	.98208	.98582	.98954
2.7	.99325	.99695	1.00063	1.00430	1.00796	1.01160	1.01523	1.01885	1.02245	1.02604
2.8	1.02962	1.03318	1.03674	1.04028	1.04380	1.04732	1.05082	1.05431	1.05779	1.06126
2.9	1.06471	1.06815	1.07158	1.07500	1.07841	1.08181	1.08519	1.08856	1.09192	1.09527
3.0	1.09861	1.10194	1.10526	1.10856	1.11186	1.11514	1.11841	1.12168	1.12493	1.12817
3.1	1.13140	1.13462	1.13783	1.14103	1.14422	1.14740	1.15057	1.15373	1.15688	1.16002
3.2	1.16315	1.16627	1.16938	1.17248	1.17557	1.17865	1.18173	1.18479	1.18784	1.19089
3.3	1.19392	1.19695	1.19996	1.20297	1.20597	1.20896	1.21194	1.21491	1.21788	1.22083
3.4	1.22378	1.22671	1.22964	1.23256	1.23547	1.23837	1.24127	1.24415	1.24703	1.24990
3.5	1.25276	1.25562	1.25846	1.26130	1.26413	1.26695	1.26976	1.27257	1.27536	1.27815
3.6	1.28093	1.28371	1.28647	1.28923	1.29198	1.29473	1.29746	1.30019	1.30291	1.30563
3.7	1.30833	1.31103	1.31372	1.31641	1.31909	1.32176	1.32442	1.32708	1.32972	1.33237
3.8	1.33500	1.33763	1.34025	1.34286	1.34547	1.34807	1.35067	1.35325	1.35584	1.35841
3.9	1.36098	1.36354	1.36609	1.36864	1.37118	1.37372	1.37624	1.37877	1.38128	1.38379
4.0	1.38629	1.38879	1.39128	1.39377	1.39624	1.39872	1.40118	1.40364	1.40610	1.40854
4.1	1.41099	1.41342	1.41585	1.41828	1.42070	1.42311	1.42552	1.42792	1.43031	1.43270
4.2	1.43508	1.43746	1.43984	1.44220	1.44456	1.44692	1.44927	1.45161	1.45395	1.45629
4.3	1.45862	1.46094	1.46326	1.46557	1.46787	1.47018	1.47247	1.47476	1.47705	1.47933
4.4	1.48160	1.48387	1.48614	1.48840	1.49065	1.49290	1.49515	1.49739	1.49962	1.50185
4.5	1.50408	1.50630	1.50851	1.51072	1.51293	1.51513	1.51732	1.51951	1.52170	1.52388
4.6	1.52606	1.52823	1.53039	1.53256	1.53471	1.53687	1.53902	1.54116	1.54330	1.54543
4.7	1.54756	1.54969	1.55181	1.55393	1.55604	1.55814	1.56025	1.56235	1.56444	1.56653
4.8	1.56862	1.57070	1.57277	1.57485	1.57691	1.57898	1.58104	1.58309	1.58515	1.58719
4.9	1.58924	1.59127	1.59331	1.59534	1.59737	1.59939	1.60141	1.60342	1.60543	1.60744

$\ln 10 = 2.30259$ $\quad$ $4 \ln 10 = 9.21034$ $\quad$ $7 \ln 10 = 16.11810$

$2 \ln 10 = 4.60517$ $\quad$ $5 \ln 10 = 11.51293$ $\quad$ $8 \ln 10 = 18.42068$

$3 \ln 10 = 6.90776$ $\quad$ $6 \ln 10 = 13.81551$ $\quad$ $9 \ln 10 = 20.72327$

7 NATURAL OR NAPIERIAN LOGARITHMS

$\log_e x$ or $\ln x$ *(Continued)*

x	0	1	2	3	4	5	6	7	8	9
5.0	1.60944	1.61144	1.61343	1.61542	1.61741	1.61939	1.62137	1.62334	1.62531	1.62728
5.1	1.62924	1.63120	1.63315	1.63511	1.63705	1.63900	1.64094	1.64287	1.64481	1.64673
5.2	1.64866	1.65058	1.65250	1.65441	1.65632	1.65823	1.66013	1.66203	1.66393	1.66582
5.3	1.66771	1.66959	1.67147	1.67335	1.67523	1.67710	1.67896	1.68083	1.68269	1.68455
5.4	1.68640	1.68825	1.69010	1.69194	1.69378	1.69562	1.69745	1.69928	1.70111	1.70293
5.5	1.70475	1.70656	1.70838	1.71019	1.71199	1.71380	1.71560	1.71740	1.71919	1.72098
5.6	1.72277	1.72455	1.72633	1.72811	1.72988	1.73166	1.73342	1.73519	1.73695	1.73871
5.7	1.74047	1.74222	1.74397	1.74572	1.74746	1.74920	1.75094	1.75267	1.75440	1.75613
5.8	1.75786	1.75958	1.76130	1.76302	1.76473	1.76644	1.76815	1.76985	1.77156	1.77326
5.9	1.77495	1.77665	1.77834	1.78002	1.78171	1.78339	1.78507	1.78675	1.78842	1.79009
6.0	1.79176	1.79342	1.79509	1.79675	1.79840	1.80006	1.80171	1.80336	1.80500	1.80665
6.1	1.80829	1.80993	1.81156	1.81319	1.81482	1.81645	1.81808	1.81970	1.82132	1.82294
6.2	1.82455	1.82616	1.82777	1.82938	1.83098	1.83258	1.83418	1.83578	1.83737	1.83896
6.3	1.84055	1.84214	1.84372	1.84530	1.84688	1.84845	1.85003	1.85160	1.85317	1.85473
6.4	1.85630	1.85786	1.85942	1.86097	1.86253	1.86408	1.86563	1.86718	1.86872	1.87026
6.5	1.87180	1.87334	1.87487	1.87641	1.87794	1.87947	1.88099	1.88251	1.88403	1.88555
6.6	1.88707	1.88858	1.89010	1.89160	1.89311	1.89462	1.89612	1.89762	1.89912	1.90061
6.7	1.90211	1.90360	1.90509	1.90658	1.90806	1.90954	1.91102	1.91250	1.91398	1.91545
6.8	1.91692	1.91839	1.91986	1.92132	1.92279	1.92425	1.92571	1.92716	1.92862	1.93007
6.9	1.93152	1.93297	1.93442	1.93586	1.93730	1.93874	1.94018	1.94162	1.94305	1.94448
7.0	1.94591	1.94734	1.94876	1.95019	1.95161	1.95303	1.95445	1.95586	1.95727	1.95869
7.1	1.96009	1.96150	1.96291	1.96431	1.96571	1.96711	1.96851	1.96991	1.97130	1.97269
7.2	1.97408	1.97547	1.97685	1.97824	1.97962	1.98100	1.98238	1.98376	1.98513	1.98650
7.3	1.98787	1.98924	1.99061	1.99198	1.99334	1.99470	1.99606	1.99742	1.99877	2.00013
7.4	2.00148	2.00283	2.00418	2.00553	2.00687	2.00821	2.00956	2.01089	2.01223	2.01357
7.5	2.01490	2.01624	2.01757	2.01890	2.02022	2.02155	2.02287	2.02419	2.02551	2.02683
7.6	2.02815	2.02946	2.03078	2.03209	2.03340	2.03471	2.03601	2.03732	2.03862	2.03992
7.7	2.04122	2.04252	2.04381	2.04511	2.04640	2.04769	2.04898	2.05027	2.05156	2.05284
7.8	2.05412	2.05540	2.05668	2.05796	2.05924	2.06051	2.06179	2.06306	2.06433	2.06560
7.9	2.06686	2.06813	2.06939	2.07065	2.07191	2.07317	2.07443	2.07568	2.07694	2.07819
8.0	2.07944	2.08069	2.08194	2.08318	2.08443	2.08567	2.08691	2.08815	2.08939	2.09063
8.1	2.09186	2.09310	2.09433	2.09556	2.09679	2.09802	2.09924	2.10047	2.10169	2.10291
8.2	2.10413	2.10535	2.10657	2.10779	2.10900	2.11021	2.11142	2.11263	2.11384	2.11505
8.3	2.11626	2.11746	2.11866	2.11986	2.12106	2.12226	2.12346	2.12465	2.12585	2.12704
8.4	2.12823	2.12942	2.13061	2.13180	2.13298	2.13417	2.13535	2.13653	2.13771	2.13889
8.5	2.14007	2.14124	2.14242	2.14359	2.14476	2.14593	2.14710	2.14827	2.14943	2.15060
8.6	2.15176	2.15292	2.15409	2.15524	2.15640	2.15756	2.15871	2.15987	2.16102	2.16217
8.7	2.16332	2.16447	2.16562	2.16677	2.16791	2.16905	2.17020	2.17134	2.17248	2.17361
8.8	2.17475	2.17589	2.17702	2.17816	2.17929	2.18042	2.18155	2.18267	2.18380	2.18493
8.9	2.18605	2.18717	2.18830	2.18942	2.19054	2.19165	2.19277	2.19389	2.19500	2.19611
9.0	2.19722	2.19834	2.19944	2.20055	2.20166	2.20276	2.20387	2.20497	2.20607	2.20717
9.1	2.20827	2.20937	2.21047	2.21157	2.21266	2.21375	2.21485	2.21594	2.21703	2.21812
9.2	2.21920	2.22029	2.22138	2.22246	2.22354	2.22462	2.22570	2.22678	2.22786	2.22894
9.3	2.23001	2.23109	2.23216	2.23324	2.23431	2.23538	2.23645	2.23751	2.23858	2.23965
9.4	2.24071	2.24177	2.24284	2.24390	2.24496	2.24601	2.24707	2.24813	2.24918	2.25024
9.5	2.25129	2.25234	2.25339	2.25444	2.25549	2.25654	2.25759	2.25863	2.25968	2.26072
9.6	2.26176	2.26280	2.26384	2.26488	2.26592	2.26696	2.26799	2.26903	2.27006	2.27109
9.7	2.27213	2.27316	2.27419	2.27521	2.27624	2.27727	2.27829	2.27932	2.28034	2.28136
9.8	2.28238	2.28340	2.28442	2.28544	2.28646	2.28747	2.28849	2.28950	2.29051	2.29152
9.9	2.29253	2.29354	2.29455	2.29556	2.29657	2.29757	2.29858	2.29958	2.30058	2.30158

EXPONENTIAL FUNCTIONS

e^x

x	0	1	2	3	4	5	6	7	8	9
.0	1.0000	1.0101	1.0202	1.0305	1.0408	1.0513	1.0618	1.0725	1.0833	1.0942
.1	1.1052	1.1163	1.1275	1.1388	1.1503	1.1618	1.1735	1.1853	1.1972	1.2092
.2	1.2214	1.2337	1.2461	1.2586	1.2712	1.2840	1.2969	1.3100	1.3231	1.3364
.3	1.3499	1.3634	1.3771	1.3910	1.4049	1.4191	1.4333	1.4477	1.4623	1.4770
.4	1.4918	1.5068	1.5220	1.5373	1.5527	1.5683	1.5841	1.6000	1.6161	1.6323
.5	1.6487	1.6653	1.6820	1.6989	1.7160	1.7333	1.7507	1.7683	1.7860	1.8040
.6	1.8221	1.8404	1.8589	1.8776	1.8965	1.9155	1.9348	1.9542	1.9739	1.9937
.7	2.0138	2.0340	2.0544	2.0751	2.0959	2.1170	2.1383	2.1598	2.1815	2.2034
.8	2.2255	2.2479	2.2705	2.2933	2.3164	2.3396	2.3632	2.3869	2.4109	2.4351
.9	2.4596	2.4843	2.5093	2.5345	2.5600	2.5857	2.6117	2.6379	2.6645	2.6912
1.0	2.7183	2.7456	2.7732	2.8011	2.8292	2.8577	2.8864	2.9154	2.9447	2.9743
1.1	3.0042	3.0344	3.0649	3.0957	3.1268	3.1582	3.1899	3.2220	3.2544	3.2871
1.2	3.3201	3.3535	3.3872	3.4212	3.4556	3.4903	3.5254	3.5609	3.5966	3.6328
1.3	3.6693	3.7062	3.7434	3.7810	3.8190	3.8574	3.8962	3.9354	3.9749	4.0149
1.4	4.0552	4.0960	4.1371	4.1787	4.2207	4.2631	4.3060	4.3492	4.3929	4.4371
1.5	4.4817	4.5267	4.5722	4.6182	4.6646	4.7115	4.7588	4.8066	4.8550	4.9037
1.6	4.9530	5.0028	5.0531	5.1039	5.1552	5.2070	5.2593	5.3122	5.3656	5.4195
1.7	5.4739	5.5290	5.5845	5.6407	5.6973	5.7546	5.8124	5.8709	5.9299	5.9895
1.8	6.0496	6.1104	6.1719	6.2339	6.2965	6.3598	6.4237	6.4883	6.5535	6.6194
1.9	6.6859	6.7531	6.8210	6.8895	6.9588	7.0287	7.0993	7.1707	7.2427	7.3155
2.0	7.3891	7.4633	7.5383	7.6141	7.6906	7.7679	7.8460	7.9248	8.0045	8.0849
2.1	8.1662	8.2482	8.3311	8.4149	8.4994	8.5849	8.6711	8.7583	8.8463	8.9352
2.2	9.0250	9.1157	9.2073	9.2999	9.3933	9.4877	9.5831	9.6794	9.7767	9.8749
2.3	9.9742	10.074	10.176	10.278	10.381	10.486	10.591	10.697	10.805	10.913
2.4	11.023	11.134	11.246	11.359	11.473	11.588	11.705	11.822	11.941	12.061
2.5	12.182	12.305	12.429	12.554	12.680	12.807	12.936	13.066	13.197	13.330
2.6	13.464	13.599	13.736	13.874	14.013	14.154	14.296	14.440	14.585	14.732
2.7	14.880	15.029	15.180	15.333	15.487	15.643	15.800	15.959	16.119	16.281
2.8	16.445	16.610	16.777	16.945	17.116	17.288	17.462	17.637	17.814	17.993
2.9	18.174	18.357	18.541	18.728	18.916	19.106	19.298	19.492	19.688	19.886
3.0	20.086	20.287	20.491	20.697	20.905	21.115	21.328	21.542	21.758	21.977
3.1	22.198	22.421	22.646	22.874	23.104	23.336	23.571	23.807	24.047	24.288
3.2	24.533	24.779	25.028	25.280	25.534	25.790	26.050	26.311	26.576	26.843
3.3	27.113	27.385	27.660	27.938	28.219	28.503	28.789	29.079	29.371	29.666
3.4	29.964	30.265	30.569	30.877	31.187	31.500	31.817	32.137	32.460	32.786
3.5	33.115	33.448	33.784	34.124	34.467	34.813	35.163	35.517	35.874	36.234
3.6	36.598	36.966	37.338	37.713	38.092	38.475	38.861	39.252	39.646	40.045
3.7	40.447	40.854	41.264	41.679	42.098	42.521	42.948	43.380	43.816	44.256
3.8	44.701	45.150	45.604	46.063	46.525	46.993	47.465	47.942	48.424	48.911
3.9	49.402	49.899	50.400	50.907	51.419	51.935	52.457	52.985	53.517	54.055
4.	54.598	60.340	66.686	73.700	81.451	90.017	99.484	109.95	121.51	134.29
5.	148.41	164.02	181.27	200.34	221.41	244.69	270.43	298.87	330.30	365.04
6.	403.43	445.86	492.75	544.57	601.85	665.14	735.10	812.41	897.85	992.27
7.	1096.6	1212.0	1339.4	1480.3	1636.0	1808.0	1998.2	2208.3	2440.6	2697.3
8.	2981.0	3294.5	3641.0	4023.9	4447.1	4914.8	5431.7	6002.9	6634.2	7332.0
9.	8103.1	8955.3	9897.1	10938	12088	13360	14765	16318	18034	19930
10.	22026									

9 EXPONENTIAL FUNCTIONS

e^{-x}

x	0	1	2	3	4	5	6	7	8	9
.0	1.00000	.99005	.98020	.97045	.96079	.95123	.94176	.93239	.92312	.91393
.1	.90484	.89583	.88692	.87810	.86936	.86071	.85214	.84366	.83527	.82696
.2	.81873	.81058	.80252	.79453	.78663	.77880	.77105	.76338	.75578	.74826
.3	.74082	.73345	.72615	.71892	.71177	.70469	.69768	.69073	.68386	.67706
.4	.67032	.66365	.65705	.65051	.64404	.63763	.63128	.62500	.61878	.61263
.5	.60653	.60050	.59452	.58860	.58275	.57695	.57121	.56553	.55990	.55433
.6	.54881	.54335	.53794	.53259	.52729	.52205	.51685	.51171	.50662	.50158
.7	.49659	.49164	.48675	.48191	.47711	.47237	.46767	.46301	.45841	.45384
.8	.44933	.44486	.44043	.43605	.43171	.42741	.42316	.41895	.41478	.41066
.9	.40657	.40252	.39852	.39455	.39063	.38674	.38289	.37908	.37531	.37158
1.0	.36788	.36422	.36060	.35701	.35345	.34994	.34646	.34301	.33960	.33622
1.1	.33287	.32956	.32628	.32303	.31982	.31664	.31349	.31037	.30728	.30422
1.2	.30119	.29820	.29523	.29229	.28938	.28650	.28365	.28083	.27804	.27527
1.3	.27253	.26982	.26714	.26448	.26185	.25924	.25666	.25411	.25158	.24908
1.4	.24660	.24414	.24171	.23931	.23693	.23457	.23224	.22993	.22764	.22537
1.5	.22313	.22091	.21871	.21654	.21438	.21225	.21014	.20805	.20598	.20393
1.6	.20190	.19989	.19790	.19593	.19398	.19205	.19014	.18825	.18637	.18452
1.7	.18268	.18087	.17907	.17728	.17552	.17377	.17204	.17033	.16864	.16696
1.8	.16530	.16365	.16203	.16041	.15882	.15724	.15567	.15412	.15259	.15107
1.9	.14957	.14808	.14661	.14515	.14370	.14227	.14086	.13946	.13807	.13670
2.0	.13534	.13399	.13266	.13134	.13003	.12873	.12745	.12619	.12493	.12369
2.1	.12246	.12124	.12003	.11884	.11765	.11648	.11533	.11418	.11304	.11192
2.2	.11080	.10970	.10861	.10753	.10646	.10540	.10435	.10331	.10228	.10127
2.3	.10026	.09926	.09827	.09730	.09633	.09537	.09442	.09348	.09255	.09163
2.4	.09072	.08982	.08892	.08804	.08716	.08629	.08543	.08458	.08374	.08291
2.5	.08208	.08127	.08046	.07966	.07887	.07808	.07730	.07654	.07577	.07502
2.6	.07427	.07353	.07280	.07208	.07136	.07065	.06995	.06925	.06856	.06788
2.7	.06721	.06654	.06587	.06522	.06457	.06393	.06329	.06266	.06204	.06142
2.8	.06081	.06020	.05961	.05901	.05843	.05784	.05727	.05670	.05613	.05558
2.9	.05502	.05448	.05393	.05340	.05287	.05234	.05182	.05130	.05079	.05029
3.0	.04979	.04929	.04880	.04832	.04783	.04736	.04689	.04642	.04596	.04550
3.1	.04505	.04460	.04416	.04372	.04328	.04285	.04243	.04200	.04159	.04117
3.2	.04076	.04036	.03996	.03956	.03916	.03877	.03839	.03801	.03763	.03725
3.3	.03688	.03652	.03615	.03579	.03544	.03508	.03474	.03439	.03405	.03371
3.4	.03337	.03304	.03271	.03239	.03206	.03175	.03143	.03112	.03081	.03050
3.5	.03020	.02990	.02960	.02930	.02901	.02872	.02844	.02816	.02788	.02760
3.6	.02732	.02705	.02678	.02652	.02625	.02599	.02573	.02548	.02522	.02497
3.7	.02472	.02448	.02423	.02399	.02375	.02352	.02328	.02305	.02282	.02260
3.8	.02237	.02215	.02193	.02171	.02149	.02128	.02107	.02086	.02065	.02045
3.9	.02024	.02004	.01984	.01964	.01945	.01925	.01906	.01887	.01869	.01850
4.	.018316	.016573	.014996	.013569	.012277	.011109	.010052	$.0^{2}90953$	$.0^{2}82297$	$.0^{2}74466$
5.	$.0^{2}67379$	$.0^{2}60967$	$.0^{2}55166$	$.0^{2}49916$	$.0^{2}45166$	$.0^{2}40868$	$.0^{2}36979$	$.0^{2}33460$	$.0^{2}30276$	$.0^{2}27394$
6.	$.0^{2}24788$	$.0^{2}22429$	$.0^{2}20294$	$.0^{2}18363$	$.0^{2}16616$	$.0^{2}15034$	$.0^{2}13604$	$.0^{2}12309$	$.0^{2}11138$	$.0^{2}10078$
7.	$.0^{3}91188$	$.0^{3}82510$	$.0^{3}74659$	$.0^{3}67554$	$.0^{3}61125$	$.0^{3}55308$	$.0^{3}50045$	$.0^{3}45283$	$.0^{3}40973$	$.0^{3}37074$
8.	$.0^{3}33546$	$.0^{3}30354$	$.0^{3}27465$	$.0^{3}24852$	$.0^{3}22487$	$.0^{3}20347$	$.0^{3}18411$	$.0^{3}16659$	$.0^{3}15073$	$.0^{3}13639$
9.	$.0^{3}12341$	$.0^{3}11167$	$.0^{3}10104$	$.0^{4}91424$	$.0^{4}82724$	$.0^{4}74852$	$.0^{4}67729$	$.0^{4}61283$	$.0^{4}55452$	$.0^{4}50175$
10.	$.0^{4}45400$									

EXPONENTIAL, SINE, AND COSINE INTEGRALS

$$\mathrm{Ei}(x) = \int_x^\infty \frac{e^{-u}}{u}\,du, \quad \mathrm{Si}(x) = \int_0^x \frac{\sin u}{u}\,du, \quad \mathrm{Ci}(x) = \int_x^\infty \frac{\cos u}{u}\,du$$

x	Ei(x)	Si(x)	Ci(x)
.0	∞	.0000	∞
.5	.5598	.4931	.1778
1.0	.2194	.9461	−.3374
1.5	.1000	1.3247	−.4704
2.0	.04890	1.6054	−.4230
2.5	.02491	1.7785	−.2859
3.0	.01305	1.8487	−.1196
3.5	$.0^{2}6970$	1.8331	.0321
4.0	$.0^{2}3779$	1.7582	.1410
4.5	$.0^{2}2073$	1.6541	.1935
5.0	$.0^{2}1148$	1.5499	.1900
5.5	$.0^{3}6409$	1.4687	.1421
6.0	$.0^{3}3601$	1.4247	.0681
6.5	$.0^{3}2034$	1.4218	−.0111
7.0	$.0^{3}1155$	1.4546	−.0767
7.5	$.0^{4}6583$	1.5107	−.1156
8.0	$.0^{4}3767$	1.5742	−.1224
8.5	$.0^{4}2162$	1.6296	−.09943
9.0	$.0^{4}1245$	1.6650	−.05535
9.5	$.0^{5}7185$	1.6745	$-.0^{2}2678$
10.0	$.0^{5}4157$	1.6583	.04546

Section II: Factorial and Gamma Function, Binomial Coefficients

FACTORIAL *n*

$$n! = 1 \cdot 2 \cdot 3 \cdot \cdots \cdot n$$

n	$n!$	n	$n!$	n	$n!$
0	1 (by definition)	40	8.15915×10^{47}	80	7.15695×10^{118}
1	1	41	3.34525×10^{49}	81	5.79713×10^{120}
2	2	42	1.40501×10^{51}	82	4.75364×10^{122}
3	6	43	6.04153×10^{52}	83	3.94552×10^{124}
4	24	44	2.65827×10^{54}	84	3.31424×10^{126}
5	120	45	1.19622×10^{56}	85	2.81710×10^{128}
6	720	46	5.50262×10^{57}	86	2.42271×10^{130}
7	5040	47	2.58623×10^{59}	87	2.10776×10^{132}
8	40,320	48	1.24139×10^{61}	88	1.85483×10^{134}
9	362,880	49	6.08282×10^{62}	89	1.65080×10^{136}
10	3,628,800	50	3.04141×10^{64}	90	1.48572×10^{138}
11	39,916,800	51	1.55112×10^{66}	91	1.35200×10^{140}
12	479,001,600	52	8.06582×10^{67}	92	1.24384×10^{142}
13	6,227,020,800	53	4.27488×10^{69}	93	1.15677×10^{144}
14	87,178,291,200	54	2.30844×10^{71}	94	1.08737×10^{146}
15	1,307,674,368,000	55	1.26964×10^{73}	95	1.03300×10^{148}
16	20,922,789,888,000	56	7.10999×10^{74}	96	9.91678×10^{149}
17	355,687,428,096,000	57	4.05269×10^{76}	97	9.61928×10^{151}
18	6,402,373,705,728,000	58	2.35056×10^{78}	98	9.42689×10^{153}
19	121,645,100,408,832,000	59	1.38683×10^{80}	99	9.33262×10^{155}
20	2,432,902,008,176,640,000	60	8.32099×10^{81}	100	9.33262×10^{157}
21	51,090,942,171,709,440,000	61	5.07580×10^{83}		
22	1,124,000,727,777,607,680,000	62	3.14700×10^{85}		
23	25,852,016,738,884,976,640,000	63	1.98261×10^{87}		
24	620,448,401,733,239,439,360,000	64	1.26887×10^{89}		
25	15,511,210,043,330,985,984,000,000	65	8.24765×10^{90}		
26	403,291,461,126,605,635,584,000,000	66	5.44345×10^{92}		
27	10,888,869,450,418,352,160,768,000,000	67	3.64711×10^{94}		
28	304,888,344,611,713,860,501,504,000,000	68	2.48004×10^{96}		
29	8,841,761,993,739,701,954,543,616,000,000	69	1.71122×10^{98}		
30	265,252,859,812,191,058,636,308,480,000,000	70	1.19786×10^{100}		
31	8.22284×10^{33}	71	8.50479×10^{101}		
32	2.63131×10^{35}	72	6.12345×10^{103}		
33	8.68332×10^{36}	73	4.47012×10^{105}		
34	2.95233×10^{38}	74	3.30789×10^{107}		
35	1.03331×10^{40}	75	2.48091×10^{109}		
36	3.71993×10^{41}	76	1.88549×10^{111}		
37	1.37638×10^{43}	77	1.45183×10^{113}		
38	5.23023×10^{44}	78	1.13243×10^{115}		
39	2.03979×10^{46}	79	8.94618×10^{116}		

GAMMA FUNCTION

$$\Gamma(x) = \int_x^\infty t^{x-1}e^{-t}\,dt \qquad \text{for } 1 \leqq x \leqq 2$$

[For other values use the formula $\Gamma(x + 1) = x\,\Gamma(x)$]

x	$\Gamma(x)$	x	$\Gamma(x)$
1.00	1.00000	1.50	.88623
1.01	.99433	1.51	.88659
1.02	.98884	1.52	.88704
1.03	.98355	1.53	.88757
1.04	.97844	1.54	.88818
1.05	.97350	1.55	.88887
1.06	.96874	1.56	.88964
1.07	.96415	1.57	.89049
1.08	.95973	1.58	.89142
1.09	.95546	1.59	.89243
1.10	.95135	1.60	.89352
1.11	.94740	1.61	.89468
1.12	.94359	1.62	.89592
1.13	.93993	1.63	.89724
1.14	.93642	1.64	.89864
1.15	.93304	1.65	.90012
1.16	.92980	1.66	.90167
1.17	.92670	1.67	.90330
1.18	.92373	1.68	.90500
1.19	.92089	1.69	.90678
1.20	.91817	1.70	.90864
1.21	.91558	1.71	.91057
1.22	.91311	1.72	.91258
1.23	.91075	1.73	.91467
1.24	.90852	1.74	.91683
1.25	.90640	1.75	.91906
1.26	.90440	1.76	.92137
1.27	.90250	1.77	.92376
1.28	.90072	1.78	.92623
1.29	.89904	1.79	.92877
1.30	.89747	1.80	.93138
1.31	.89600	1.81	.93408
1.32	.89464	1.82	.93685
1.33	.89338	1.83	.93969
1.34	.89222	1.84	.94261
1.35	.89115	1.85	.94561
1.36	.89018	1.86	.94869
1.37	.88931	1.87	.95184
1.38	.88854	1.88	.95507
1.39	.88785	1.89	.95838
1.40	.88726	1.90	.96177
1.41	.88676	1.91	.96523
1.42	.88636	1.92	.96877
1.43	.88604	1.93	.97240
1.44	.88581	1.94	.97610
1.45	.88566	1.95	.97988
1.46	.88560	1.96	.98374
1.47	.88563	1.97	.98768
1.48	.88575	1.98	.99171
1.49	.88595	1.99	.99581
1.50	.88623	2.00	1.00000

13 BINOMIAL COEFFICIENTS

$$\binom{n}{k} = \frac{n!}{k!(n-k)!} = \frac{n(n-1)\cdots(n-k+1)}{k!} = \binom{n}{n-k}, \quad 0! = 1$$

Note that each number is the sum of two numbers in the row above; one of these numbers is in the same column and the other is in the preceding column (e.g., 56 = 35 + 21). The arrangement is often called *Pascal's triangle* (see 3.6, page 8).

n \ k	0	1	2	3	4	5	6	7	8	9
1	1	1								
2	1	2	1							
3	1	3	3	1						
4	1	4	6	4	1					
5	1	5	10	10	5	1				
6	1	6	15	20	15	6	1			
7	1	7	21	35	35	21	7	1		
8	1	8	28	56	70	56	28	8	1	
9	1	9	36	84	126	126	84	36	9	1
10	1	10	45	120	210	252	210	120	45	10
11	1	11	55	165	330	462	462	330	165	55
12	1	12	66	220	495	792	924	792	495	220
13	1	13	78	286	715	1287	1716	1716	1287	715
14	1	14	91	364	1001	2002	3003	3432	3003	2002
15	1	15	105	455	1365	3003	5005	6435	6435	5005
16	1	16	120	560	1820	4368	8008	11440	12870	11440
17	1	17	136	680	2380	6188	12376	19448	24310	24310
18	1	18	153	816	3060	8568	18564	31824	43758	48620
19	1	19	171	969	3876	11628	27132	50388	75582	92378
20	1	20	190	1140	4845	15504	38760	77520	125970	167960
21	1	21	210	1330	5985	20349	54264	116280	203490	293930
22	1	22	231	1540	7315	26334	74613	170544	319770	497420
23	1	23	253	1771	8855	33649	100947	245157	490314	817190
24	1	24	276	2024	10626	42504	134596	346104	735471	1307504
25	1	25	300	2300	12650	53130	177100	480700	1081575	2042975
26	1	26	325	2600	14950	65780	230230	657800	1562275	3124550
27	1	27	351	2925	17550	80730	296010	888030	2220075	4686825
28	1	28	378	3276	20475	98280	376740	1184040	3108105	6906900
29	1	29	406	3654	23751	118755	475020	1560780	4292145	10015005
30	1	30	435	4060	27405	142506	593775	2035800	5852925	14307150

13 BINOMIAL COEFFICIENTS

$$\binom{n}{k} = \frac{n!}{k!(n-k)!} = \frac{n(n-1)\cdots(n-k+1)}{k!} = \binom{n}{n-k}, \quad 0! = 1 \quad \textit{(Continued)}$$

n \ k	10	11	12	13	14	15
10	1					
11	11	1				
12	66	12	1			
13	286	78	13	1		
14	1001	364	91	14	1	
15	3003	1365	455	105	15	1
16	8008	4368	1820	560	120	16
17	19448	12376	6188	2380	680	136
18	43758	31824	18564	8568	3060	816
19	92378	75582	50388	27132	11628	3876
20	184756	167960	125970	77520	38760	15504
21	352716	352716	293930	203490	116280	54264
22	646646	705432	646646	497420	319770	170544
23	1144066	1352078	1352078	1144066	817190	490314
24	1961256	2496144	2704156	2496144	1961256	1307504
25	3268760	4457400	5200300	5200300	4457400	3268760
26	5311735	7726160	9657700	10400600	9657700	7726160
27	8436285	13037895	17383860	20058300	20058300	17383860
28	13123110	21474180	30421755	37442160	40116600	37442160
29	20030010	34597290	51895935	67863915	77558760	77558760
30	30045015	54627300	86493225	119759850	145422675	155117520

For $k > 15$ use the fact that $\binom{n}{k} = \binom{n}{n-k}$.

Section III: Bessel Functions

BESSEL FUNCTIONS

$J_0(x)$

x	0	1	2	3	4	5	6	7	8	9
0.	1.0000	.9975	.9900	.9776	.9604	.9385	.9120	.8812	.8463	.8075
1.	.7652	.7196	.6711	.6201	.5669	.5118	.4554	.3980	.3400	.2818
2.	.2239	.1666	.1104	.0555	.0025	−.0484	−.0968	−.1424	−.1850	−.2243
3.	−.2601	−.2921	−.3202	−.3443	−.3643	−.3801	−.3918	−.3992	−.4026	−.4018
4.	−.3971	−.3887	−.3766	−.3610	−.3423	−.3205	−.2961	−.2693	−.2404	−.2097
5.	−.1776	−.1443	−.1103	−.0758	−.0412	−.0068	.0270	.0599	.0917	.1220
6.	.1506	.1773	.2017	.2238	.2433	.2601	.2740	.2851	.2931	.2981
7.	.3001	.2991	.2951	.2882	.2786	.2663	.2516	.2346	.2154	.1944
8.	.1717	.1475	.1222	.0960	.0692	.0419	.0146	−.0125	−.0392	−.0653
9.	−.0903	−.1142	−.1367	−.1577	−.1768	−.1939	−.2090	−.2218	−.2323	−.2403

15 BESSEL FUNCTIONS

$J_1(x)$

x	0	1	2	3	4	5	6	7	8	9
0.	.0000	.0499	.0995	.1483	.1960	.2423	.2867	.3290	.3688	.4059
1.	.4401	.4709	.4983	.5220	.5419	.5579	.5699	.5778	.5815	.5812
2.	.5767	.5683	.5560	.5399	.5202	.4971	.4708	.4416	.4097	.3754
3.	.3391	.3009	.2613	.2207	.1792	.1374	.0955	.0538	.0128	−.0272
4.	−.0660	−.1033	−.1386	−.1719	−.2028	−.2311	−.2566	−.2791	−.2985	−.3147
5.	−.3276	−.3371	−.3432	−.3460	−.3453	−.3414	−.3343	−.3241	−.3110	−.2951
6.	−.2767	−.2559	−.2329	−.2081	−.1816	−.1538	−.1250	−.0953	−.0652	−.0349
7.	−.0047	.0252	.0543	.0826	.1096	.1352	.1592	.1813	.2014	.2192
8.	.2346	.2476	.2580	.2657	.2708	.2731	.2728	.2697	.2641	.2559
9.	.2453	.2324	.2174	.2004	.1816	.1613	.1395	.1166	.0928	.0684

16 BESSEL FUNCTIONS

$Y_0(x)$

x	0	1	2	3	4	5	6	7	8	9
0.	−∞	−1.5342	−1.0811	−.8073	−.6060	−.4445	−.3085	−.1907	−.0868	.0056
1.	.0883	.1622	.2281	.2865	.3379	.3824	.4204	.4520	.4774	.4968
2.	.5104	.5183	.5208	.5181	.5104	.4981	.4813	.4605	.4359	.4079
3.	.3769	.3431	.3071	.2691	.2296	.1890	.1477	.1061	.0645	.0234
4.	−.0169	−.0561	−.0938	−.1296	−.1633	−.1947	−.2235	−.2494	−.2723	−.2921
5.	−.3085	−.3216	−.3313	−.3374	−.3402	−.3395	−.3354	−.3282	−.3177	−.3044
6.	−.2882	−.2694	−.2483	−.2251	−.1999	−.1732	−.1452	−.1162	−.0864	−.0563
7.	−.0259	.0042	.0339	.0628	.0907	.1173	.1424	.1658	.1872	.2065
8.	.2235	.2381	.2501	.2595	.2662	.2702	.2715	.2700	.2659	.2592
9.	.2499	.2383	.2245	.2086	.1907	.1712	.1502	.1279	.1045	.0804

17 BESSEL FUNCTIONS

$Y_1(x)$

x	0	1	2	3	4	5	6	7	8	9
0.	−∞	−6.4590	−3.3238	−2.2931	−1.7809	−1.4715	−1.2604	−1.1032	−.9781	−.8731
1.	−.7812	−.6981	−.6211	−.5485	−.4791	−.4123	−.3476	−.2847	−.2237	−.1644
2.	−.1070	−.0517	.0015	.0523	.1005	.1459	.1884	.2276	.2635	.2959
3.	.3247	.3496	.3707	.3879	.4010	.4102	.4154	.4167	.4141	.4078
4.	.3979	.3846	.3680	.3484	.3260	.3010	.2737	.2445	.2136	.1812
5.	.1479	.1137	.0792	.0445	.0101	−.0238	−.0568	−.0887	−.1192	−.1481
6.	−.1750	−.1998	−.2223	−.2422	−.2596	−.2741	−.2857	−.2945	−.3002	−.3029
7.	−.3027	−.2995	−.2934	−.2846	−.2731	−.2591	−.2428	−.2243	−.2039	−.1817
8.	−.1581	−.1331	−.1072	−.0806	−.0535	−.0262	.0011	.0280	.0544	.0799
9.	.1043	.1275	.1491	.1691	.1871	.2032	.2171	.2287	.2379	.2447

BESSEL FUNCTIONS

$I_0(x)$

x	0	1	2	3	4	5	6	7	8	9
0.	1.000	1.003	1.010	1.023	1.040	1.063	1.092	1.126	1.167	1.213
1.	1.266	1.326	1.394	1.469	1.553	1.647	1.750	1.864	1.990	2.128
2.	2.280	2.446	2.629	2.830	3.049	3.290	3.553	3.842	4.157	4.503
3.	4.881	5.294	5.747	6.243	6.785	7.378	8.028	8.739	9.517	10.37
4.	11.30	12.32	13.44	14.67	16.01	17.48	19.09	20.86	22.79	24.91
5.	27.24	29.79	32.58	35.65	39.01	42.69	46.74	51.17	56.04	61.38
6.	67.23	73.66	80.72	88.46	96.96	106.3	116.5	127.8	140.1	153.7
7.	168.6	185.0	202.9	222.7	244.3	268.2	294.3	323.1	354.7	389.4
8.	427.6	469.5	515.6	566.3	621.9	683.2	750.5	824.4	905.8	995.2
9.	1094	1202	1321	1451	1595	1753	1927	2119	2329	2561

19

BESSEL FUNCTIONS

$I_1(x)$

x	0	1	2	3	4	5	6	7	8	9
0.	.0000	.0501	.1005	.1517	.2040	.2579	.3137	.3719	.4329	.4971
1.	.5652	.6375	.7147	.7973	.8861	.9817	1.085	1.196	1.317	1.448
2.	1.591	1.745	1.914	2.098	2.298	2.517	2.755	3.016	3.301	3.613
3.	3.953	4.326	4.734	5.181	5.670	6.206	6.793	7.436	8.140	8.913
4.	9.759	10.69	11.71	12.82	14.05	15.39	16.86	18.48	20.25	22.20
5.	24.34	26.68	29.25	32.08	35.18	38.59	42.33	46.44	50.95	55.90
6.	61.34	67.32	73.89	81.10	89.03	97.74	107.3	117.8	129.4	142.1
7.	156.0	171.4	188.3	206.8	227.2	249.6	274.2	301.3	331.1	363.9
8.	399.9	439.5	483.0	531.0	583.7	641.6	705.4	775.5	852.7	937.5
9.	1031	1134	1247	1371	1508	1658	1824	2006	2207	2428

20

BESSEL FUNCTIONS

$K_0(x)$

x	0	1	2	3	4	5	6	7	8	9
0.	∞	2.4271	1.7527	1.3725	1.1145	.9244	.7775	.6605	.5653	.4867
1.	.4210	.3656	.3185	.2782	.2437	.2138	.1880	.1655	.1459	.1288
2.	.1139	.1008	.08927	.07914	.07022	.06235	.05540	.04926	.04382	.03901
3.	.03474	.03095	.02759	.02461	.02196	.01960	.01750	.01563	.01397	.01248
4.	.01116	$.0^{2}9980$	$.0^{2}8927$	$.0^{2}7988$	$.0^{2}7149$	$.0^{2}6400$	$.0^{2}5730$	$.0^{2}5132$	$.0^{2}4597$	$.0^{2}4119$
5.	$.0^{2}3691$	$.0^{2}3308$	$.0^{2}2966$	$.0^{2}2659$	$.0^{2}2385$	$.0^{2}2139$	$.0^{2}1918$	$.0^{2}1721$	$.0^{2}1544$	$.0^{2}1386$
6.	$.0^{2}1244$	$.0^{2}1117$	$.0^{2}1003$	$.0^{3}9001$	$.0^{3}8083$	$.0^{3}7259$	$.0^{3}6520$	$.0^{3}5857$	$.0^{3}5262$	$.0^{3}4728$
7.	$.0^{3}4248$	$.0^{3}3817$	$.0^{3}3431$	$.0^{3}3084$	$.0^{3}2772$	$.0^{3}2492$	$.0^{3}2240$	$.0^{3}2014$	$.0^{3}1811$	$.0^{3}1629$
8.	$.0^{3}1465$	$.0^{3}1317$	$.0^{3}1185$	$.0^{3}1066$	$.0^{4}9588$	$.0^{4}8626$	$.0^{4}7761$	$.0^{4}6983$	$.0^{4}6283$	$.0^{4}5654$
9.	$.0^{4}5088$	$.0^{4}4579$	$.0^{4}4121$	$.0^{4}3710$	$.0^{4}3339$	$.0^{4}3006$	$.0^{4}2706$	$.0^{4}2436$	$.0^{4}2193$	$.0^{4}1975$

BESSEL FUNCTIONS

$K_1(x)$

x	0	1	2	3	4	5	6	7	8	9
0.	∞	9.8538	4.7760	3.0560	2.1844	1.6564	1.3028	1.0503	.8618	.7165
1.	.6019	.5098	.4346	.3725	.3208	.2774	.2406	.2094	.1826	.1597
2.	.1399	.1227	.1079	.09498	.08372	.07389	.06528	.05774	.05111	.04529
3.	.04016	.03563	.03164	.02812	.02500	.02224	.01979	.01763	.01571	.01400
4.	.01248	.01114	$.0^{2}9938$	$.0^{2}8872$	$.0^{2}7923$	$.0^{2}7078$	$.0^{2}6325$	$.0^{2}5654$	$.0^{2}5055$	$.0^{2}4521$
5.	$.0^{2}4045$	$.0^{2}3619$	$.0^{2}3239$	$.0^{2}2900$	$.0^{2}2597$	$.0^{2}2326$	$.0^{2}2083$	$.0^{2}1866$	$.0^{2}1673$	$.0^{2}1499$
6.	$.0^{2}1344$	$.0^{2}1205$	$.0^{2}1081$	$.0^{3}9691$	$.0^{3}8693$	$.0^{3}7799$	$.0^{3}6998$	$.0^{3}6280$	$.0^{3}5636$	$.0^{3}5059$
7.	$.0^{3}4542$	$.0^{3}4078$	$.0^{3}3662$	$.0^{3}3288$	$.0^{3}2953$	$.0^{3}2653$	$.0^{3}2383$	$.0^{3}2141$	$.0^{3}1924$	$.0^{3}1729$
8.	$.0^{3}1554$	$.0^{3}1396$	$.0^{3}1255$	$.0^{3}1128$	$.0^{3}1014$	$.0^{4}9120$	$.0^{4}8200$	$.0^{4}7374$	$.0^{4}6631$	$.0^{4}5964$
9.	$.0^{4}5364$	$.0^{4}4825$	$.0^{4}4340$	$.0^{4}3904$	$.0^{4}3512$	$.0^{4}3160$	$.0^{4}2843$	$.0^{4}2559$	$.0^{4}2302$	$.0^{4}2072$

BESSEL FUNCTIONS
Ber(x)

x	0	1	2	3	4	5	6	7	8	9
0.	1.0000	1.0000	1.0000	.9999	.9996	.9990	.9980	.9962	.9936	.9898
1.	.9844	.9771	.9676	.9554	.9401	.9211	.8979	.8700	.8367	.7975
2.	.7517	.6987	.6377	.5680	.4890	.4000	.3001	.1887	.06511	−.07137
3.	−.2214	−.3855	−.5644	−.7584	−.9680	−1.1936	−1.4353	−1.6933	−1.9674	−2.2576
4.	−2.5634	−2.8843	−3.2195	−3.5679	−3.9283	−4.2991	−4.6784	−5.0639	−5.4531	−5.8429
5.	−6.2301	−6.6107	−6.9803	−7.3344	−7.6674	−7.9736	−8.2466	−8.4794	−8.6644	−8.7937
6.	−8.8583	−8.8491	−8.7561	−8.5688	−8.2762	−7.8669	−7.3287	−6.6492	−5.8155	−4.8146
7.	−3.6329	−2.2571	−.6737	1.1308	3.1695	5.4550	7.9994	10.814	13.909	17.293
8.	20.974	24.957	29.245	33.840	38.738	43.936	49.423	55.187	61.210	67.469
9.	73.936	80.576	87.350	94.208	101.10	107.95	114.70	121.26	127.54	133.43

BESSEL FUNCTIONS
Bei(x)

x	0	1	2	3	4	5	6	7	8	9
0.	.0000	$.0^{2}2500$	.01000	.02250	.04000	.06249	.08998	.1224	.1599	.2023
1.	.2496	.3017	.3587	.4204	.4867	.5576	.6327	.7120	.7953	.8821
2.	.9723	1.0654	1.1610	1.2585	1.3575	1.4572	1.5569	1.6557	1.7529	1.8472
3.	1.9376	2.0228	2.1016	2.1723	2.2334	2.2832	2.3199	2.3413	2.3454	2.3300
4.	2.2927	2.2309	2.1422	2.0236	1.8726	1.6860	1.4610	1.1946	.8837	.5251
5.	.1160	−.3467	−.8658	−1.4443	−2.0845	−2.7890	−3.5597	−4.3986	−5.3068	−6.2854
6.	−7.3347	−8.4545	−9.6437	−10.901	−12.223	−13.607	−15.047	−16.538	−18.074	−19.644
7.	−21.239	−22.848	−24.456	−26.049	−27.609	−29.116	−30.548	−31.882	−33.092	−34.147
8.	−35.017	−35.667	−36.061	−36.159	−35.920	−35.298	−34.246	−32.714	−30.651	−28.003
9.	−24.713	−20.724	−15.976	−10.412	−3.9693	3.4106	11.787	21.218	31.758	43.459

24 BESSEL FUNCTIONS

Ker(x)

x	0	1	2	3	4	5	6	7	8	9
0.	∞	2.4205	1.7331	1.3372	1.0626	.8559	.6931	.5614	.4529	.3625
1.	.2867	.2228	.1689	.1235	.08513	.05293	.02603	$.0^{2}3691$	−.01470	−.02966
2.	−.04166	−.05111	−.05834	−.06367	−.06737	−.06969	−.07083	−.07097	−.07030	−.06894
3.	−.06703	−.06468	−.06198	−.05903	−.05590	−.05264	−.04932	−.04597	−.04265	−.03937
4.	−.03618	−.03308	−.03011	−.02726	−.02456	−.02200	−.01960	−.01734	−.01525	−.01330
5.	−.01151	$-.0^{2}9865$	$-.0^{2}8359$	$-.0^{2}6989$	$-.0^{2}5749$	$-.0^{2}4632$	$-.0^{2}3632$	$-.0^{2}2740$	$-.0^{2}1952$	$-.0^{2}1258$
6.	$-.0^{3}6530$	$-.0^{3}1295$	$.0^{3}3191$	$.0^{3}6991$	$.0^{2}1017$	$.0^{2}1278$	$.0^{2}1488$	$.0^{2}1653$	$.0^{2}1777$	$.0^{2}1866$
7.	$.0^{2}1922$	$.0^{2}1951$	$.0^{2}1956$	$.0^{2}1940$	$.0^{2}1907$	$.0^{2}1860$	$.0^{2}1800$	$.0^{2}1731$	$.0^{2}1655$	$.0^{2}1572$
8.	$.0^{2}1486$	$.0^{2}1397$	$.0^{2}1306$	$.0^{2}1216$	$.0^{2}1126$	$.0^{2}1037$	$.0^{3}9511$	$.0^{3}8675$	$.0^{3}7871$	$.0^{3}7102$
9.	$.0^{3}6372$	$.0^{3}5681$	$.0^{3}5030$	$.0^{3}4422$	$.0^{3}3855$	$.0^{3}3330$	$.0^{3}2846$	$.0^{3}2402$	$.0^{3}1996$	$.0^{3}1628$

25 BESSEL FUNCTIONS

Kei(x)

x	0	1	2	3	4	5	6	7	8	9
0.	−.7854	−.7769	−.7581	−.7331	−.7038	−.6716	−.6374	−.6022	−.5664	−.5305
1.	−.4950	−.4601	−.4262	−.3933	−.3617	−.3314	−.3026	−.2752	−.2494	−.2251
2.	−.2024	−.1812	−.1614	−.1431	−.1262	−.1107	−.09644	−.08342	−.07157	−.06083
3.	−.05112	−.04240	−.03458	−.02762	−.02145	−.01600	−.01123	$-.0^{2}7077$	$-.0^{2}3487$	$-.0^{3}4108$
4.	$.0^{2}2198$	$.0^{2}4386$	$.0^{2}6194$	$.0^{2}7661$	$.0^{2}8826$	$.0^{2}9721$	.01038	.01083	.01110	.01121
5.	.01119	.01105	.01082	.01051	.01014	$.0^{2}9716$	$.0^{2}9255$	$.0^{2}8766$	$.0^{2}8258$	$.0^{2}7739$
6.	$.0^{2}7216$	$.0^{2}6696$	$.0^{2}6183$	$.0^{2}5681$	$.0^{2}5194$	$.0^{2}4724$	$.0^{2}4274$	$.0^{2}3846$	$.0^{2}3440$	$.0^{2}3058$
7.	$.0^{2}2700$	$.0^{2}2366$	$.0^{2}2057$	$.0^{2}1770$	$.0^{2}1507$	$.0^{2}1267$	$.0^{2}1048$	$.0^{3}8498$	$.0^{3}6714$	$.0^{3}5117$
8.	$.0^{3}3696$	$.0^{3}2440$	$.0^{3}1339$	$.0^{4}3809$	$-.0^{4}4449$	$-.0^{3}1149$	$-.0^{3}1742$	$-.0^{3}2233$	$-.0^{3}2632$	$-.0^{3}2949$
9.	$-.0^{3}3192$	$-.0^{3}3368$	$-.0^{3}3486$	$-.0^{3}3552$	$-.0^{3}3574$	$-.0^{3}3557$	$-.0^{3}3508$	$-.0^{3}3430$	$-.0^{3}3329$	$-.0^{3}3210$

26 VALUES FOR APPROXIMATE ZEROS OF BESSEL FUNCTIONS

The following table lists the first few positive roots of various equations. Note that for all cases listed the successive large roots differ approximately by $\pi = 3.14159\ldots$

	$n = 0$	$n = 1$	$n = 2$	$n = 3$	$n = 4$	$n = 5$	$n = 6$
$J_n(x) = 0$	2.4048	3.8317	5.1356	6.3802	7.5883	8.7715	9.9361
	5.5201	7.0156	8.4172	9.7610	11.0647	12.3386	13.5893
	8.6537	10.1735	11.6198	13.0152	14.3725	15.7002	17.0038
	11.7915	13.3237	14.7960	16.2235	17.6160	18.9801	20.3208
	14.9309	16.4706	17.9598	19.4094	20.8269	22.2178	23.5861
	18.0711	19.6159	21.1170	22.5827	24.0190	25.4303	26.8202
$Y_n(x) = 0$	0.8936	2.1971	3.3842	4.5270	5.6452	6.7472	7.8377
	3.9577	5.4297	6.7938	8.0976	9.3616	10.5972	11.8110
	7.0861	8.5960	10.0235	11.3965	12.7301	14.0338	15.3136
	10.2223	11.7492	13.2100	14.6231	15.9996	17.3471	18.6707
	13.3611	14.8974	16.3790	17.8185	19.2244	20.6029	21.9583
	16.5009	18.0434	19.5390	20.9973	22.4248	23.8265	25.2062
$J'_n(x) = 0$	0.0000	1.8412	3.0542	4.2012	5.3176	6.4156	7.5013
	3.8317	5.3314	6.7061	8.0152	9.2824	10.5199	11.7349
	7.0156	8.5363	9.9695	11.3459	12.6819	13.9872	15.2682
	10.1735	11.7060	13.1704	14.5859	15.9641	17.3128	18.6374
	13.3237	14.8636	16.3475	17.7888	19.1960	20.5755	21.9317
	16.4706	18.0155	19.5129	20.9725	22.4010	23.8036	25.1839
$Y'_n(x) = 0$	2.1971	3.6830	5.0026	6.2536	7.4649	8.6496	9.8148
	5.4297	6.9415	8.3507	9.6988	11.0052	12.2809	13.5328
	8.5960	10.1234	11.5742	12.9724	14.3317	15.6608	16.9655
	11.7492	13.2858	14.7609	16.1905	17.5844	18.9497	20.2913
	14.8974	16.4401	17.9313	19.3824	20.8011	22.1928	23.5619
	18.0434	19.5902	21.0929	22.5598	23.9970	25.4091	26.7995

LEGENDRE POLYNOMIALS $P_n(x)$

$[P_0(x)=1,\ P_1(x)=x]$

x	$P_2(x)$	$P_3(x)$	$P_4(x)$	$P_5(x)$
.00	−.5000	.0000	.3750	.0000
.05	−.4963	−.0747	.3657	.0927
.10	−.4850	−.1475	.3379	.1788
.15	−.4663	−.2166	.2928	.2523
.20	−.4400	−.2800	.2320	.3075
.25	−.4063	−.3359	.1577	.3397
.30	−.3650	−.3825	.0729	.3454
.35	−.3163	−.4178	−.0187	.3225
.40	−.2600	−.4400	−.1130	.2706
.45	−.1963	−.4472	−.2050	.1917
.50	−.1250	−.4375	−.2891	.0898
.55	−.0463	−.4091	−.3590	−.0282
.60	.0400	−.3600	−.4080	−.1526
.65	.1338	−.2884	−.4284	−.2705
.70	.2350	−.1925	−.4121	−.3652
.75	.3438	−.0703	−.3501	−.4164
.80	.4600	.0800	−.2330	−.3995
.85	.5838	.2603	−.0506	−.2857
.90	.7150	.4725	.2079	−.0411
.95	.8538	.7184	.5541	.3727
1.00	1.0000	1.0000	1.0000	1.0000

28 LEGENDRE POLYNOMIALS $P_n(\cos\theta)$

$[P_0(\cos\theta)=1]$

θ	$P_1(\cos\theta)$	$P_2(\cos\theta)$	$P_3(\cos\theta)$	$P_4(\cos\theta)$	$P_5(\cos\theta)$
0°	1.0000	1.0000	1.0000	1.0000	1.0000
5°	.9962	.9886	.9773	.9623	.9437
10°	.9848	.9548	.9106	.8532	.7840
15°	.9659	.8995	.8042	.6847	.5471
20°	.9397	.8245	.6649	.4750	.2715
25°	.9063	.7321	.5016	.2465	.0009
30°	.8660	.6250	.3248	.0234	−.2233
35°	.8192	.5065	.1454	−.1714	−.3691
40°	.7660	.3802	−.0252	−.3190	−.4197
45°	.7071	.2500	−.1768	−.4063	−.3757
50°	.6428	.1198	−.3002	−.4275	−.2545
55°	.5736	−.0065	−.3886	−.3852	−.0868
60°	.5000	−.1250	−.4375	−.2891	.0898
65°	.4226	−.2321	−.4452	−.1552	.2381
70°	.3420	−.3245	−.4130	−.0038	.3281
75°	.2588	−.3995	−.3449	.1434	.3427
80°	.1737	−.4548	−.2474	.2659	.2810
85°	.0872	−.4886	−.1291	.3468	.1577
90°	.0000	−.5000	.0000	.3750	.0000

Section V: Elliptic Integrals

COMPLETE ELLIPTIC INTEGRALS OF FIRST AND SECOND KINDS

$$K = \int_0^{\pi/2} \frac{d\theta}{\sqrt{1 - k^2 \sin^2 \theta}}, \quad E = \int_0^{\pi/2} \sqrt{1 - k^2 \sin^2 \theta}\, d\theta, \quad k = \sin \psi$$

ψ	K	E
0°	1.5708	1.5708
1	1.5709	1.5707
2	1.5713	1.5703
3	1.5719	1.5697
4	1.5727	1.5689
5	1.5738	1.5678
6	1.5751	1.5665
7	1.5767	1.5649
8	1.5785	1.5632
9	1.5805	1.5611
10	1.5828	1.5589
11	1.5854	1.5564
12	1.5882	1.5537
13	1.5913	1.5507
14	1.5946	1.5476
15	1.5981	1.5442
16	1.6020	1.5405
17	1.6061	1.5367
18	1.6105	1.5326
19	1.6151	1.5283
20	1.6200	1.5238
21	1.6252	1.5191
22	1.6307	1.5141
23	1.6365	1.5090
24	1.6426	1.5037
25	1.6490	1.4981
26	1.6557	1.4924
27	1.6627	1.4864
28	1.6701	1.4803
29	1.6777	1.4740
30	1.6858	1.4675

ψ	K	E
30°	1.6858	1.4675
31	1.6941	1.4608
32	1.7028	1.4539
33	1.7119	1.4469
34	1.7214	1.4397
35	1.7312	1.4323
36	1.7415	1.4248
37	1.7522	1.4171
38	1.7633	1.4092
39	1.7748	1.4013
40	1.7868	1.3931
41	1.7992	1.3849
42	1.8122	1.3765
43	1.8256	1.3680
44	1.8396	1.3594
45	1.8541	1.3506
46	1.8691	1.3418
47	1.8848	1.3329
48	1.9011	1.3238
49	1.9180	1.3147
50	1.9356	1.3055
51	1.9539	1.2963
52	1.9729	1.2870
53	1.9927	1.2776
54	2.0133	1.2681
55	2.0347	1.2587
56	2.0571	1.2492
57	2.0804	1.2397
58	2.1047	1.2301
59	2.1300	1.2206
60	2.1565	1.2111

ψ	K	E
60°	2.1565	1.2111
61	2.1842	1.2015
62	2.2132	1.1920
63	2.2435	1.1826
64	2.2754	1.1732
65	2.3088	1.1638
66	2.3439	1.1545
67	2.3809	1.1453
68	2.4198	1.1362
69	2.4610	1.1272
70	2.5046	1.1184
71	2.5507	1.1096
72	2.5998	1.1011
73	2.6521	1.0927
74	2.7081	1.0844
75	2.7681	1.0764
76	2.8327	1.0686
77	2.9026	1.0611
78	2.9786	1.0538
79	3.0617	1.0468
80	3.1534	1.0401
81	3.2553	1.0338
82	3.3699	1.0278
83	3.5004	1.0223
84	3.6519	1.0172
85	3.8317	1.0127
86	4.0528	1.0086
87	4.3387	1.0053
88	4.7427	1.0026
89	5.4349	1.0008
90	∞	1.0000

30 INCOMPLETE ELLIPTIC INTEGRAL OF THE FIRST KIND

$$F(k,\phi)=\int_0^{\phi}\frac{d\theta}{\sqrt{1-k^2\sin^2\theta}},\quad k=\sin\psi$$

ϕ \ ψ	0°	10°	20°	30°	40°	50°	60°	70°	80°	90°
0°	0.0000	0.0000	0.0000	0.0000	0.0000	0.0000	0.0000	0.0000	0.0000	0.0000
10°	0.1745	0.1746	0.1746	0.1748	0.1749	0.1751	0.1752	0.1753	0.1754	0.1754
20°	0.3491	0.3493	0.3499	0.3508	0.3520	0.3533	0.3545	0.3555	0.3561	0.3564
30°	0.5236	0.5243	0.5263	0.5294	0.5334	0.5379	0.5422	0.5459	0.5484	0.5493
40°	0.6981	0.6997	0.7043	0.7116	0.7213	0.7323	0.7436	0.7535	0.7604	0.7629
50°	0.8727	0.8756	0.8842	0.8982	0.9173	0.9401	0.9647	0.9876	1.0044	1.0107
60°	1.0472	1.0519	1.0660	1.0896	1.1226	1.1643	1.2126	1.2619	1.3014	1.3170
70°	1.2217	1.2286	1.2495	1.2853	1.3372	1.4068	1.4944	1.5959	1.6918	1.7354
80°	1.3963	1.4056	1.4344	1.4846	1.5597	1.6660	1.8125	2.0119	2.2653	2.4362
90°	1.5708	1.5828	1.6200	1.6858	1.7868	1.9356	2.1565	2.5046	3.1534	∞

31 INCOMPLETE ELLIPTIC INTEGRAL OF THE SECOND KIND

$$E(k,\phi)=\int_0^{\phi}\sqrt{1-k^2\sin^2\theta}\,d\theta,\quad k=\sin\psi$$

ϕ \ ψ	0°	10°	20°	30°	40°	50°	60°	70°	80°	90°
0°	0.0000	0.0000	0.0000	0.0000	0.0000	0.0000	0.0000	0.0000	0.0000	0.0000
10°	0.1745	0.1745	0.1744	0.1743	0.1742	0.1740	0.1739	0.1738	0.1737	0.1736
20°	0.3491	0.3489	0.3483	0.3473	0.3462	0.3450	0.3438	0.3429	0.3422	0.3420
30°	0.5236	0.5229	0.5209	0.5179	0.5141	0.5100	0.5061	0.5029	0.5007	0.5000
40°	0.6981	0.6966	0.6921	0.6851	0.6763	0.6667	0.6575	0.6497	0.6446	0.6428
50°	0.8727	0.8698	0.8614	0.8483	0.8317	0.8134	0.7954	0.7801	0.7697	0.7660
60°	1.0472	1.0426	1.0290	1.0076	0.9801	0.9493	0.9184	0.8914	0.8728	0.8660
70°	1.2217	1.2149	1.1949	1.1632	1.1221	1.0750	1.0266	0.9830	0.9514	0.9397
80°	1.3963	1.3870	1.3597	1.3161	1.2590	1.1926	1.1225	1.0565	1.0054	0.9848
90°	1.5708	1.5589	1.5238	1.4675	1.3931	1.3055	1.2111	1.1184	1.0401	1.0000

Section VI: Financial Tables

COMPOUND AMOUNT: $(1 + r)^n$

If a principal P is deposited at interest rate r (in decimals) compounded annually, then at the end of n years the accumulated amount $A = P(1 + r)^n$.

n \ r	1%	$1\frac{1}{4}$%	$1\frac{1}{2}$%	2%	$2\frac{1}{2}$%	3%	4%	5%	6%
1	1.0100	1.0125	1.0150	1.0200	1.0250	1.0300	1.0400	1.0500	1.0600
2	1.0201	1.0252	1.0302	1.0404	1.0506	1.0609	1.0816	1.1025	1.1236
3	1.0303	1.0380	1.0457	1.0612	1.0769	1.0927	1.1249	1.1576	1.1910
4	1.0406	1.0509	1.0614	1.0824	1.1038	1.1255	1.1699	1.2155	1.2625
5	1.0510	1.0641	1.0773	1.1041	1.1314	1.1593	1.2167	1.2763	1.3382
6	1.0615	1.0774	1.0934	1.1262	1.1597	1.1941	1.2653	1.3401	1.4185
7	1.0721	1.0909	1.1098	1.1487	1.1887	1.2299	1.3159	1.4071	1.5036
8	1.0829	1.1045	1.1265	1.1717	1.2184	1.2668	1.3688	1.4775	1.5938
9	1.0937	1.1183	1.1434	1.1951	1.2489	1.3048	1.4233	1.5513	1.6895
10	1.1046	1.1323	1.1605	1.2190	1.2801	1.3439	1.4802	1.6289	1.7908
11	1.1157	1.1464	1.1779	1.2434	1.3121	1.3842	1.5395	1.7103	1.8983
12	1.1268	1.1608	1.1956	1.2682	1.3449	1.4258	1.6010	1.7959	2.0122
13	1.1381	1.1753	1.2136	1.2936	1.3785	1.4685	1.6651	1.8856	2.1329
14	1.1495	1.1900	1.2318	1.3195	1.4130	1.5126	1.7317	1.9799	2.2609
15	1.1610	1.2048	1.2502	1.3459	1.4483	1.5580	1.8009	2.0789	2.3966
16	1.1726	1.2199	1.2690	1.3728	1.4845	1.6047	1.8730	2.1829	2.5404
17	1.1843	1.2351	1.2880	1.4002	1.5216	1.6528	1.9479	2.2920	2.6928
18	1.1961	1.2506	1.3073	1.4282	1.5597	1.7024	2.0258	2.4066	2.8543
19	1.2081	1.2662	1.3270	1.4568	1.5987	1.7535	2.1068	2.5270	3.0256
20	1.2202	1.2820	1.3469	1.4859	1.6386	1.8061	2.1911	2.6533	3.2071
21	1.2324	1.2981	1.3671	1.5157	1.6796	1.8603	2.2788	2.7860	3.3996
22	1.2447	1.3143	1.3876	1.5460	1.7216	1.9161	2.3699	2.9253	3.6035
23	1.2572	1.3307	1.4084	1.5769	1.7646	1.9736	2.4647	3.0715	3.8197
24	1.2697	1.3474	1.4295	1.6084	1.8087	2.0328	2.5633	3.2251	4.0489
25	1.2824	1.3642	1.4509	1.6406	1.8539	2.0938	2.6658	3.3864	4.2919
26	1.2953	1.3812	1.4727	1.6734	1.9003	2.1566	2.7725	3.5557	4.5494
27	1.3082	1.3985	1.4948	1.7069	1.9478	2.2213	2.8834	3.7335	4.8223
28	1.3213	1.4160	1.5172	1.7410	1.9965	2.2879	2.9987	3.9201	5.1117
29	1.3345	1.4337	1.5400	1.7758	2.0464	2.3566	3.1187	4.1161	5.4184
30	1.3478	1.4516	1.5631	1.8114	2.0976	2.4273	3.2434	4.3219	5.7435
31	1.3613	1.4698	1.5865	1.8476	2.1500	2.5001	3.3731	4.5380	6.0881
32	1.3749	1.4881	1.6103	1.8845	2.2038	2.5751	3.5081	4.7649	6.4534
33	1.3887	1.5067	1.6345	1.9222	2.2589	2.6523	3.6484	5.0032	6.8406
34	1.4026	1.5256	1.6590	1.9607	2.3153	2.7319	3.7943	5.2533	7.2510
35	1.4166	1.5446	1.6839	1.9999	2.3732	2.8139	3.9461	5.5160	7.6861
36	1.4308	1.5639	1.7091	2.0399	2.4325	2.8983	4.1039	5.7918	8.1473
37	1.4451	1.5835	1.7348	2.0807	2.4933	2.9852	4.2681	6.0814	8.6361
38	1.4595	1.6033	1.7608	2.1223	2.5557	3.0748	4.4388	6.3855	9.1543
39	1.4741	1.6233	1.7872	2.1647	2.6196	3.1670	4.6164	6.7048	9.7035
40	1.4889	1.6436	1.8140	2.2080	2.6851	3.2620	4.8010	7.0400	10.2857
41	1.5038	1.6642	1.8412	2.2522	2.7522	3.3599	4.9931	7.3920	10.9029
42	1.5188	1.6850	1.8688	2.2972	2.8210	3.4607	5.1928	7.7616	11.5570
43	1.5340	1.7060	1.8969	2.3432	2.8915	3.5645	5.4005	8.1497	12.2505
44	1.5493	1.7274	1.9253	2.3901	2.9638	3.6715	5.6165	8.5572	12.9855
45	1.5648	1.7489	1.9542	2.4379	3.0379	3.7816	5.8412	8.9850	13.7646
46	1.5805	1.7708	1.9835	2.4866	3.1139	3.8950	6.0748	9.4343	14.5905
47	1.5963	1.7929	2.0133	2.5363	3.1917	4.0119	6.3178	9.9060	15.4659
48	1.6122	1.8154	2.0435	2.5871	3.2715	4.1323	6.5705	10.4013	16.3939
49	1.6283	1.8380	2.0741	2.6388	3.3533	4.2562	6.8333	10.9213	17.3775
50	1.6446	1.8610	2.1052	2.6916	3.4371	4.3839	7.1067	11.4674	18.4202

PRESENT VALUE OF AN AMOUNT: $(1 + r)^{-n}$

The present value P which will amount to A in n years at an interest rate of r (in decimals) compounded annually is $P = A(1 + r)^{-n}$.

n \ r	1%	$1\frac{1}{4}$%	$1\frac{1}{2}$%	2%	$2\frac{1}{2}$%	3%	4%	5%	6%
1	.99010	.98765	.98522	.98039	.97561	.97087	.96154	.95238	.94340
2	.98030	.97546	.97066	.96117	.95181	.94260	.92456	.90703	.89000
3	.97059	.96342	.95632	.94232	.92860	.91514	.88900	.86384	.83962
4	.96098	.95152	.94218	.92385	.90595	.88849	.85480	.82270	.79209
5	.95147	.93978	.92826	.90573	.88385	.86261	.82193	.78353	.74726
6	.94205	.92817	.91454	.88797	.86230	.83748	.79031	.74622	.70496
7	.93272	.91672	.90103	.87056	.84127	.81309	.75992	.71068	.66506
8	.92348	.90540	.88771	.85349	.82075	.78941	.73069	.67684	.62741
9	.91434	.89422	.87459	.83676	.80073	.76642	.70259	.64461	.59190
10	.90529	.88318	.86167	.82035	.78120	.74409	.67556	.61391	.55839
11	.89632	.87228	.84893	.80426	.76214	.72242	.64958	.58468	.52679
12	.88745	.86151	.83639	.78849	.74356	.70138	.62460	.55684	.49697
13	.87866	.85087	.82403	.77303	.72542	.68095	.60057	.53032	.46884
14	.86996	.84037	.81185	.75788	.70773	.66112	.57748	.50507	.44230
15	.86135	.82999	.79985	.74301	.69047	.64186	.55526	.48102	.41727
16	.85282	.81975	.78803	.72845	.67362	.62317	.53391	.45811	.39365
17	.84438	.80963	.77639	.71416	.65720	.60502	.51337	.43630	.37136
18	.83602	.79963	.76491	.70016	.64117	.58739	.49363	.41552	.35034
19	.82774	.78976	.75361	.68643	.62553	.57029	.47464	.39573	.33051
20	.81954	.78001	.74247	.67297	.61027	.55368	.45639	.37689	.31180
21	.81143	.77038	.73150	.65978	.59539	.53755	.43883	.35894	.29416
22	.80340	.76087	.72069	.64684	.58086	.52189	.42196	.34185	.27751
23	.79544	.75147	.71004	.63416	.56670	.50669	.40573	.32557	.26180
24	.78757	.74220	.69954	.62172	.55288	.49193	.39012	.31007	.24698
25	.77977	.73303	.68921	.60953	.53939	.47761	.37512	.29530	.23300
26	.77205	.72398	.67902	.59758	.52623	.46369	.36069	.28124	.21981
27	.76440	.71505	.66899	.58586	.51340	.45019	.34682	.26785	.20737
28	.75684	.70622	.65910	.57437	.50088	.43708	.33348	.25509	.19563
29	.74934	.69750	.64936	.56311	.48866	.42435	.32065	.24295	.18456
30	.74192	.68889	.63976	.55207	.47674	.41199	.30832	.23138	.17411
31	.73458	.68038	.63031	.54125	.46511	.39999	.29646	.22036	.16425
32	.72730	.67198	.62099	.53063	.45377	.38834	.28506	.20987	.15496
33	.72010	.66369	.61182	.52023	.44270	.37703	.27409	.19987	.14619
34	.71297	.65549	.60277	.51003	.43191	.36604	.26355	.19035	.13791
35	.70591	.64740	.59387	.50003	.42137	.35538	.25342	.18129	.13011
36	.69892	.63941	.58509	.49022	.41109	.34503	.24367	.17266	.12274
37	.69200	.63152	.57644	.48061	.40107	.33498	.23430	.16444	.11579
38	.68515	.62372	.56792	.47119	.39128	.32523	.22529	.15661	.10924
39	.67837	.61602	.55953	.46195	.38174	.31575	.21662	.14915	.10306
40	.67165	.60841	.55126	.45289	.37243	.30656	.20829	.14205	.09722
41	.66500	.60090	.54312	.44401	.36335	.29763	.20028	.13528	.09172
42	.65842	.59348	.53509	.43530	.35448	.28896	.19257	.12884	.08653
43	.65190	.58616	.52718	.42677	.34584	.28054	.18517	.12270	.08163
44	.64545	.57892	.51939	.41840	.33740	.27237	.17805	.11686	.07701
45	.63905	.57177	.51171	.41020	.32917	.26444	.17120	.11130	.07265
46	.63273	.56471	.50415	.40215	.32115	.25674	.16461	.10600	.06854
47	.62646	.55774	.49670	.39427	.31331	.24926	.15828	.10095	.06466
48	.62026	.55086	.48936	.38654	.30567	.24200	.15219	.09614	.06100
49	.61412	.54406	.48213	.37896	.29822	.23495	.14634	.09156	.05755
50	.60804	.53734	.47500	.37153	.29094	.22811	.14071	.08720	.05429

AMOUNT OF AN ANNUITY: $\dfrac{(1+r)^n-1}{r}$

If a principal P is deposited at the end of each year at interest rate r (in decimals) compounded annually, then at the end of n years the accumulated amount is $P\left[\dfrac{(1-r)^n-1}{r}\right]$. The process is often called an *annuity*.

n \ r	1%	$1\frac{1}{4}$%	$1\frac{1}{2}$%	2%	$2\frac{1}{2}$%	3%	4%	5%	6%
1	1.0000	1.0000	1.0000	1.0000	1.0000	1.0000	1.0000	1.0000	1.0000
2	2.0100	2.0125	2.0150	2.0200	2.0250	2.0300	2.0400	2.0500	2.0600
3	3.0301	3.0377	3.0452	3.0604	3.0756	3.0909	3.1216	3.1525	3.1836
4	4.0604	4.0756	4.0909	4.1216	4.1525	4.1836	4.2465	4.3101	4.3746
5	5.1010	5.1266	5.1523	5.2040	5.2563	5.3091	5.4163	5.5256	5.6371
6	6.1520	6.1907	6.2296	6.3081	6.3877	6.4684	6.6330	6.8019	6.9753
7	7.2135	7.2680	7.3230	7.4343	7.5474	7.6625	7.8983	8.1420	8.3938
8	8.2857	8.3589	8.4328	8.5830	8.7361	8.8923	9.2142	9.5491	9.8975
9	9.3685	9.4634	9.5593	9.7546	9.9545	10.1591	10.5828	11.0266	11.4913
10	10.4622	10.5817	10.7027	10.9497	11.2034	11.4639	12.0061	12.5779	13.1808
11	11.5668	11.7139	11.8633	12.1687	12.4835	12.8078	13.4864	14.2068	14.9716
12	12.6825	12.8604	13.0412	13.4121	13.7956	14.1920	15.0258	15.9171	16.8699
13	13.8093	14.0211	14.2368	14.6803	15.1404	15.6178	16.6268	17.7130	18.8821
14	14.9474	15.1964	15.4504	15.9739	16.5190	17.0863	18.2919	19.5986	21.0151
15	16.0969	16.3863	16.6821	17.2934	17.9319	18.5989	20.0236	21.5786	23.2760
16	17.2579	17.5912	17.9324	18.6393	19.3802	20.1569	21.8245	23.6575	25.6725
17	18.4304	18.8111	19.2014	20.0121	20.8647	21.7616	23.6975	25.8404	28.2129
18	19.6147	20.0462	20.4894	21.4123	22.3863	23.4144	25.6454	28.1324	30.9057
19	20.8109	21.2968	21.7967	22.8406	23.9460	25.1169	27.6712	30.5390	33.7600
20	22.0190	22.5630	23.1237	24.2974	25.5447	26.8704	29.7781	33.0660	36.7856
21	23.2392	23.8450	24.4705	25.7833	27.1833	28.6765	31.9692	35.7193	39.9927
22	24.4716	25.1431	25.8376	27.2990	28.8629	30.5368	34.2480	38.5052	43.3923
23	25.7163	26.4574	27.2251	28.8450	30.5844	32.4529	36.6179	41.4305	46.9958
24	26.9735	27.7881	28.6335	30.4219	32.3490	34.4265	39.0826	44.5020	50.8156
25	28.2432	29.1354	30.0630	32.0303	34.1578	36.4593	41.6459	47.7271	54.8645
26	29.5256	30.4996	31.5140	33.6709	36.0117	38.5530	44.3117	51.1135	59.1564
27	30.8209	31.8809	32.9867	35.3443	37.9120	40.7096	47.0842	54.6691	63.7058
28	32.1291	33.2794	34.4815	37.0512	39.8598	42.9309	49.9676	58.4026	68.5281
29	33.4504	34.6954	35.9987	38.7922	41.8563	45.2189	52.9663	62.3227	73.6398
30	34.7849	36.1291	37.5387	40.5681	43.9027	47.5754	56.0849	66.4388	79.0582
31	36.1327	37.5807	39.1018	42.3794	46.0003	50.0027	59.3283	70.7608	84.8017
32	37.4941	39.0504	40.6883	44.2270	48.1503	52.5028	62.7015	75.2988	90.8898
33	38.8690	40.5386	42.2986	46.1116	50.3540	55.0778	66.2095	80.0638	97.3432
34	40.2577	42.0453	43.9331	48.0338	52.6129	57.7302	69.8579	85.0670	104.1838
35	41.6603	43.5709	45.5921	49.9945	54.9282	60.4621	73.6522	90.3203	111.4348
36	43.0769	45.1155	47.2760	51.9944	57.3014	63.2759	77.5983	95.8363	119.1209
37	44.5076	46.6794	48.9851	54.0343	59.7339	66.1742	81.7022	101.6281	127.2681
38	45.9527	48.2629	50.7199	56.1149	62.2273	69.1594	85.9703	107.7095	135.9042
39	47.4123	49.8662	52.4807	58.2372	64.7830	72.2342	90.4091	114.0950	145.0585
40	48.8864	51.4896	54.2679	60.4020	67.4026	75.4013	95.0255	120.7998	154.7620
41	50.3752	53.1332	56.0819	62.6100	70.0876	78.6633	99.8265	127.8398	165.0477
42	51.8790	54.7973	57.9231	64.8622	72.8398	82.0232	104.8196	135.2318	175.9505
43	53.3978	56.4823	59.7920	67.1595	75.6608	85.4839	110.0124	142.9933	187.5076
44	54.9318	58.1883	61.6889	69.5027	78.5523	89.0484	115.4129	151.1430	199.7580
45	56.4811	59.9157	63.6142	71.8927	81.5161	92.7199	121.0294	159.7002	212.7435
46	58.0459	61.6646	65.5684	74.3306	84.5540	96.5015	126.8706	168.6852	226.5081
47	59.6263	63.4354	67.5519	76.8172	87.6679	100.3965	132.9454	178.1194	241.0986
48	61.2226	65.2284	69.5652	79.3535	90.8596	104.4084	139.2632	188.0254	256.5645
49	62.8348	67.0437	71.6087	81.9406	94.1311	108.5406	145.8337	198.4267	272.9584
50	64.4632	68.8818	73.6828	84.5794	97.4843	112.7969	152.6671	209.3480	290.3359

35

PRESENT VALUE OF AN ANNUITY: $\dfrac{1-(1+r)^{-n}}{r}$

An annuity in which the yearly payment at the end of each of n years is A at an interest rate r (in decimals) compounded annually has present value $A\left[\dfrac{1-(1+r)^{-n}}{r}\right]$.

n \ r	1%	$1\frac{1}{4}$%	$1\frac{1}{2}$%	2%	$2\frac{1}{2}$%	3%	4%	5%	6%
1	0.9901	0.9877	0.9852	0.9804	0.9756	0.9709	0.9615	0.9524	0.9434
2	1.9704	1.9631	1.9559	1.9416	1.9274	1.9135	1.8861	1.8594	1.8334
3	2.9410	2.9265	2.9122	2.8839	2.8560	2.8286	2.7751	2.7232	2.6730
4	3.9020	3.8781	3.8544	3.8077	3.7620	3.7171	3.6299	3.5460	3.4651
5	4.8534	4.8178	4.7826	4.7135	4.6458	4.5797	4.4518	4.3295	4.2124
6	5.7955	5.7460	5.6972	5.6014	5.5081	5.4172	5.2421	5.0757	4.9173
7	6.7282	6.6627	6.5982	6.4720	6.3494	6.2303	6.0021	5.7864	5.5824
8	7.6517	7.5681	7.4859	7.3255	7.1701	7.0197	6.7327	6.4632	6.2098
9	8.5660	8.4623	8.3605	8.1622	7.9709	7.7861	7.4353	7.1078	6.8017
10	9.4713	9.3455	9.2222	8.9826	8.7521	8.5302	8.1109	7.7217	7.3601
11	10.3676	10.2178	10.0711	9.7868	9.5142	9.2526	8.7605	8.3064	7.8869
12	11.2551	11.0793	10.9075	10.5753	10.2578	9.9540	9.3851	8.8633	8.3838
13	12.1337	11.9302	11.7315	11.3484	10.9832	10.6350	9.9856	9.3936	8.8527
14	13.0037	12.7706	12.5434	12.1062	11.6909	11.2961	10.5631	9.8986	9.2950
15	13.8651	13.6005	13.3432	12.8493	12.3814	11.9379	11.1184	10.3797	9.7122
16	14.7179	14.4203	14.1313	13.5777	13.0550	12.5611	11.6523	10.8378	10.1059
17	15.5623	15.2299	14.9076	14.2919	13.7122	13.1661	12.1657	11.2741	10.4773
18	16.3983	16.0295	15.6726	14.9920	14.3534	13.7535	12.6593	11.6896	10.8276
19	17.2260	16.8193	16.4262	15.6785	14.9789	14.3238	13.1339	12.0853	11.1581
20	18.0456	17.5993	17.1686	16.3514	15.5892	14.8775	13.5903	12.4622	11.4699
21	18.8570	18.3697	17.9001	17.0112	16.1845	15.4150	14.0292	12.8212	11.7641
22	19.6604	19.1306	18.6208	17.6580	16.7654	15.9369	14.4511	13.1630	12.0416
23	20.4558	19.8820	19.3309	18.2922	17.3321	16.4436	14.8568	13.4886	12.3034
24	21.2434	20.6242	20.0304	18.9139	17.8850	16.9355	15.2470	13.7986	12.5504
25	22.0232	21.3573	20.7196	19.5235	18.4244	17.4131	15.6221	14.0939	12.7834
26	22.7952	22.0813	21.3986	20.1210	18.9506	17.8768	15.9828	14.3752	13.0032
27	23.5596	22.7963	22.0676	20.7069	19.4640	18.3270	16.3296	14.6430	13.2105
28	24.3164	23.5025	22.7267	21.2813	19.9649	18.7641	16.6631	14.8981	13.4062
29	25.0658	24.2000	23.3761	21.8444	20.4535	19.1885	16.9837	15.1411	13.5907
30	25.8077	24.8889	24.0158	22.3965	20.9303	19.6004	17.2920	15.3725	13.7648
31	26.5423	25.5693	24.6461	22.9377	21.3954	20.0004	17.5885	15.5928	13.9291
32	27.2696	26.2413	25.2671	23.4683	21.8492	20.3888	17.8736	15.8027	14.0840
33	27.9897	26.9050	25.8790	23.9886	22.2919	20.7658	18.1476	16.0025	14.2302
34	28.7027	27.5605	26.4817	24.4986	22.7238	21.1318	18.4112	16.1929	14.3681
35	29.4086	28.2079	27.0756	24.9986	23.1452	21.4872	18.6646	16.3742	14.4982
36	30.1075	28.8473	27.6607	25.4888	23.5563	21.8323	18.9083	16.5469	14.6210
37	30.7995	29.4788	28.2371	25.9695	23.9573	22.1672	19.1426	16.7113	14.7368
38	31.4847	30.1025	28.8051	26.4406	24.3486	22.4925	19.3679	16.8679	14.8460
39	32.1630	30.7185	29.3646	26.9026	24.7303	22.8082	19.5845	17.0170	14.9491
40	32.8347	31.3269	29.9158	27.3555	25.1028	23.1148	19.7928	17.1591	15.0463
41	33.4997	31.9278	30.4590	27.7995	25.4661	23.4124	19.9931	17.2944	15.1380
42	34.1581	32.5213	30.9941	28.2348	25.8206	23.7014	20.1856	17.4232	15.2245
43	34.8100	33.1075	31.5212	28.6616	26.1664	23.9819	20.3708	17.5459	15.3062
44	35.4555	33.6864	32.0406	29.0800	26.5038	24.2543	20.5488	17.6628	15.3832
45	36.0945	34.2582	32.5523	29.4902	26.8330	24.5187	20.7200	17.7741	15.4558
46	36.7272	34.8229	33.0565	29.8923	27.1542	24.7754	20.8847	17.8801	15.5244
47	37.3537	35.3806	33.5532	30.2866	27.4675	25.0247	21.0429	17.9810	15.5890
48	37.9740	35.9315	34.0426	30.6731	27.7732	25.2667	21.1951	18.0772	15.6500
49	38.5881	36.4755	34.5247	31.0521	28.0714	25.5017	21.3415	18.1687	15.7076
50	39.1961	37.0129	34.9997	31.4236	28.3623	25.7298	21.4822	18.2559	15.7619

Section VII: Probability and Statistics

AREAS UNDER THE STANDARD NORMAL CURVE

from $-\infty$ to x

$$\Phi(x) = \frac{1}{\sqrt{2\pi}} \int_{-\infty}^{x} e^{-t^2/2}\,dt$$

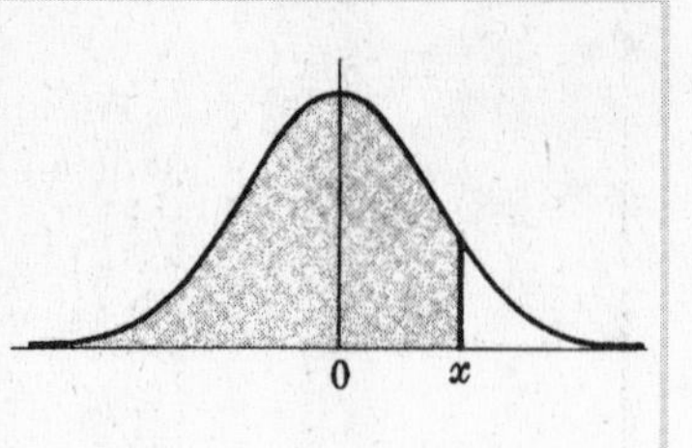

NOTE: $\operatorname{erf}(x) = 2\Phi(x\sqrt{2}) - 1$

x	0	1	2	3	4	5	6	7	8	9
0.0	.5000	.5040	.5080	.5120	.5160	.5199	.5239	.5279	.5319	.5359
0.1	.5398	.5438	.5478	.5517	.5557	.5596	.5636	.5675	.5714	.5754
0.2	.5793	.5832	.5871	.5910	.5948	.5987	.6026	.6064	.6103	.6141
0.3	.6179	.6217	.6255	.6293	.6331	.6368	.6406	.6443	.6480	.6517
0.4	.6554	.6591	.6628	.6664	.6700	.6736	.6772	.6808	.6844	.6879
0.5	.6915	.6950	.6985	.7019	.7054	.7088	.7123	.7157	.7190	.7224
0.6	.7258	.7291	.7324	.7357	.7389	.7422	.7454	.7486	.7518	.7549
0.7	.7580	.7612	.7642	.7673	.7704	.7734	.7764	.7794	.7823	.7852
0.8	.7881	.7910	.7939	.7967	.7996	.8023	.8051	.8078	.8106	.8133
0.9	.8159	.8186	.8212	.8238	.8264	.8289	.8315	.8340	.8365	.8389
1.0	.8413	.8438	.8461	.8485	.8508	.8531	.8554	.8577	.8599	.8621
1.1	.8643	.8665	.8686	.8708	.8729	.8749	.8770	.8790	.8810	.8830
1.2	.8849	.8869	.8888	.8907	.8925	.8944	.8962	.8980	.8997	.9015
1.3	.9032	.9049	.9066	.9082	.9099	.9115	.9131	.9147	.9162	.9177
1.4	.9192	.9207	.9222	.9236	.9251	.9265	.9279	.9292	.9306	.9319
1.5	.9332	.9345	.9357	.9370	.9382	.9394	.9406	.9418	.9429	.9441
1.6	.9452	.9463	.9474	.9484	.9495	.9505	.9515	.9525	.9535	.9545
1.7	.9554	.9564	.9573	.9582	.9591	.9599	.9608	.9616	.9625	.9633
1.8	.9641	.9649	.9656	.9664	.9671	.9678	.9686	.9693	.9699	.9706
1.9	.9713	.9719	.9726	.9732	.9738	.9744	.9750	.9756	.9761	.9767
2.0	.9772	.9778	.9783	.9788	.9793	.9798	.9803	.9808	.9812	.9817
2.1	.9821	.9826	.9830	.9834	.9838	.9842	.9846	.9850	.9854	.9857
2.2	.9861	.9864	.9868	.9871	.9875	.9878	.9881	.9884	.9887	.9890
2.3	.9893	.9896	.9898	.9901	.9904	.9906	.9909	.9911	.9913	.9916
2.4	.9918	.9920	.9922	.9925	.9927	.9929	.9931	.9932	.9934	.9936
2.5	.9938	.9940	.9941	.9943	.9945	.9946	.9948	.9949	.9951	.9952
2.6	.9953	.9955	.9956	.9957	.9959	.9960	.9961	.9962	.9963	.9964
2.7	.9965	.9966	.9967	.9968	.9969	.9970	.9971	.9972	.9973	.9974
2.8	.9974	.9975	.9976	.9977	.9977	.9978	.9979	.9979	.9980	.9981
2.9	.9981	.9982	.9982	.9983	.9984	.9984	.9985	.9985	.9986	.9986
3.0	.9987	.9987	.9987	.9988	.9988	.9989	.9989	.9989	.9990	.9990
3.1	.9990	.9991	.9991	.9991	.9992	.9992	.9992	.9992	.9993	.9993
3.2	.9993	.9993	.9994	.9994	.9994	.9994	.9994	.9995	.9995	.9995
3.3	.9995	.9995	.9995	.9996	.9996	.9996	.9996	.9996	.9996	.9997
3.4	.9997	.9997	.9997	.9997	.9997	.9997	.9997	.9997	.9997	.9998
3.5	.9998	.9998	.9998	.9998	.9998	.9998	.9998	.9998	.9998	.9998
3.6	.9998	.9998	.9999	.9999	.9999	.9999	.9999	.9999	.9999	.9999
3.7	.9999	.9999	.9999	.9999	.9999	.9999	.9999	.9999	.9999	.9999
3.8	.9999	.9999	.9999	.9999	.9999	.9999	.9999	.9999	.9999	.9999
3.9	1.0000	1.0000	1.0000	1.0000	1.0000	1.0000	1.0000	1.0000	1.0000	1.0000

37 ORDINATES OF THE STANDARD NORMAL CURVE

$$y = \frac{1}{\sqrt{2\pi}} e^{-x^2/2}$$

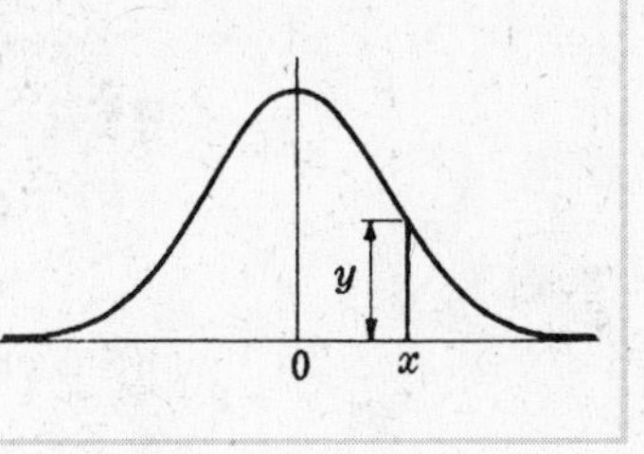

x	0	1	2	3	4	5	6	7	8	9
0.0	.3989	.3989	.3989	.3988	.3986	.3984	.3982	.3980	.3977	.3973
0.1	.3970	.3965	.3961	.3956	.3951	.3945	.3939	.3932	.3925	.3918
0.2	.3910	.3902	.3894	.3885	.3876	.3867	.3857	.3847	.3836	.3825
0.3	.3814	.3802	.3790	.3778	.3765	.3752	.3739	.3725	.3712	.3697
0.4	.3683	.3668	.3653	.3637	.3621	.3605	.3589	.3572	.3555	.3538
0.5	.3521	.3503	.3485	.3467	.3448	.3429	.3410	.3391	.3372	.3352
0.6	.3332	.3312	.3292	.3271	.3251	.3230	.3209	.3187	.3166	.3144
0.7	.3123	.3101	.3079	.3056	.3034	.3011	.2989	.2966	.2943	.2920
0.8	.2897	.2874	.2850	.2827	.2803	.2780	.2756	.2732	.2709	.2685
0.9	.2661	.2637	.2613	.2589	.2565	.2541	.2516	.2492	.2468	.2444
1.0	.2420	.2396	.2371	.2347	.2323	.2299	.2275	.2251	.2227	.2203
1.1	.2179	.2155	.2131	.2107	.2083	.2059	.2036	.2012	.1989	.1965
1.2	.1942	.1919	.1895	.1872	.1849	.1826	.1804	.1781	.1758	.1736
1.3	.1714	.1691	.1669	.1647	.1626	.1604	.1582	.1561	.1539	.1518
1.4	.1497	.1476	.1456	.1435	.1415	.1394	.1374	.1354	.1334	.1315
1.5	.1295	.1276	.1257	.1238	.1219	.1200	.1182	.1163	.1145	.1127
1.6	.1109	.1092	.1074	.1057	.1040	.1023	.1006	.0989	.0973	.0957
1.7	.0940	.0925	.0909	.0893	.0878	.0863	.0848	.0833	.0818	.0804
1.8	.0790	.0775	.0761	.0748	.0734	.0721	.0707	.0694	.0681	.0669
1.9	.0656	.0644	.0632	.0620	.0608	.0596	.0584	.0573	.0562	.0551
2.0	.0540	.0529	.0519	.0508	.0498	.0488	.0478	.0468	.0459	.0449
2.1	.0440	.0431	.0422	.0413	.0404	.0396	.0387	.0379	.0371	.0363
2.2	.0355	.0347	.0339	.0332	.0325	.0317	.0310	.0303	.0297	.0290
2.3	.0283	.0277	.0270	.0264	.0258	.0252	.0246	.0241	.0235	.0229
2.4	.0224	.0219	.0213	.0208	.0203	.0198	.0194	.0189	.0184	.0180
2.5	.0175	.0171	.0167	.0163	.0158	.0154	.0151	.0147	.0143	.0139
2.6	.0136	.0132	.0129	.0126	.0122	.0119	.0116	.0113	.0110	.0107
2.7	.0104	.0101	.0099	.0096	.0093	.0091	.0088	.0086	.0084	.0081
2.8	.0079	.0077	.0075	.0073	.0071	.0069	.0067	.0065	.0063	.0061
2.9	.0060	.0058	.0056	.0055	.0053	.0051	.0050	.0048	.0047	.0046
3.0	.0044	.0043	.0042	.0040	.0039	.0038	.0037	.0036	.0035	.0034
3.1	.0033	.0032	.0031	.0030	.0029	.0028	.0027	.0026	.0025	.0025
3.2	.0024	.0023	.0022	.0022	.0021	.0020	.0020	.0019	.0018	.0018
3.3	.0017	.0017	.0016	.0016	.0015	.0015	.0014	.0014	.0013	.0013
3.4	.0012	.0012	.0012	.0011	.0011	.0010	.0010	.0010	.0009	.0009
3.5	.0009	.0008	.0008	.0008	.0008	.0007	.0007	.0007	.0007	.0006
3.6	.0006	.0006	.0006	.0005	.0005	.0005	.0005	.0005	.0005	.0004
3.7	.0004	.0004	.0004	.0004	.0004	.0004	.0003	.0003	.0003	.0003
3.8	.0003	.0003	.0003	.0003	.0003	.0002	.0002	.0002	.0002	.0002
3.9	.0002	.0002	.0002	.0002	.0002	.0002	.0002	.0002	.0001	.0001

38 PERCENTILE VALUES (t_p) FOR STUDENT'S t DISTRIBUTION

with n degrees of freedom (shaded area = p)

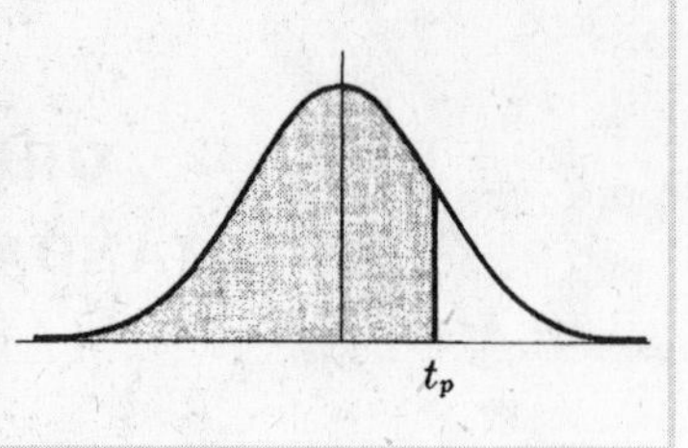

n	$t_{.995}$	$t_{.99}$	$t_{.975}$	$t_{.95}$	$t_{.90}$	$t_{.80}$	$t_{.75}$	$t_{.70}$	$t_{.60}$	$t_{.55}$
1	63.66	31.82	12.71	6.31	3.08	1.376	1.000	.727	.325	.158
2	9.92	6.96	4.30	2.92	1.89	1.061	.816	.617	.289	.142
3	5.84	4.54	3.18	2.35	1.64	.978	.765	.584	.277	.137
4	4.60	3.75	2.78	2.13	1.53	.941	.741	.569	.271	.134
5	4.03	3.36	2.57	2.02	1.48	.920	.727	.559	.267	.132
6	3.71	3.14	2.45	1.94	1.44	.906	.718	.553	.265	.131
7	3.50	3.00	2.36	1.90	1.42	.896	.711	.549	.263	.130
8	3.36	2.90	2.31	1.86	1.40	.889	.706	.546	.262	.130
9	3.25	2.82	2.26	1.83	1.38	.883	.703	.543	.261	.129
10	3.17	2.76	2.23	1.81	1.37	.879	.700	.542	.260	.129
11	3.11	2.72	2.20	1.80	1.36	.876	.697	.540	.260	.129
12	3.06	2.68	2.18	1.78	1.36	.873	.695	.539	.259	.128
13	3.01	2.65	2.16	1.77	1.35	.870	.694	.538	.259	.128
14	2.98	2.62	2.14	1.76	1.34	.868	.692	.537	.258	.128
15	2.95	2.60	2.13	1.75	1.34	.866	.691	.536	.258	.128
16	2.92	2.58	2.12	1.75	1.34	.865	.690	.535	.258	.128
17	2.90	2.57	2.11	1.74	1.33	.863	.689	.534	.257	.128
18	2.88	2.55	2.10	1.73	1.33	.862	.688	.534	.257	.127
19	2.86	2.54	2.09	1.73	1.33	.861	.688	.533	.257	.127
20	2.84	2.53	2.09	1.72	1.32	.860	.687	.533	.257	.127
21	2.83	2.52	2.08	1.72	1.32	.859	.686	.532	.257	.127
22	2.82	2.51	2.07	1.72	1.32	.858	.686	.532	.256	.127
23	2.81	2.50	2.07	1.71	1.32	.858	.685	.532	.256	.127
24	2.80	2.49	2.06	1.71	1.32	.857	.685	.531	.256	.127
25	2.79	2.48	2.06	1.71	1.32	.856	.684	.531	.256	.127
26	2.78	2.48	2.06	1.71	1.32	.856	.684	.531	.256	.127
27	2.77	2.47	2.05	1.70	1.31	.855	.684	.531	.256	.127
28	2.76	2.47	2.05	1.70	1.31	.855	.683	.530	.256	.127
29	2.76	2.46	2.04	1.70	1.31	.854	.683	.530	.256	.127
30	2.75	2.46	2.04	1.70	1.31	.854	.683	.530	.256	.127
40	2.70	2.42	2.02	1.68	1.30	.851	.681	.529	.255	.126
60	2.66	2.39	2.00	1.67	1.30	.848	.679	.527	.254	.126
120	2.62	2.36	1.98	1.66	1.29	.845	.677	.526	.254	.126
∞	2.58	2.33	1.96	1.645	1.28	.842	.674	.524	.253	.126

Source: R. A. Fisher and F. Yates, *Statistical Tables for Biological, Agricultural and Medical Research* (6th edition, 1963), Table III, Oliver and Boyd Ltd., Edinburgh, by permission of the authors and publishers.

39 PERCENTILE VALUES (χ^2_p) FOR χ^2 (CHI-SQUARE) DISTRIBUTION

with n degrees of freedom (shaded area = p)

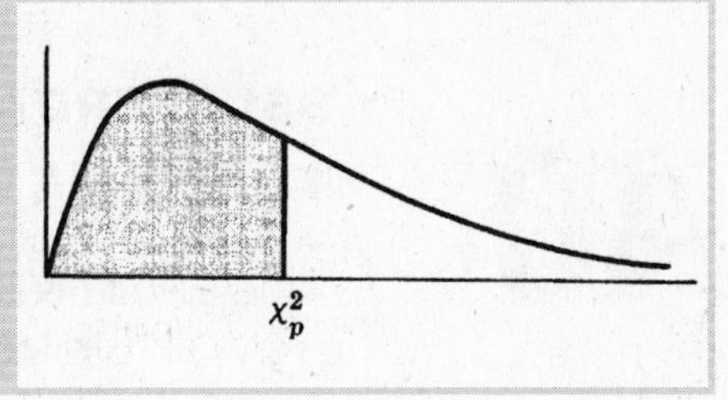

n	$\chi^2_{.995}$	$\chi^2_{.99}$	$\chi^2_{.975}$	$\chi^2_{.95}$	$\chi^2_{.90}$	$\chi^2_{.75}$	$\chi^2_{.50}$	$\chi^2_{.25}$	$\chi^2_{.10}$	$\chi^2_{.05}$	$\chi^2_{.025}$	$\chi^2_{.01}$	$\chi^2_{.005}$
1	7.88	6.63	5.02	3.84	2.71	1.32	.455	.102	.0158	.0039	.0010	.0002	.0000
2	10.6	9.21	7.38	5.99	4.61	2.77	1.39	.575	.211	.103	.0506	.0201	.0100
3	12.8	11.3	9.35	7.81	6.25	4.11	2.37	1.21	.584	.352	.216	.115	.072
4	14.9	13.3	11.1	9.49	7.78	5.39	3.36	1.92	1.06	.711	.484	.297	.207
5	16.7	15.1	12.8	11.1	9.24	6.63	4.35	2.67	1.61	1.15	.831	.554	.412
6	18.5	16.8	14.4	12.6	10.6	7.84	5.35	3.45	2.20	1.64	1.24	.872	.676
7	20.3	18.5	16.0	14.1	12.0	9.04	6.35	4.25	2.83	2.17	1.69	1.24	.989
8	22.0	20.1	17.5	15.5	13.4	10.2	7.34	5.07	3.49	2.73	2.18	1.65	1.34
9	23.6	21.7	19.0	16.9	14.7	11.4	8.34	5.90	4.17	3.33	2.70	2.09	1.73
10	25.2	23.2	20.5	18.3	16.0	12.5	9.34	6.74	4.87	3.94	3.25	2.56	2.16
11	26.8	24.7	21.9	19.7	17.3	13.7	10.3	7.58	5.58	4.57	3.82	3.05	2.60
12	28.3	26.2	23.3	21.0	18.5	14.8	11.3	8.44	6.30	5.23	4.40	3.57	3.07
13	29.8	27.7	24.7	22.4	19.8	16.0	12.3	9.30	7.04	5.89	5.01	4.11	3.57
14	31.3	29.1	26.1	23.7	21.1	17.1	13.3	10.2	7.79	6.57	5.63	4.66	4.07
15	32.8	30.6	27.5	25.0	22.3	18.2	14.3	11.0	8.55	7.26	6.26	5.23	4.60
16	34.3	32.0	28.8	26.3	23.5	19.4	15.3	11.9	9.31	7.96	6.91	5.81	5.14
17	35.7	33.4	30.2	27.6	24.8	20.5	16.3	12.8	10.1	8.67	7.56	6.41	5.70
18	37.2	34.8	31.5	28.9	26.0	21.6	17.3	13.7	10.9	9.39	8.23	7.01	6.26
19	38.6	36.2	32.9	30.1	27.2	22.7	18.3	14.6	11.7	10.1	8.91	7.63	6.84
20	40.0	37.6	34.2	31.4	28.4	23.8	19.3	15.5	12.4	10.9	9.59	8.26	7.43
21	41.4	38.9	35.5	32.7	29.6	24.9	20.3	16.3	13.2	11.6	10.3	8.90	8.03
22	42.8	40.3	36.8	33.9	30.8	26.0	21.3	17.2	14.0	12.3	11.0	9.54	8.64
23	44.2	41.6	38.1	35.2	32.0	27.1	22.3	18.1	14.8	13.1	11.7	10.2	9.26
24	45.6	43.0	39.4	36.4	33.2	28.2	23.3	19.0	15.7	13.8	12.4	10.9	9.89
25	46.9	44.3	40.6	37.7	34.4	29.3	24.3	19.9	16.5	14.6	13.1	11.5	10.5
26	48.3	45.6	41.9	38.9	35.6	30.4	25.3	20.8	17.3	15.4	13.8	12.2	11.2
27	49.6	47.0	43.2	40.1	36.7	31.5	26.3	21.7	18.1	16.2	14.6	12.9	11.8
28	51.0	48.3	44.5	41.3	37.9	32.6	27.3	22.7	18.9	16.9	15.3	13.6	12.5
29	52.3	49.6	45.7	42.6	39.1	33.7	28.3	23.6	19.8	17.7	16.0	14.3	13.1
30	53.7	50.9	47.0	43.8	40.3	34.8	29.3	24.5	20.6	18.5	16.8	15.0	13.8
40	66.8	63.7	59.3	55.8	51.8	45.6	39.3	33.7	29.1	26.5	24.4	22.2	20.7
50	79.5	76.2	71.4	67.5	63.2	56.3	49.3	42.9	37.7	34.8	32.4	29.7	28.0
60	92.0	88.4	83.3	79.1	74.4	67.0	59.3	52.3	46.5	43.2	40.5	37.5	35.5
70	104.2	100.4	95.0	90.5	85.5	77.6	69.3	61.7	55.3	51.7	48.8	45.4	43.3
80	116.3	112.3	106.6	101.9	96.6	88.1	79.3	71.1	64.3	60.4	57.2	53.5	51.2
90	128.3	124.1	118.1	113.1	107.6	98.6	89.3	80.6	73.3	69.1	65.6	61.8	59.2
100	140.2	135.8	129.6	124.3	118.5	109.1	99.3	90.1	82.4	77.9	74.2	70.1	67.3

Source: Catherine M. Thompson, *Table of percentage points of the χ^2 distribution*, Biometrika, Vol. 32 (1941), by permission of the author and publisher.

40 95th PERCENTILE VALUES FOR THE *F* DISTRIBUTION

n_1 = degrees of freedom for numerator
n_2 = degrees of freedom for denominator
(shaded area = .95)

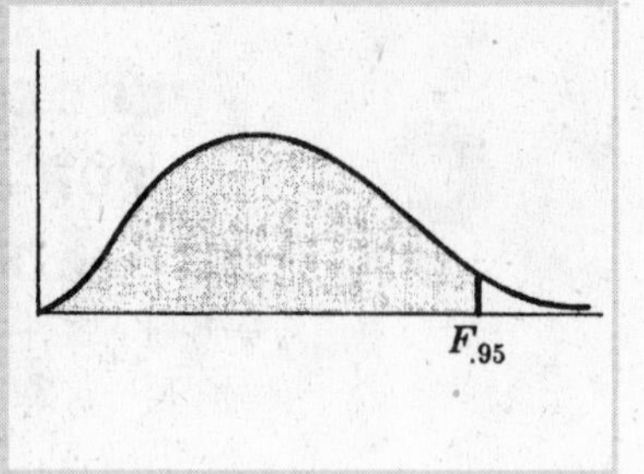

n_2 \ n_1	1	2	3	4	5	6	8	12	16	20	30	40	50	100	∞
1	161.4	199.5	215.7	224.6	230.2	234.0	238.9	243.9	246.3	248.0	250.1	251.1	252.2	253.0	254.3
2	18.51	19.00	19.16	19.25	19.30	19.33	19.37	19.41	19.43	19.45	19.46	19.46	19.47	19.49	19.50
3	10.13	9.55	9.28	9.12	9.01	8.94	8.85	8.74	8.69	8.66	8.62	8.60	8.58	8.56	8.53
4	7.71	6.94	6.59	6.39	6.26	6.16	6.04	5.91	5.84	5.80	5.75	5.71	5.70	5.66	5.63
5	6.61	5.79	5.41	5.19	5.05	4.95	4.82	4.68	4.60	4.56	4.50	4.46	4.44	4.40	4.36
6	5.99	5.14	4.76	4.53	4.39	4.28	4.15	4.00	3.92	3.87	3.81	3.77	3.75	3.71	3.67
7	5.59	4.74	4.35	4.12	3.97	3.87	3.73	3.57	3.49	3.44	3.38	3.34	3.32	3.28	3.23
8	5.32	4.46	4.07	3.84	3.69	3.58	3.44	3.28	3.20	3.15	3.08	3.05	3.03	2.98	2.93
9	5.12	4.26	3.86	3.63	3.48	3.37	3.23	3.07	2.98	2.93	2.86	2.82	2.80	2.76	2.71
10	4.96	4.10	3.71	3.48	3.33	3.22	3.07	2.91	2.82	2.77	2.70	2.67	2.64	2.59	2.54
11	4.84	3.98	3.59	3.36	3.20	3.09	2.95	2.79	2.70	2.65	2.57	2.53	2.50	2.45	2.40
12	4.75	3.89	3.49	3.26	3.11	3.00	2.85	2.69	2.60	2.54	2.46	2.42	2.40	2.35	2.30
13	4.67	3.81	3.41	3.18	3.03	2.92	2.77	2.60	2.51	2.46	2.38	2.34	2.32	2.26	2.21
14	4.60	3.74	3.34	3.11	2.96	2.85	2.70	2.53	2.44	2.39	2.31	2.27	2.24	2.19	2.13
15	4.54	3.68	3.29	3.06	2.90	2.79	2.64	2.48	2.39	2.33	2.25	2.21	2.18	2.12	2.07
16	4.49	3.63	3.24	3.01	2.85	2.74	2.59	2.42	2.33	2.28	2.20	2.16	2.13	2.07	2.01
17	4.45	3.59	3.20	2.96	2.81	2.70	2.55	2.38	2.29	2.23	2.15	2.11	2.08	2.02	1.96
18	4.41	3.55	3.16	2.93	2.77	2.66	2.51	2.34	2.25	2.19	2.11	2.07	2.04	1.98	1.92
19	4.38	3.52	3.13	2.90	2.74	2.63	2.48	2.31	2.21	2.15	2.07	2.02	2.00	1.94	1.88
20	4.35	3.49	3.10	2.87	2.71	2.60	2.45	2.28	2.18	2.12	2.04	1.99	1.96	1.90	1.84
22	4.30	3.44	3.05	2.82	2.66	2.55	2.40	2.23	2.13	2.07	1.98	1.93	1.91	1.84	1.78
24	4.26	3.40	3.01	2.78	2.62	2.51	2.36	2.18	2.09	2.03	1.94	1.89	1.86	1.80	1.73
26	4.23	3.37	2.98	2.74	2.59	2.47	2.32	2.15	2.05	1.99	1.90	1.85	1.82	1.76	1.69
28	4.20	3.34	2.95	2.71	2.56	2.45	2.29	2.12	2.02	1.96	1.87	1.81	1.78	1.72	1.65
30	4.17	3.32	2.92	2.69	2.53	2.42	2.27	2.09	1.99	1.93	1.84	1.79	1.76	1.69	1.62
40	4.08	3.23	2.84	2.61	2.45	2.34	2.18	2.00	1.90	1.84	1.74	1.69	1.66	1.59	1.51
50	4.03	3.18	2.79	2.56	2.40	2.29	2.13	1.95	1.85	1.78	1.69	1.63	1.60	1.52	1.44
60	4.00	3.15	2.76	2.53	2.37	2.25	2.10	1.92	1.81	1.75	1.65	1.59	1.56	1.48	1.39
70	3.98	3.13	2.74	2.50	2.35	2.23	2.07	1.89	1.79	1.72	1.62	1.56	1.53	1.45	1.35
80	3.96	3.11	2.72	2.48	2.33	2.21	2.05	1.88	1.77	1.70	1.60	1.54	1.51	1.42	1.32
100	3.94	3.09	2.70	2.46	2.30	2.19	2.03	1.85	1.75	1.68	1.57	1.51	1.48	1.39	1.28
150	3.91	3.06	2.67	2.43	2.27	2.16	2.00	1.82	1.71	1.64	1.54	1.47	1.44	1.34	1.22
200	3.89	3.04	2.65	2.41	2.26	2.14	1.98	1.80	1.69	1.62	1.52	1.45	1.42	1.32	1.19
400	3.86	3.02	2.62	2.39	2.23	2.12	1.96	1.78	1.67	1.60	1.49	1.42	1.38	1.28	1.13
∞	3.84	2.99	2.60	2.37	2.21	2.09	1.94	1.75	1.64	1.57	1.46	1.40	1.32	1.24	1.00

Source: **G. W. Snedecor and W. G. Cochran, *Statistical Methods* (6th edition, 1967), Iowa State University Press, Ames, Iowa, by permission of the authors and publisher.**

41

99th PERCENTILE VALUES FOR THE *F* DISTRIBUTION

n_1 = degrees of freedom for numerator
n_2 = degrees of freedom for denominator
(shaded area = .99)

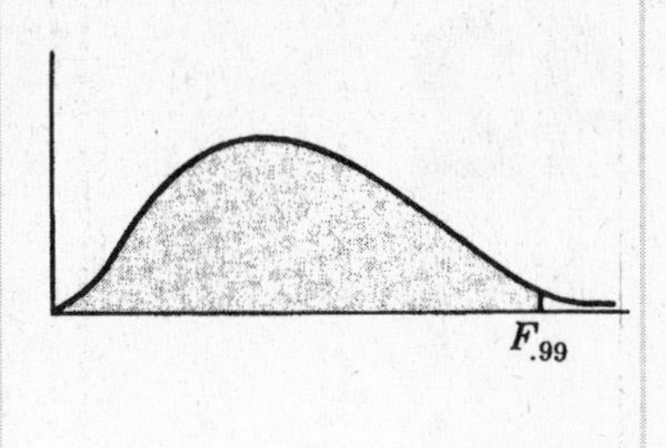

n_2 \ n_1	1	2	3	4	5	6	8	12	16	20	30	40	50	100	∞
1	4052	4999	5403	5625	5764	5859	5981	6106	6169	6208	6258	6286	6302	6334	6366
2	98.49	99.01	99.17	99.25	99.30	99.33	99.36	99.42	99.44	99.45	99.47	99.48	99.48	99.49	99.50
3	34.12	30.81	29.46	28.71	28.24	27.41	27.49	27.05	28.63	26.69	26.50	26.41	26.35	26.23	26.12
4	21.20	18.00	16.69	15.98	15.52	15.21	14.80	14.37	14.15	14.02	13.83	13.74	13.69	13.57	13.46
5	16.26	13.27	12.06	11.39	10.97	10.67	10.27	9.89	9.68	9.55	9.38	9.29	9.24	9.13	9.02
6	13.74	10.92	9.78	9.15	8.75	8.47	8.10	7.72	7.52	7.39	7.23	7.14	7.09	6.99	6.88
7	12.25	9.55	8.45	7.85	7.46	7.19	6.84	6.47	6.27	6.15	5.98	5.90	5.85	5.75	5.65
8	11.26	8.65	7.59	7.01	6.63	6.37	6.03	5.67	5.48	5.36	5.20	5.11	5.06	4.96	4.86
9	10.56	8.02	6.99	6.42	6.06	5.80	5.47	5.11	4.92	4.80	4.64	4.56	4.51	4.41	4.31
10	10.04	7.56	6.55	5.99	5.64	5.39	5.06	4.71	4.52	4.41	4.25	4.17	4.12	4.01	3.91
11	9.05	7.20	6.22	5.67	5.32	5.07	4.74	4.40	4.21	4.10	3.94	3.86	3.80	3.70	3.60
12	9.33	6.93	5.95	5.41	5.06	4.82	4.50	4.16	3.98	3.86	3.70	3.61	3.56	3.46	3.36
13	9.07	6.70	5.74	5.20	4.86	4.62	4.30	3.96	3.78	3.67	3.51	3.42	3.37	3.27	3.16
14	8.86	6.51	5.56	5.03	4.69	4.46	4.14	3.80	3.62	3.51	3.34	3.26	3.21	3.11	3.00
15	8.68	6.36	5.42	4.89	4.56	4.32	4.00	3.67	3.48	3.36	3.20	3.12	3.07	2.97	2.87
16	8.53	6.23	5.29	4.77	4.44	4.20	3.89	3.55	3.37	3.25	3.10	3.01	2.96	2.86	2.75
17	8.40	6.11	5.18	4.67	4.34	4.10	3.79	3.45	3.27	3.16	3.00	2.92	2.86	2.76	2.65
18	8.28	6.01	5.09	4.58	4.25	4.01	3.71	3.37	3.19	3.07	2.91	2.83	2.78	2.68	2.57
19	8.18	5.93	5.01	4.50	4.17	3.94	3.63	3.30	3.12	3.00	2.84	2.76	2.70	2.60	2.49
20	8.10	5.85	4.94	4.43	4.10	3.87	3.56	3.23	3.05	2.94	2.77	2.69	2.63	2.53	2.42
22	7.94	5.72	4.82	4.31	3.99	3.76	3.45	3.12	2.94	2.83	2.67	2.58	2.53	2.42	2.31
24	7.82	5.61	4.72	4.22	3.90	3.67	3.36	3.03	2.85	2.74	2.58	2.49	2.44	2.33	2.21
26	7.72	5.53	4.64	4.14	3.82	3.59	3.29	2.96	2.77	2.66	2.50	2.41	2.36	2.25	2.13
28	7.64	5.45	4.57	4.07	3.76	3.53	3.23	2.90	2.71	2.60	2.44	2.35	2.30	2.18	2.06
30	7.56	5.39	4.51	4.02	3.70	3.47	3.17	2.84	2.66	2.55	2.38	2.29	2.24	2.13	2.01
40	7.31	5.18	4.31	3.83	3.51	3.29	2.99	2.66	2.49	2.37	2.20	2.11	2.05	1.94	1.81
50	7.17	5.06	4.20	3.72	3.41	3.18	2.88	2.56	2.39	2.26	2.10	2.00	1.94	1.82	1.68
60	7.08	4.98	4.13	3.65	3.34	3.12	2.82	2.50	2.32	2.20	2.03	1.93	1.87	1.74	1.60
70	7.01	4.92	4.08	3.60	3.29	3.07	2.77	2.45	2.28	2.15	1.98	1.88	1.82	1.69	1.53
80	6.96	4.88	4.04	3.56	3.25	3.04	2.74	2.41	2.24	2.11	1.94	1.84	1.78	1.65	1.49
100	6.90	4.82	3.98	3.51	3.20	2.99	2.69	2.36	2.19	2.06	1.89	1.79	1.73	1.59	1.43
150	6.81	4.75	3.91	3.44	3.14	2.92	2.62	2.30	2.12	2.00	1.83	1.72	1.66	1.51	1.33
200	6.76	4.71	3.88	3.41	3.11	2.90	2.60	2.28	2.09	1.97	1.79	1.69	1.62	1.48	1.28
400	6.70	4.66	3.83	3.36	3.06	2.85	2.55	2.23	2.04	1.92	1.74	1.64	1.57	1.42	1.19
∞	6.64	4.60	3.78	3.32	3.02	2.80	2.51	2.18	1.99	1.87	1.69	1.59	1.52	1.36	1.00

Source: G. W. Snedecor and W. G. Cochran, *Statistical Methods* (6th edition, 1967), Iowa State University Press, Ames, Iowa, by permission of the authors and publisher.

42 RANDOM NUMBERS

51772	74640	42331	29044	46621	62898	93582	04186	19640	87056
24033	23491	83587	06568	21960	21387	76105	10863	97453	90581
45939	60173	52078	25424	11645	55870	56974	37428	93507	94271
30586	02133	75797	45406	31041	86707	12973	17169	88116	42187
03585	79353	81938	82322	96799	85659	36081	50884	14070	74950
64937	03355	95863	20790	65304	55189	00745	65253	11822	15804
15630	64759	51135	98527	62586	41889	25439	88036	24034	67283
09448	56301	57683	30277	94623	85418	68829	06652	41982	49159
21631	91157	77331	60710	52290	16835	48653	71590	16159	14676
91097	17480	29414	06829	87843	28195	27279	47152	35683	47280
50532	25496	95652	42457	73547	76552	50020	24819	52984	76168
07136	40876	79971	54195	25708	51817	36732	72484	94923	75936
27989	64728	10744	08396	56242	90985	28868	99431	50995	20507
85184	73949	36601	46253	00477	25234	09908	36574	72139	70185
54398	21154	97810	36764	32869	11785	55261	59009	38714	38723
65544	34371	09591	07839	58892	92843	72828	91341	84821	63886
08263	65952	85762	64236	39238	18776	84303	99247	46149	03229
39817	67906	48236	16057	81812	15815	63700	85915	19219	45943
62257	04077	79443	95203	02479	30763	92486	54083	23631	05825
53298	90276	62545	21944	16530	03878	07516	95715	02526	33537

Index of Special Symbols and Notations

The following list show special symbols and notations together with pages on which they are defined or first appear. Cases where a symbol has more than one meaning will be clear from the context.

Symbols

$\mathrm{Ber}_n(x)$, $\mathrm{Bei}_n(x)$	Ber and Bei functions, 157
$B(m, n)$	beta function, 152
B_b	Bernoulli numbers, 142
$C(x)$	Fresnel cosine integral, 204
$\mathrm{C_i}(x)$	cosine integral, 204
$\mathrm{e}_1, \mathrm{e}_2, \mathrm{e}_3$	unit vectors in curvilinear coordinates, 127
$\mathrm{erf}(x)$	error function, 203
$\mathrm{erfc}(x)$	complementary error function, 203
$E = E(k, \pi/2)$	complete elliptic integral of the second kind, 198
$E = E(k, \phi)$	incomplete elliptic integral of the second kind, 198
$\mathrm{Ei}(x)$	exponential integral, 203
E_n	Euler number, 142
$E(X)$	mean or expectation of random variable X, 223
$f[x_0, x_1, ..., x_k]$	divided distance formula, 287, 288
$F(a)$, $F(x)$	cumulative distribution function, 209
$F(a, b; c; x)$	hypergeometric function, 178
$F(k, \phi)$	incomplete elliptic integral of the first kind, 198
$\mathcal{G}$, $\mathcal{G}^{-1}$	Fourier transform and inverse Fourier transform, 194
G. M.	geometric mean, 209
h_1, h_2, h_3	scale factors in curvilinear coordinates, 127
$H_n(x)$	Hermite polynomial, 169
$H_n^{(1)}(x)$, $H_n^{(2)}(x)$	Hankel functions of the first and second kind, 155
H. M.	harmonic mean, 210
i, j, k	unit vectors in rectangular coordinates, 120
$I_n(x)$	modified Bessel function of the first kind, 155
$J_n(x)$	Bessel function of the first kind, 153
$K = F(k, \pi/2)$	complete elliptic integral of the first kind, 198
$\mathrm{Ker}_n(x)$, $\mathrm{Kei}_n(x)$	Ker and Kei functions, 158
$K_n(x)$	modified Bessel function of the second kind, 156
$\ln x$ or $\log_e x$	natural logarithm of x, 53
$\log x$ or $\log_{10} x$	common logarithm of x, 53
$L_n(x)$	Laguerre polynomials, 171
$L_n^m(x)$	associated Laguerre polynomials, 173
$\mathcal{L}$, $\mathcal{L}^{-1}$	Laplace transform and inverse Laplace transform, 180
M.D.	mean deviation
$P(A/E)$	conditional probability of A given E, 219
$P_n(x)$	Legendre polynomials, 164
$P_n^m(x)$	associated Legendre polynomials, 173
Q_U, M, Q_L	quartiles, 211
$Q_n(x)$	Legendre functions of second kind, 167
$Q_n^m(x)$	associated Legendre functions of second kind, 168
r	sample correlation coefficient, 213
R.M.S.	root-mean-square, 211
s	sample standard deviation, 208
s^2	sample variance, 210
s_{xy}	sample covariance, 213
$\mathrm{Si}(x)$	Sine integral, 203
$S(x)$	Fresnel sine integral, 204
$T_n(x)$	Chebyshev polynomials of first kind, 175

$U_n(x)$ Chebyshev polynomials of second kind, 176
Var(X) variance of random variable X, 224
$\bar{x}$, $\bar{\bar{x}}$ sample mean, grand mean, 208, 209
$x_k^{(n)}$ kth zero of Legendre polynomial $P_n(x)$, 232
$Y_n(x)$ Bessel function of second kind, 153
Z standardized random variable, 226

Greek Symbols

α_r rth moment in standard units, 212
γ Euler's constant, 4
$\Gamma(x)$ gamma function, 149
$\zeta(x)$ Rieman zeta function, 204
μ population mean, 208
θ coordinate: cylindrical 37,
polar, 11, 24; spherical, 38
π pi, 3
ϕ spherical coordinate, 38
$\Phi(p)$ sum $1+\frac{1}{2}+\frac{1}{3}+\cdots+\frac{1}{p}$, $\Phi(0)=0$, 154
$\Phi(x)$ probability distribution function, 226
σ population standard deviation, 223
σ^2 population variance, 223

Notations

$A \sim B$ A is asymptotic to B or A/B approaches 1, 151
$|A|$ absolute value of $A = \begin{cases} A \text{ if } A \geq 0 \\ -A \text{ if } A<0 \end{cases}$
$n!$ factorial n, 7
$\binom{n}{k}$ binomial coefficients, 8
$y'=\frac{dy}{dx}=f'(x)$, $y''=\frac{d^2y}{dx^2}=f''(x)$, etc. derivatives of y or $f(x)$ with respect to x, 62
$D^p=\frac{d^p}{dx^p}$ pth derivative with respect to x, 64
$\frac{\partial f}{\partial x}, \frac{\partial f}{\partial x}, \frac{\partial^2 f}{\partial x \partial y}$, etc. partial derivatives, 65
$\frac{\partial(x,y,z)}{\partial(u_1,u_2,u_3)}$ Jacobian, 128
$\int f(x)dx$ indefinite integral, 67
$\int_a^b f(x)dx$ definite integral, 108
$\int_C \mathbf{A} \cdot dr$ line integral of **A** along C, 124
$\mathbf{A} \cdot \mathbf{B}$ dot product of **A** and **B**, 120
$\mathbf{A} \times \mathbf{B}$ cross product of **A** and **B**, 121
∇ del operator, 122
$\nabla^2 = \nabla \cdot \nabla$ Laplacian operator, 123
$\nabla^4 = \nabla^2(\nabla^2)$ biharmonic operator, 123

Index

Adams-Bashforth methods, 236
Adams-Moulton methods, 236
Addition formula:
  Bessel functions, 163
  Hermite polynomials, 170
Addition rule (probability) 208
Addition of vectors, 119
Algebra of sets, 217
Algebraic equations, solutions of, 13
Alphabet, Greek, 3
Analytic geometry, plane, 22–33
  solid, 34–40
Annuity table, 274
Anti-derivative, 67
Anti-logarithms, 53
Arithmetic:
  mean, 208
  series, 134
Arithmetic-geometric series, 134
Associated Laguerre polynomials, 173
  (*See also* Laguerre polynomials)
Associated Legendre functions, 164
  (*See also* Legendre functions)
  of the first kind, 168
  of the second kind, 168
Asymptotic expansions or formulas:

Bernoulli numbers, 143
Bessel functions, 160
Backward difference formulas, 228
  Her and Bei functions, 157
Bayes formula, 220
Bernoulli numbers, 142
  asymptotic formula, 143
  series, 143
Bernoulli's differential equation, 116
Bessel functions, 153–164
  graphs, 159
  integral representation, 161
  modified, 155
  recurrence formulas, 154, 157
  series, orthogonal, 161
  tables, 261–267
Bessel's differential equation,
    118, 153
  general solution, 154
  modified differential equation, 155
Best fit, line of, 214
Beta function, 152
Biharmonic operator, 123
Binomial:
  coefficients, 7, 228, 259
  distribution, 226
  formula, 7
  series, 136
Bipolar coordinates, 131
Bisection method, 223
Bivariate data, 212

Carioid, 29
Cassini, ovals of, 32
Catalan's constant, 200
Catenary, 29
Cauchy or Euler differential equation, 117
Cauchy's form of remainder in Taylor series, 134
Cauchy-Schwarz inequality, 205
  for integrals, 206
Central tendency, 208
Chain rule for derivatives, 67
Chebyshev polynomials, 175
  of the first kind, 175
  of the second kind, 176
  recurrence formula, 175
Chebyshev's differential equation, 175
  general solution, 177
Chebyshev's inequality, 206
Chi-square distribution, 226
  table of values, 279
Circle, 17, 25
Coefficient:
  of excess (kurtosis), 212
  of skewness, 212
Coefficients:
  binomial, 7
  multinomial, 9
Complementary error function, 203
Complex:
  conjugate, 10
  numbers, 10
  logarithm of, 55
  plane, 10
Components of a vector, 120
Compound amount, 262
Confocal:
  ellipsdoidal coordinates, 133
  paraboloidal coordinates, 133
Conical coordinates, 129
Conics, 25 (*See also* Ellipse, Parabola, Hyperbola)
Conjugate, complex, 10
Constant of integration, 67
Constants, 3
  series of, 134
Continuous random variable, 224
Convergence, interval of, 138.
Conversion factors, 15
Convolution theorem, Fourier transform, 194
Coordinates, 127
  bipolar, 131
  confocal ellipsoidal, 133
  confocal paraboloidal, 133
  conical, 132
  curvilinear, 127
  cylindrical, 129
  elliptic cylindrical, 130
  oblate spheroidal, 131
  paraboloidal, 130
  prolate spheroidal, 131
  spherical, 129
  toroidal, 132
Correlation coefficient, 213
Cosine, 43
  graph of, 46
  table of values, 245
Cosine integral, 203, 256
Cosines, law of, 51
Covariance, 213
Cross or vector product, 121
Cubic equation, solution of, 13

Cumulative distribution function, 225
Curl, 123
Curve fitting, 215
Curvilinear coordinates, 134
Cycloid, 28
Cylindrical coordinates, 37, 129

Definite integrals, 108–116
  approximate formula, 109
  definition of, 108
Degrees, conversion to radians, 251
Del operator, 122
DeMoivre's theorem, 11
Derivatives, 62–66
  chain rule for, 62
  higher, 64
  Leibniz's rule, 64
  of vectors, 122
Deviation:
  mean, 210
  standard, 210
Differential equations, numerical methods for solution:
  ordinary, 235–236
  partial, 237–240
Differentials, 65, 66
Differentiation, 62–66 (*See also* Derivatives)
Direction numbers, 34
  cosines, 34
Discrete random variable, 223
Distributions, probability, 226
Divergence, 122, 128
  theorem, 126
Divided-difference formula (general), 228
Dot or scalar product, 120
Double integrals, 125

Eccentricity, 25
Ellipse, 18, 25
Ellipsoid, 39
Elliptic cylinder, 41
Elliptic cylindrical coordinates, 130
Elliptic functions, 198–202
  Jacobi's, 199
  series expansion, 200
Elliptic integrals, 198–199
  table of values, 270–271
Epicycloid, 30
Equality of vectors, 119
Equations, algebraic, 13
Error functions, 203
Euler:
  constant, 4
  differential equation, 117
  methods, 235
  numbers, 142
Euler-Maclaurin summation formula,
    137
Exact differential equation, 116
Excess, coefficient of kurtosis, 212
Exponential curve (least-squares), 215
Exponential function, 53–54
  series for, 139
  table of values, 254–255
Exponential integral, 203, 256
Exponents, 53

F distribution, 226
  table of values, 280–281
Factorial n, 7
  table of values, 257
Factors, special, 5
Financial tables, 272–275

Finite-difference methods for solution of:
  heat equation, 237
  Poisson equation, 237
  wave equation, 238
First-order divided-difference formula, 227
Five number summary [L, Q_L, M, Q_H, H], 211
Fixed-point iteration, 234
Folium of Descartes, 31
Forward difference formulas, 228
Fourier series, 144–146
Fourier transform, 193
  convolution of, 194
  cosine, 194, 197
  Parseval's identity for, 193
  sine, 194, 196
  tables, 195–199
Fourier's integral theorem, 193
Fresnel sine and cosine integral, 204
Frullani's integral, 115

Gamma function, 149, 150
  relation to beta function, 152
  table of values, 258
Gauss' theorem, 126
Gauss-Legendre formula, 232
Gauss-Seidel method, 230
Gaussian quadrature formula, 231
Generating functions, 157, 165, 168, 169, 171, 173, 175, 176
Geometric:
  mean (G.M.), 209
  series, 134
Geometry, 16–21
  analytic, 22–40
Gradient, 122, 128
Grand mean, 209
Greek alphabet, 3
Green's theorem, 126
Griggsian logarithms, 53

Half angle formulas, 48
Half rectified sine wave function, 191
Hankel functions, 155
Harmonic mean, 209
Heat equation, 237
Heaviside's unit function, 192
Hermite:
  interpolation,229
  polynomials, 169–170
Hermite's differential equation, 169
Heun's method, 235
Holder's inequality, 205
  for integrals, 206
Homogeneous differential equation, 116
  linear second order, 117
Hyperbola, 25
Hyperbolic functions, 56–61
  graphs of, 59
  inverse, 59–61
  series for, 140
Hyperboloid, 39
Hypergeometric:
  differential equation, 178
  distribution, 226
  functions, 178
Hypocycloid, 28, 30

Imaginary part of a complex number, 10
Indefinite integrals, 67–107
  definition of, 67
  tables of, 71–107
  transformation of, 69
Independent events, 221

Inequalities, 205
Infinite products, 207
Integral calculus, fundamental theorem, 108
Integrals:
  definite (*see* Definite integrals)
  improper, 108
  indefinite (*see* Indefinite integrals)
  line, 124
  multiple, 125
  surface, 125
Integration, 64 (*See also* Integrals)
  constant of, 67
  general rules, 67–69
Integration by parts, 67
  generalized, 69
Intercepts, 22
Interest, 272–275
Intermediate Value Theorem, 233
Interpolation, 227
  Hermite, 229
Interpolatory formula (general), 228
Interquartile range, 211
Interval of convergence, 138
Inverse:
  hyperbolic functions, 59–61
  Laplace transforms, 180
  trigonometric functions, 49–51
Iteration methods, 240
  for general linear systems, 240
  for Poisson equation, 240

Jacobi method, 240
Jacobi's elliptic functions, 199
Jacobian, 128

Ker and Kei functions, 158–159
Kurtosis, 212

Lagrange:
  form of remainder, 138
  interpolation, 227
Laguerre polynomials, 172
  generating function for, 173
  recurrence formula, 192
Laguerre's associated differential equation, 170
Laguerre's differential equation, 172
Landen's transformation, 199
Laplace transform, 180–192
  complex inversion formula for, 180
  definition of, 180
  inverse, 180
  tables of, 181–192
Laplacian, 123, 128
Least-squares:
  curve, 215
  line, 214
Legendre functions, 164–168
  of the second kind, 166
Legendre polynomial, 164–165, 232
  generating function for, 164
  recurrence formula for, 166
  tables of values for, 269
Legendre's associated differential equation, 168
Legendre's differential equation, 118, 164
Leibniz's rule, 64
Lemniscate, 28
Limacon of Pascal, 32
Line, 22, 35
  of best fit, 214
  regression, 214
Line integral, 124

Logarithmic functions, 53–55 (*See also* Logarithms)
  series for, 139
  table of values, 245–246, 252–253
Logarithms, 53–55
  of complex numbers, 55
  Griggsian, 53

Maclaurin series, 138
Mean, 208
  continuous random variable, 224
  deviation (M.D.), 211
  discrete random variable, 223
  geometric, 209
  grand, 209
  harmonic, 209
  population, 212
  weighted. 209
Mean value theorem,
  for definite integrals, 108
  generalized, 109
Median, 208
Midpoint rule, 231, 235
Midrange, 210
Milne's method, 236
Minkowski's inequality, 206
  for integrals, 206
Mode, 209

Modified Bessel functions, 155–157
  generating function for, 157
  graphs of, 159
  recurrence formulas for, 157
Modulus of a complex number, 11
Moment, rth, 212
Momental skewness, 212
Moments of inertia, 41
Monoticity Rule (Probability), 218
Mutinomial coefficients, 9
Multiple integrals, 125

Napier's rules, 52
Natural logarithms and antilogarithms, 53
  tables of, 252–253
Neumann's function, 153
Newton's:
  backward-difference formula, 228
  forward-difference formula, 228
  interpolation, 227
  method, 233
Nonhomogeneous differential equation, linear second order, 117
Nonlinear equations, solution of, 233
Normal curve, 276–277
  distribution, 226
Normal equations for least-squares line, 214
Null function, 189
Numbers:
  Bernoulli, 142
  Euler, 142
Numerical methods for partial differential equations, 237–239

Oblate spheroidal coordinates, 131
Orthogonal curvilinear coordinates, 127–128
  formulas involving, 128
Orthogonality:
  Chebyshev's polynomials, 176
  Laguerre polynomials, 172
  Legendre polynomials, 165
Ovals of Cassini, 32

Parabola, 25
segment of, 18
Parabolic cylindrical coordinates, 129
Paraboloid, 40
Paraboloidal coordinates, 130
Parallelepiped, 19
Parallelogram, 7
Parameter, 208
Parseval's identity for:
Fourier series, 144
Fourier transform, 194
Partial:
derivatives, 65
differential equations, numerical methods, 237
Pascal's triangle, 8
Percentile, kth, 211
Periods of elliptic functions, 200
Plane analytic geometry, formulas from, 22–27
Plane, complex, 10
Poisson:
distribution, 226
equation, 237
summation formula, 137
Polar:
coordinates, 24
form of a complex number, 11
Polygon, regular, 17
Polynomial function (least-squares), 214
Polynomials:
Chebyshev's, 175
Laguerre, 171
Legendre, 164
Population, 208
mean 210
standard deviation, 212
variance, 212
Power function (least-squares), 214
Power series, 138–141
reversion of 141
Powers, sums of, 134
Present value, of an amount, 273
of an annuity, 275
Probability, 217
distribution, 223
function, 218
tables, 276
Products, infinite, 207
special, 5
Pulse function, 192
Pyramid, volume of, 20

Quadrants, 43
Quadratic convergence, 233
Quadratic equation, solution of, 103
Quadrature, 231–232
Quartic equation, solution of, 13
Quartile coefficient of skewness, 212
Quartiles [Q_L, M, Q_U], 211

Radians, 4, 44
table of conversion to degrees, 250
Random numbers table, 282
Random variable, 223–226
standardized, 226
Range, sample, 210
Real part of a complex number, 10
Reciprocals of powers, sums of, 135
Rectangle, 13
Rectangular coordinate system, 120
Rectangular coordinates, 24
transformation to polar coordinates, 24
Rectangular formula, 109
Rectified sine wave function, 191
Recurrence or recursion formulas:
Bessel functions, 154
Chebyshev's polynomials, 175
gamma function, 149
Hermite polynomials, 169
Laguerre polynomials, 171
Legendre polynomials, 165
Regression line, 214
Regular polygon, 17
Remainder:
Cauchy's form, 13
Lagrange form, 138
Remainder formula:
Gauss-Legendre interpolation, 232
Hermite interpolation, 230
Lagrange interpolation, 227
Reversion of power series, 141
Richardson method, 240
Riemann zeta function, 204
Right circular cone, 20
Rochigue's formula:
Laguerre polynomials, 171
Legendre's polynomials, 164
Root mean square (R.M.S.), 211
Roots of complex numbers, 11
Rose, 29
Rotation, 24, 37
Runge-Kutta method, 236

Sample, 208
covariance, 213
Saw tooth wave function, 191
Scalar, 119
multiplication of vectors, 119
Scalar or dot product, 120
Scale factors, 127
Scatterplot, 212
Schwarz (Cauchy-Schwarz) inequality, 205
for integrals, 206
Secant method, 233
Second-order differential equation, 117
Second-order divided-difference formula, 228
Sector of a circle, 17
Segment:
of circle, 18
of parabola, 18
Semi-interquartile range, 211
Separation of variables, 116
Series, arithmetic, 134
arithmetic-geometric, 134
binomial, 188
of constants, 134
Fourier, 144–148
geometric, 134
power, 138
of sums of powers, 134
Taylor, 138–141
Simpson's formula, 109, 231
Sine, 43
graph of, 46
table of values, 247
Sine integral, 88
table of values, 264
Sines, law of, 51
Skewness, 212
Solid analytic geometry, 34–40
Solutions of algebraic equations, 13–14
SOR (successive-overrelaxation) method, 240
Sphere, equations of, 38
surface area, 19
volume, 21
Spherical coordinates, 38, 129
Spherical triangle, 51

Spiral of Archimedes, 33
Square wave function, 191
Squares error, 215
Standard deviation, 210
  continuous random variable, 225
  discrete random variable, 224
  population, 212
  sample, 210
Standardized random variable, 215
Statistics, 208–216
  tables, 276–281
Step function, 192
Stirling's formula, 150
Stochastic process, 219
Stokes' theorem, 126
Student's *t* distribution, 226
  table of, 298
Successive-overrelaxation (SOR) method, 240
Summation formula:
  Euler-Maclaurin, 137
  Poisson, 137
Surface integrals, 125

Tangent function, 43
  graph of, 46
  table of values, 249
Tangents, law of, 51, 52
Taylor series, 138–141
  two variables, 141
Three-point interpolatory formula, 228
Toroidal coordinates, 132
Torus, surface area, volume, 18
Total probability, Law of, 220
Tractrix, 31
Transformation:
  Jacobian of, 128
  of coordinates, 24, 36–37, 128
  of integrals, 70, 128
Translation of coordinates:
  in a plane, 24
  in space, 36
Trapezoid, area, perimeter, 16
Trapezoidal rule (formula), 109, 231, 235
Tree diagrams, Probability, 219
Triangle inequality, 205
Triangular wave function, 191
Trigonometric functions, 43–52
  definition of, 43
  graphs of, 46
  inverse, 49–50
  series for, 139
  tables of, 247–249
Triple integrals, 125
Trochoid, 30
Two-point formula, 228
Two-point interpolatory
    formula, 228

Unit function, Heaviside's, 192
Unit normal to the surface, 125
Unit vector, 120

Variance, 210
  continuous random variable, 225
  discrete random variable, 224
  population, 210
  sample, 210
Vector analysis, 119–133
Vector or cross-product, 121
Vectors, 119
  derivatives of, 122
  integrals involving, 124
  unit, 119
Volume integrals, 125

Wallis' product, 207
Wave equation, 238
Weber's function, 153
Weighted mean, 209
Witch of Agnesi, 31

x-intercept, 22

y-intercept, 22

Zero vector, 119
Zeros of Bessel functions, 267
Zeta function of Riemann, 204